Benjamin Gal-Or

Vectored Propulsion, Supermaneuverability and Robot Aircraft

With 190 Figures

Springer-Verlag
New York Berlin Heidelberg
London Paris Tokyo Hong Kong

Professor Benjamin Gal-Or
Technion–Israel Institute of Technology
Department of Aeronautical Engineering
32000 Haifa
Israel

Library of Congress Cataloging-in-Publication Data
Gal-Or, Benjamin.
Vectored propulsion, supermaneuverability, and robot aircraft / by Benjamin Gal-Or.
P. cm.
1. Airplanes – Jet propulsion.
2. Drone aircraft.
3. Jet planes, Military.
4. stealth aircraft. I. Title.
TL685.3.G23 1989
629.134′353—dc20 89-21797

Typeset, Printed, and Bound by Keterpress, Jerusalem, Israel

ISBN 0-387-97161-0 Springer-Verlag New York Berlin Heidelberg
ISBN 3-540-97161-0 Springer-Verlag Berlin Heidelberg New York

Dedicated to my family;
Leah, Amir and Gillad,
who, by their respective professions,
and by their encouragement,
have given me invaluable help
in preparing this volume,

and to the members of the
Jet Propulsion Laboratory Team,
Azar, B.;
Cohen, Z.;
Dekel, E.;
Friedman, E.;
Golijow, E.;
Mashiach, E.;
Rasputnis, A.;
Soreq, I.;
Turgemann M.;
and
Vorobeichik, S.,

to whom I am greatly indebted.

Mistakes!, Who may pretend to comprehend them? And for the unknown ones, please forgive me."

Based on Psalms 19:12

"Our whole problem is to make the mistakes as fast as possible..."

John Archibald Wheeler

"All our knowledge grows only through the correcting of our mistakes."

The Philosopher of Science
Sir Karl R. Popper

(Conjectures and Refutations; The Growth of Scientific Knowledge, K.R. Popper, Routledge and Kegan Paul, London, 1972).

Foreword

With the advent of digital flight control and digital engine control technologies, airframe and propulsion systems designers can now consider a much higher degree of coupling between the aircraft and its engine than ever before to achieve revolutionary new capabilities for high performance aircraft.

Professor Gal-Or's landmark book is, in my opinion, the most complete and definitive treatment to date of the complex aerodynamic and control integration associated with vectored-thrust propulsion, aircraft agility enhanced by multi-axis thrust vectoring and reaction control systems, and provides valuable insight into applications for piloted and robotic aircraft.

Although the military utility of enhanced agility using vectored propulsion is not fully understood, it is clear that a better experimental and analytical technology base is necessary to evaluate the concepts in realistic air combat scenarios. Professor Gal-Or's book is a valuable reference in this regard, although the reader should be cautioned that a text on such a rapidly-developing field may need to be updated as new information becomes available. Professor Gal-Or addresses the key questions in the Introduction, which are the subject of active research and development programs in all major aerospace establishments.

The above comments should not be interpreted as an endorsement of this book, or its contents, by the United States Air Force.

Dr. G. KEITH RICHEY
Technical Director of the Wright Research and Development Center,
Wright-Patterson Air Force Base,
Ohio, USA

TABLE OF CONTENTS

PREFACE

> The hypotheses with which (a new development) starts, become steadily more abstract and remote from experience... Meanwhile the train of thought leading from the axioms to the empirical facts of verifiable consequences gets longer and more subtle.
>
> **Albert Einstein**

This book is designed to fill a professional vacuum in the new field of advanced, high-α, *vectored-stealth aircraft.* It contains a core of *unclassified* knowledge that, according to the latest trends in military aviation, should be mastered by all aeronautical engineers and students, as well as by advanced pilots and various R&D people.

The subject matter of this book has never before been investigated, or presented, as a *unified* field of study, partially because it covers an entirely new field of study, and partially because specialized fragments of this unified field are scattered throughout the literature on specific problems.

As a result, different engineers have approached the new *integrated* systems from widely varying, and sometimes misleading viewpoints, employing disjointed concepts to what should be a unified R&D-design methodology.

Consequently, this book departs from traditional, isolated texts on aerodynamics, thermodynamics, propulsion, electrodynamics, radar, control, materials, flight mechanics and design, in its emphasis on the *interconnectedness* of up-dated information in vectored, agile, STOL, or V/STOL, manned or unmanned aircraft.

Thus, in trying to develop a new, *integrated methodology*, we refrain from repeating various, well-known, isolated, mathematical-computational analyses, and, instead, stress the physical and performance consequences from the *end-user point-of-view.*

For the sake of unification and simplicity, the physico-performance options are first introduced in as simple a manner as possible, using many illustrated graphs and diagrams.

Part of the text is an expanded version of *three university courses on "Jet-Engine Design"*, "Fluid Dynamics, Heat Transfer and the Performance of Jet Engines", and "Technology of Jet Engines", given at this institute over a period of years. The other part, and perhaps the major part of the text, is the exposition of the new technology with every argument that can lead to testable consequences in the laboratory and in flight testing.

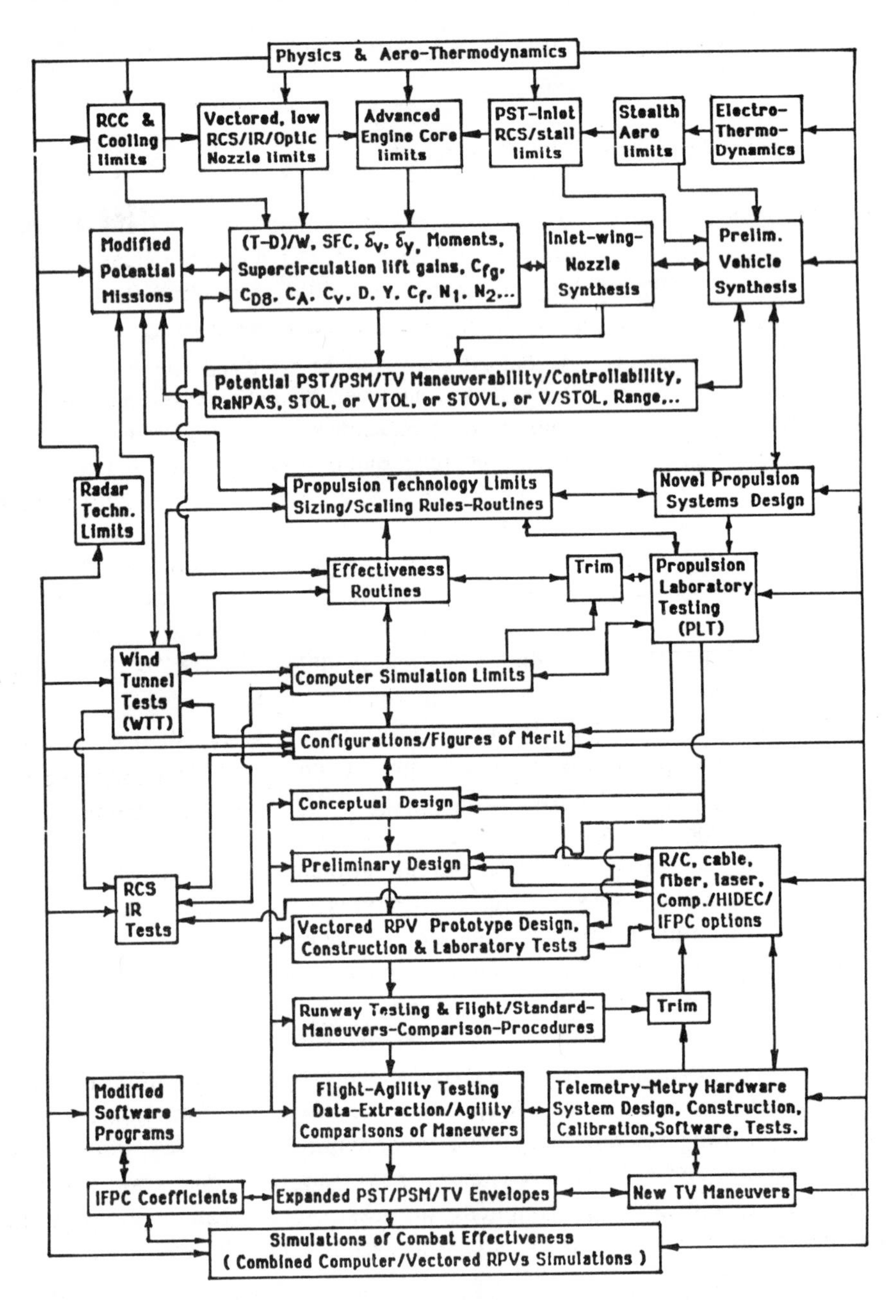

Fig. I. **Proper use of *vectored RPVs* emerges as a *highly-effective research tool*, replacing some of the more traditional roles of wind-tunnel simulations.** The actual, *feedback mechanisms* employed in this type of *highly-integrated design methodology*, are marked here as a preliminary definition of *integrated vectored aircraft*.

This scheme demonstrates that an advanced propulsion/airframe system, as a whole, cannot be optimized by improving each variable separately, and, then, superimposing the results on each other to get the whole optimized system. It is, thus, imperative to strive for the overall design, i.e., to strive for an

Another viewpoint on the need of this technology is presented below.

* * *

All advanced fighter aircraft will be based on the new technology called vectored propulsion.

This technology is the key element in enabling fighters to survive and win in battles both beyond and within visual range of the enemy. It also provides the best design elements for highly agile, stealth, supermaneuverability aircraft and Remote Piloted Vehicles (RPVs) of the future, including some vectored helicopters.

The first vectored flights in "the open history of aviation" were conducted in 1987 by our team, using various *vectored RPVs.* These flights were the final results of unpublished work conducted during the past eight years.

The fundamental concepts and the design methodology required to understand this field of engineering are detailed and explained here, using original drawings and figures which have never been published before.

Most of the experimental work conducted in this laboratory on vectored engines, and in flying "pure" vectored RPVs, is classified as the *proprietary* of our financial sources, and is, consequently, *unpublishable.* However, the fundamental concepts and the various methodologies described in this book stand out as a generic, non-sponsored, academic investigation.

Another way to describe the aim and scope of this book is to examine the *integrated design of vectored (manned and unmanned) aircraft* as depicted in the following figure on R&D cycles-feedbacks between theory and lab-flight testing and analysis. *This figure, somewhat loosely, defines the work cycles of many R&D and industrial development programs, as expected for the 90s* and beyond. It also defines the actual work-cycles practiced in this laboratory over the past few years.

* * *

A few preliminary concepts for vectored aircraft R&D may be introduced now:

- *Vectored RPVs* emerge as highly-effective *R&D tools*, replacing some of the more traditional roles of wind-tunnel testing and computational simulations.
- Mission definition and synthesis is a time-variant, testing-variant matrix, that should *not* be specified *prematurely* as a "final cause".
- *Design for low signatures* should be of prime importance in conceptual and preliminary designs.
- Propulsion-laboratory testing of *integrated wing-nozzle-engine-inlet systems* is of prime importance.

---

interconnected design philosophy in which the various variables are complementary within the framework of a reliable, whole-system-engineering methodology. However, such methodology does not exist yet. Instead, one may use the feedback methodology depicted here and detailed in the text.

- The fundamental concepts of physics, and especially of *aero-electro-thermo-dynamics*, must constantly serve as *general guidelines* for the emergence and testing of novel ideas.
- As the complexity of highly-integrated vectored aircraft increases, the gap between extant *mathematical-computational tools* and actual design needs widens, thereby making the use of the former more and more ineffective, and, at times, even misleading (except for the analysis of specific components). Consequently, during this early evolutionary stage of the technology, one should avoid the introduction of *premature* mathematical tools into advanced design courses of highly integrated systems. Instead, *physico-engineering insight*, combined with *proper integration methods* between preliminary test results and tentative mission definitions should be given the first priority.

* * *

A Few Notes For Pilots, Engineers and Teachers.

This Volume contains a core of basic material that both the advanced researcher and the desinger or test-pilot should master; namely, redefined and newly defined *fundamental* concepts, components, variables, experimental data, basic unit operations, and *partially integrated systems.*

It supplies *two tracks through the subjects.*

The *first track* focuses on general, *"interdisciplinary"* material, which is so ordered as to help those readers who are either unfamiliar with, or uninterested in detailed formulations and empirical data of a specific "discipline", but would rather consider the general arguments involed without interruption. Hence, a substantial part of this Volume is available as a *second track.*

All "track two" material is marked with ●, or included in the *Appendices.* It includes the more specific *data base* and various experimental results of earlier designs, and of some unorthodox subsystems. It also includes a few notes on the *latest evolutionary trends* in this technology, including brief summaries on the expected *PST/Vectoring technology limits in the 90s and beyond.*

The selection of the material for the appendices is intended to provide a greater degree of competence in a few sub-systems. Nevertheless, researchers, teachers, designers and pilots, especially those in need of further enrichment material on experimental data and limiting testing conditions, or on some highly specific designs and missions that have recently gained special importance, are invited to read the original references to those subjects that interest them the most.

All "track two" material can be understood by *undergraduate students* who have taken at least one-semester courses on (compressible) fluid dynamics and on jet propulsion.

With a few exceptions, "track one" material can be easily understood by readers who have studied only *"Freshman" physics*, and, in addition, have participated in basic *pilot training courses.*

Volume II, to be published later, is to stress the fundamental design methodology and the flight-propulsion control of *fully integrated stealth-vectored systems.* Consequently, it is designed to concentrate on new "figures of merit" and on *integrated propulsion/flight control theories* and components of pure vectored aircraft.

Revisiting the fundamentals of electro-aero-thermodynamics is included in Volume II, for proper component integration in pure vectored/stealth aircraft design, in terms of new materials, control systems, inlets and powerplants, while simultaneously *maximizing supermaneuverability, STOL, or V/STOL performance. All in all, it shows that vectored aircraft design is not separable into disciplines.*

* * *

The Lectures presented in Volume I may be adapted for a *one-semester*, undergraduate course on *"Design"*, or on *"Advanced Flight/Jet Propulsion"*. Volume II, when available, is designed to become useful in a combined, *two-semester graduate/undergraduate course*, or as an independent, one-semester, *graduate course.*

* * *

It was in 1980 that I started to lecture on these topics.

Ever since I have been working on this book, which reproduces, if not the letter, then the spirit of my lectures.

No series of lectures was ever a set-piece; they have remained in a state of flux, until now, when the final writing and printing has "frozen" them. This "freezing" is very apprehensive to me, but it cannot be helped. My hope is to complete the 2nd volume within the next few years. Meanwhile, I hope that at least some of the readers will "thaw out" the figures and printed lines, and give them greater dynamical force through their own critical attention.

* * *

Indeed, it was the work on the present volume that gave me the opportunity to develop more fully than in my earlier research projects, the general principles of this field of study, as well as the integrated methodology of laboratory/flight testings of post-stall, purely-vectored, research RPVs. Joining the test results obtained from the jet-propulsion laboratory with the design and flight-testing results of vectored RPVs, seemed, at times, like the generation of a vast jigsaw puzzle from scattered and (apparently) unrelated pieces of test results. But the eventual emergence of new regularities and unconventional design principles, which, at the beginning, I had not even suspected to be part of the overall research picture, was often a rewarding surprise. I can only invite the interested reader to join me in further work on some of these excursions along trails leading to new fields of inquiry.

* * *

Verifiable evidence cannot yet be admitted to a number of topics presented in this book. Therefore, if engineers (and pilots) are counted as "moderate skeptics", as no doubt they should, they must adopt the practice of "suspense of judgment". By this we do not mean radical empiricism, nor the denial of any rational addition to the advancement of high-performance aircraft design.

ACKNOWLEDGEMENTS

Part of the unclassified research work published here, as well as the **integrated methodology** of laboratory/flight testing developed by the author, has been financially sponsored by

Teledyne CAE,

The General Electric Co.,

The United States Air Force,

General Dynamics.

I wish to thank my research sponsors as well as the editors, publishers and boards of the various periodicals, reports and books referenced in this volume for allowing me, directly and indirectly, by policy or word, to include figures and pictures appearing in this volume, as quoted below and in the respective references. Indeed, I am especially indebted to AIAA, ASME, SAE, AHS, IDR, NASA, USAF-WPAFB, US Army, IDF, IAF, Journal of Propulsion, International Journal of Turbo and Jet Engines, Journal of Power and Gas Turbines, Journal of Aircraft, the General Electric Co., Pratt & Whitney, R.R., Teledyne, Allison Gas Turbines of GM, Garrett, Bet Shemesh Engines, Turbomeca, MBB, MTU, General Dynamics, Northrop, Lockheed, IAI, NPT, TAT, IGTA, KHD, Micro-Turbo, Donaldson Co., Avco-Lycoming and the McGraw Hill Co.

Most important, the evolution of this important field of aviation has been made possible by the significant contributions of the many individuals whose works have been selected in the preparation process of this book. A few of these pioneers are mentioned by name in Appendix A. However, our greatest gratitude is with the long list of contributors as reflected in the References, and in the text itself.

In revising the text, I have been helped, in various ways, by Doron Bar Annann, Dr. A. Rasputnis, Erez Friedman, Sara Voroveitchik, Mike Turgemann, Dan Erez, Eduardo Golijow, Eran Liron, Prof. Yoram Tambour, and, indirectly, by General David Ivry, the General Director of Israel's MoD, Dr. W. B. Herbst of MBB, Mr. T. P. McAtee, Mr. Jerry Murff, Mr. Trey Durham, Mr. Fran Ketter, Mr. Tom Barret, and Mr. C. Porcher of General Dynamics, Mr. F. Ehric, Mr. Ed Rogala and Mr. Don Dunbar of General Electric, Mr. E. Benstein of Teledyne, Mr. E. Albert and Mr. S. Levy of I.A.I., and Mr. D. Bowers and Mr. W. Lindsay, of the US Air Force. Indirectly, I have also been helped by my students, who frequently contributed comments and questions, especially by those who had selected to participate in my courses "Fluid Dynamics, Heat Transfer, and the Performance of Jet Engines", "Design of Propulsion Systems", "Jet-Engines Technology" and "Jet Engines-1". Erez Friedman has given me invaluable, first-class professional help in constructing the first Pure Vectored RPVs, and the first vectored F-15/F-16 RPVs, and in flight testing them in the spring of 1987, and in the summer of 1989, respectively. Similar acknowledgement and gratitude go to Mike Turgemann, who constructed our very first vectored RPV and flight tested it in April and May 1987.

My deepest gratitude goes to my two sons, Amir Gal-Or and Gillad Gal-Or, who, during our long, enthusiastic, professional discussions, have contributed many ideas, and have also encouraged me to pursue these ideas in laboratory and RPVs flight testing.

Finally I want to express my greatest indebtedness to the Publisher, Springer Verlag, and especially to the Engineering Editors Dr. Z.Ruder, and A. Von Hagen, and to Zvi Weller, Yaacov Polak, Ami Green and Hannah Lipschitz of Keter Press, Jerusalem.

B. Gal-Or
Haifa, 1989

GLOSSARY AND NOTATION

A – Dimension defined in Fig. III-10
$A_e - A_9$ – Exhaust nozzle exit area.
$A_t - A_8$ – Exhaust nozzle throat area.
AC – Aerodynamic Center.
$ACC–D$ – Acceleration–Deceleration dynamic response of Trim-IFPC-engine.
ACM – Advanced Cruise Missiles.
ADEN – A thrust-vectoring nozzle illustated in Appendix A.
AoA – Angle of Attack
As – Surface area of a wing with jet vectoring issuing from the airfoil's trailing edge.
AB – After-burning
ATA – Advanced Tactical Aircraft (A-12).
ATB – Advanced Tactical Bomber (B-2).
ATF – Advanced tactical fighter.
AVCM – Advanced Vectoring Cruise Missile.
b – wing span.
B – Two-dimensional nozzle height
BLS – Boundary Layer Separation/Thickness at each propulsion station.
BPR – By-Pass Ratio.
C – Defined in Fig. III-10. An upper bar refers to average values.
C' – Defined in Eq. III-10 (Superscript h means "cos δ_v-component", Fig. III-10).
C_A – Angularity Coefficient, Eq. III-13, accounting for Thrust Angularity Losses.
C_D – Drag coefficient (cf. Fig. 2, Introduction), also = C_d
C_{D8} – Nozzle Flow Coefficient, Eq. III-7
C_f – $2D$ vectoring-nozzle flap width, (cf. Fig. 1, Introduction and Fig. III-10).
C_{fg} – Nozzle Thrust Coefficient, Eq. III-6
C.G. – Center of gravity.
C_L – Wing sectional lift coefficient. See also ΔC_L and $(\Delta C_L)_v$
$C–R$ – Circular-to-Rectangular transition ducts in vectoring nozzles (cf. Fig. III-17 and Figs. I-13, 14).
C_v – Velocity Coefficient
C_v^* – Equivalent Flap Length – see Fig. III-10
C_p – Pressure Coefficient.
CP – Center of Pressure.
C_n – Yawing moment coefficient.
C_{ny} – Increment in yawing moment per degree of deflection of both yawing vectoring nozzles (or of both rudders).
C_T – Aerodynamic thrust coefficient, F_g/qSw

C_μ – Exhaust gross thrust blowing coefficient; Refers to the flight dynamic pressure and wing area s_j; see Fig. III-18 and Eqs. III-29, 30.
D – Dimension of arm in pitching moment (Fig. II-10 and Fig. 1, Introduction). Drag.
D_y – Drag in the y-direction.
$2D$–CD – Two Dimensional–Converging–Diverging, vectoring exhaust nozzle. (Sometimes written as 2-D/C–D).
Degr. – Engine Degradation Status/History.
DTV – Directional Thrust Vector
D_Γ – Induced drag force due to supercirculation
E – Dimension defined in Fig. I, Introduction.
EGT – Exhaust Gas Temperature.
EW – Electronic Warfare.
ETV – External Thrust Vectoring.
F_g – Nozzle gross thrust.
FBL – Fly-By-Light.
FBW – Fly-By-Wire.
FCE – Canard Effects.
G – Supercirculation gain factor defined by Eqs. III-18, 19
G' – Adjusted supercirculation gain factor – $G/(S_j/S_w)$. (Fig. III-18).
G.E. – Ground Effects.
FQ – Fuel quality and caloric value and fuel-air ratio.
H – Altitude.
HIDEC – Highly Integrated Digital Engine Control.
IFPC – Integrated Flight-Propulsion Control.
Inlet – Aircraft air inlet geometric characterization + dynamic mode of operation and maximum distortion limits. It also depends on engine stall margins, $t°$, $\alpha_{\max}$, etc. (see Appendix F).
IR – Infra Red
ITV – Internal Thrust Vectoring.
LTV – Longitudinal Thrust Vector
M – Gas Mass Flow Rate. Also Mach Number.
M_a – Aerodynamic yawing moment
NAR – Nozzle Aspect Ratio – $A^2/4A_8$ [nominally defined with A_8 in S.L., static, "dry" or "*AB*" engine conditions].
NG – Nozzle Geometry
NPR – Nozzle Pressure Ratio
Pa – Ambient Pressure.
P_{T8} – Total pressure at station 8 (i.e., at the nozzle throat).
PJV – Partial Jet Vectoring
PSM – Pure Sideslip Maneuvers
PST – Post-Stall Technology
PVA – Pure Vectored Aircraft
PVR – Pure Vectored RPV
P_w – Static pressure at the wall.
P_{T2} – Total pressure at compressor's inlet.
q_0 – Free-stream dynamic pressure
q – Jet dynamic pressure.
RA – Robot Aircraft
RaNPAS – Rapid-Nose-Pointing-and-Shooting.
R/C – Radio Control.
RCC – Reinforced Carbon–Carbon (Appendix B).
RCS – Radar Cross Section.

Re – Reynolds Number.
RPM – Rounds Per Minute.
RPV – Remotely Piloted Vehicle, or, more generally, RA.
R_t – Radius of nozzle flap surface at nozzle throat.
SAM – Surface to Air Missile
S_w – Wing reference, or total area.
S_j – Supercirculation affected wing area (cf. Figs. III-18, 19, and II-1).
$SERN$ – A type of thrust vectoring nozzle. see Fig. A-3
S/MTD – STOL and Maneuver Technology Demonstrator
SSW – Supersonic Waves interactions
STOL – Short Takeoff and Landing.
STOVL – Short Takeoff and Vertical Landing
t – time
t° – Dynamic Distortion & Turbulence, especially at station 2 (pressure and velocity).
T – Actual (total) Net Thrust.
Ta – Ambient Temperature.
T_i – Ideal isentropic thrust.
T_x, T_y, T_z – Thrust-vectored components in the x-, y-, z-directions
TR – Thrust Reversal mode of operation (cf. Fig. 14, Intr., Figs. I.1,4 and I.7).
T_t – Total Temperature.
Trim – Engine Trim and Compressor stall margin (depend on altitude, M, Re, etc.).
Turb – Local Turbulence Degrees.
TV – Thrust Vectoring.
T_v – Vertical Thrust Component (normally equals to T_z).
U_∞ – Free-stream velocity
V – Aircraft velocity.
V/STOL – Vertical and Short Takeoff and Landing.
VTOL – Vertical Takeoff and Landing.
VRA – Vectored Robot Aircraft.
VRT – Vectoring, Reversing, Targeting (or simultaneous use of pitch-yaw-reversal thrust-vectoring).
V_s – Isentropic, or fully expanded, jet velocity.
V_{9i} – Ideal velocity based on A_9/A_8.
WVR – Within Visual Range

Greek

$\bar{\alpha}$ – Divergence angle of nozzle flap surfaces (cf. Fig. I-1).
α^* – Engine throttle angle.
α – Angle of attack (AoA).
$\dot{\alpha}$ – Time Rate of Change of α.
β – Angle of sideslip.
$\dot{\beta}$ – Time Rate of Change of β.
ε – A_e/A_t – Nozzle expansion ratio = A_9/A_8.
$\gamma_{\text{dist.}}$ – Circulation distribution over the wing, including wing's twist, etc.
δ_c – Canard angle of attack.
δ_r – Rudder deflection angle.
δ_e – Elevator deflection angle.
δ_v – Pitch Thrust Vectoring Angle (counted positive for downward flap/jet rotation)
δ_y – Yaw thrust Vectoring Angle, see Fig. 1, Introduction.
ΔC_L – Increased C_L due to "down" thrust-vectored-induced lift and $(\Delta C_L)_v$

$(\Delta C_L)_v$ – Increased C_L due to direct engine lift T_v *(or* T_z).
Δ_H – Horizontal tail deflection angle.
ρ – Gas or air density. Also, in a few references, equals to $\bar{\alpha}$ or θ.
θ – Cant angle of vertical tails/stabilizers
μ – Missile off-boresight angle. Viscosity.

Subscripts/Superscripts

The subscript T refers to total conditions, C – to critical values, SC to supercirculation and *'a'* to ambient conditions. Superscript c to refers to canard. For the definitions of a few other parameters see the figures and the text. Engine station numbers follow the conventional thermodynamic definitions.

INTRODUCTION

"Powered lift is the biggest contributor to military aviation today – and may be civil aviation tomorrow – since the invention of the jet engine."

DR. JOHN FOZARD
British Aerospace Division Director (204)

THE MAIN PROBLEMS OF THRUST-VECTORED MANEUVERABILITY

– Is thrust vectoring becoming the standard technology of fighter aircraft? Indeed, how important it is to enhance maneuverability and controllability?
– Are the roads to thrust vectoring also the roads to Post-Stall Technology (PST), and to low observability?
– What are the fundamental concepts of "pure" and "partial" vectored aircraft?; or of "internal" and "external" thrust vectoring?
– What are the technology limits, and what is the state-of-the-art of vectored propulsion/vectored aircraft?
– What is the lowest thrust-to-weight-ratio above which one can extract clear-cut advantages of vectored fighters over conventional ones? I.e., which of the existing fighters can be upgraded to become PST-vectored fighters?
– What are the most promising designs of PST-Stealth/vectored propulsion/vectored aircraft? Do we have the proper design philosophy to handle these problems? Are Soviet and Western thrust-vectoring methodologies similar?
– Does yaw-pitch-roll thrust vectoring constitute a basic requirement for survivability and winning in the air combat arena? How can it contribute to the aircraft's STOL and agility characteristics?
– Can an efficient PST-inlet be developed? Can such inlets be installed on, and flight-tested by PST-vectored RPVs? Indeed, what should be the R&D tools for the evaluation of vectored aircraft?
– How should PST-fighter agility be defined? What are the measurable parameters, or "metrics", which define vectored aircraft agility?
– How can we identify maneuvers, missions, and flight regimes in which vectored aircraft demonstrate advantages over conventional ones? Can one express these advantages in terms of killing-ratios, or other measurable metrics?
– What should be the new flight/propulsion control rules for PST-vectored aircraft?
– What specific maneuvers, and what new pilot tactics are associated with "partial" and "pure" PST-vectored aircraft?

– What are the expected g-loads, and other limitations, associated with thrust-vectored, PST-maneuvers?
– And, most important, how should vectorable engine nozzle and vectorable inlet geometric design, and aspect ratio, be modified to meet a given set of mission needs, such as low signatures, STOL-VTOL, air-to-air, or air-to-ground supermaneuverability and supercontrollability? In particular, what are the engine/nozzle/inlet efficiency variations and limitations associated with these new design trends?

These are some of the main problems dealt with in this book.

1. VECTORED AIRCRAFT: BREAKING THE "STALL BARRIER"

Assessing these problems first in this introduction we will assert that in future aerial combat, pointing the nose/weapon of the aircraft at the adversary first will be required to win, since pointing first may mean having the first opportunity to shoot (§3.5). It is also, under proper (post-stall) maneuverability rules, the required technology to dramatically increase survivability.

The availability of PST vectored fighters, helmet-sight-aiming systems, all-aspect missiles and the new generation of EW systems, require reassessment of the optimal balance between aircraft agility and effectiveness, and the agility and effectiveness of missile/helmet-sight-aiming systems (§3.5).

Whatever is the aforementioned balance, high-performance fighter aircraft will gradually be based on improved thurst vectored propulsion/maneuverability/controllability.

Since future fighter aircraft would be thrust-vectored, and since thrust-vectoring engines would be used for enhanced maneuverability and controllability, as well as for conventional propulsion, one must first define new concepts and new measurable "metrics", which would be employed in a realistic comparison of vectored aircraft maneuverability-controllability with that of conventional fighter aircraft. For this to be properly done one needs a new, highly-integrated methodology – a methodology which does not exist yet.

THE NEW STANDARD TECHNOLOGY

Nevertheless, as will be demonstrated in this book, thrust vectoring is now becoming the basic requirement and the standard technology for all future fighter aircraft. Yet, as it stands now, this technology is still in its embryonic state. While the pitch/thrust-reversal technology (see below) appears now to be maturing, the most critical technology of simultaneous yaw-pitch-roll thrust vectoring (see below), is still far from this stage.

In light of the prolonged time inherently associated with the advancement and maturity of such an engineering field, one may expect its full exploitation only in the

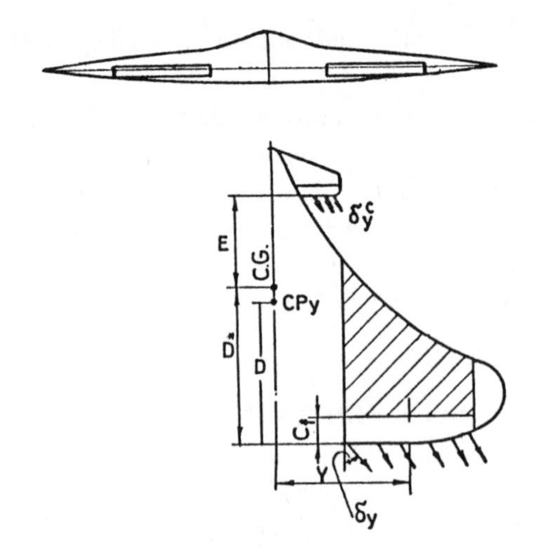

Fig. 1. **A Schematic Representation of Pure Vectored Aircraft**
Why the needs for Rapid-Nose-Pointing-And-Shooting dictate the need for vectored fighter aircraft?

The roles played by all-aspect missiles, and other, newer, combat-effectiveness factors, are evaluated in detail in the main text. Altogether, this evaluation demonstrates why simultaneous yaw-pitch-roll thrust vectoring is the basic requirement for all future fighter aircraft.

Vectored aircraft may be divided into those that are "pure" or "partial".

In pure thrust vectoring the flight-control forces generated by the conventional, aerodynamically-affected, external control surfaces of the aircraft, are replaced by the stronger, internal, thrust forces of the jet engine(s). These forces may be simultaneously, or separately, directed in all directions; i.e., in the yaw, pitch, roll, thrust-reversal, and forward thrust coordinates of the aircraft.

Since engine forces (for PST-tailored inlets) are essentially independent of the external-flow angle-of-attack and slip angle, the flight-control forces of PVA remain effective even beyond the maximum-lift Angle-of-Attack (AoA), i.e., PVA are fully controllable even in the PST domain. Thus, vectored flight provides the highest payoffs at the weakest domains of aerodynamically-controlled aircraft (i.e., at high-alpha-beta values, at very low (or zero) speeds, high altitude, high-rate spins, very-short run-ways, and during all PST, Rapid Nose-Pointing-and-Shooting (RaNPAS) maneuvers).

Consequently, no rudders, ailerons, flaps, elevators and flaperons are required in pure vectored flight, and even the vertical stabilizers may become redundant. Thus, by employing FBW, FBL and Integrated Flight/Propulsion Control (IFPC) systems, vectored aircraft may resemble "flying wings", or variously-shaped "body lift objects" which can provide Pure Sideslip Maneuvers (PSM), or even PSM/RaNPAS maneuvers (cf. Lecture II).

In pure-vectored-takeoff-methodology one first turns the jet upward, and rotates (nose liftup) the aircraft at a much lower speed (i.e., using a much shorter runway), than with conventional-technology aircraft. Then, and under the automatic guidance of IFPC, the jets are turned down, adding direct lift and supercirculation to proper wing sections (cf. Fig. II-5). Approach and landing methodologies of PVA (Fig. II-6) also afford the use of very short runways. Thus, STOL, V/STOL, STOVL and VTOL performance become the "natural" domains of PVA. These terms are defined in Appendix B and in Lecture II.

NOTES: This figure is limited to a schematic representation of this class of aircraft. The canards, or the vectored canards, are non-essential options (cf. Fig. II-1). The shaded area represents supercirculation-affected wing sections (cf. Parag. III-7.5 for the definition of supercirculation effects in vectored aircraft). PST is defined in Fig. 2. Pure sideslip, RaNPAS maneuvers are not consuming as much energy as pure PST-maneuvers (cf. Fig. 3). Thrust reversal may be less effective in flight than PST-maneuvers, cf. Fig. 21.

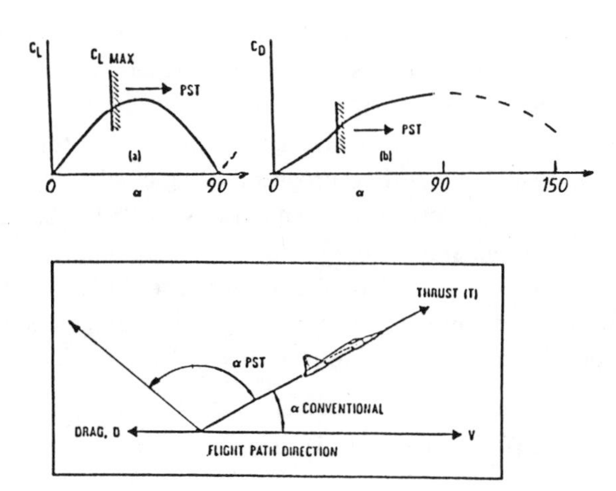

Fig. 2. **The Definition of the PST Domain in which Rapid Nose-Pointing-and-Shooting (RaNPAS) Provides the Highest Combat Payoffs.** (See also Figs. 3 to 6).

By employing the change of the lift and drag coefficients with AoA (alpha), one may define the domain of Post-Stall Technology (PST) as that marked in the upper two figures. In addition, one may split alpha into alpha-conventional and alpha-PST, as shown in the lower figure. It should be stressed that, in practice, AoA may be greater than 90 degrees. Vectored aircraft will provide combat effectiveness in the STOL, PST and PSM/RaNPAS domains (cf. Fig. 3).

post-ATF era. Nevertheless, some of its proven elements may be gradually incorporated in such upgrading/transforming designs as those feasible for the new vectored F-15 STOL Demonstrator, or for the proposed vectored F-16 and F-18 upgrading programs.

Attempting the integration of some advanced propulsion concepts with vectored aircraft's superagility concepts, is one of the central goals of this book. No doubt it will also become the central goal of well-integrated engineering and pilot education, and of research strategies for advanced propulsion/airframe programs.

PRELIMINARY REMARKS ON SAFETY AND CIVIL APPLICATIONS

This introduction defines and depicts the fundamental concepts associated with the aforementioned new technologies. It is demonstrated, during this process, that advanced propulsion research must be expanded now to include aircraft's thrust-vectored agility, and, in particular, thrust-vectored maneuverability and controllability. Thus, student instruction, as well as research and development in jet propulsion are gradually emerging as a highly integrated field of engineering.

At this early stage of the course one may also stress three points:

(i) – Thrust-vectored controllability of pure vectored aircraft is much safer and easier, as well as less complex and costly, than that of partially vectored aircraft (see the fundamental definitions below). Furthermore, maneuverability, or post-stall supermaneuverability of pure vectored aircraft is considerably superior to that obtainable with partially vectored systems, or with conventional fighter aircraft.

(ii) – Vectored propulsion systems may be developed and tested both in the jet-propulsion laboratory and in flight-test programs using small, vectored RPVs (Cf. Lecture IV). Moreover, various missions/design options for PST-vectored RPVs are now emerging as a highly promising field of study.

(iii) – The cargo and civil aircraft industries may also exploit the methodologies of vectored propulsion, for instance, by introducing low-drag, cost-effective, STOL, pure-vectoring propulsion/flight-control systems.

1.2 Vectored Aircraft: Soviet and Western Concepts

Since thrust vectoring is now emerging as the pace-maker technology of all superagile fighter aircraft, it is imperative to examine first the different design philosophies of this important field of air defence. The growing importance of this field is well reflected by the increasing number of heavily-financed R&D projects conducted now by various governmental, industrial and academic bodies, as well as by the increasing number of recent publications on this topic (see the two hundred and more references at the end of this book). These papers reflect different international methodologies.

For instance, the Central Institute of Aviation Motors in Moscow, has recently published computer simulations of thrust-vectored aircraft (188). To start with, one may note that the Soviet scientists present their analysis for aircraft propelled and controlled by simultaneous yaw-pitch thrust vectoring.

Unlike the Soviets, who appear to be newcomers to this field, the American designers had previously adopted a more conservative design philosophy, concentrating their main R&D efforts only on pitch, or pitch/reversal, thrust-vectored aircraft. (E.g., the GE/PW-thrust-vectoring engines for the new F-15 STOL and Maneuver Technology Demonstrator (S/MTD) and for the Advanced Tactical Fighter – the ATF.)

There is, nevertheless, a new US-German program to flight-test [see below] an experimental yaw/pitch vectored plane – the X-31, as well as a new, comprehensive NASA program (270–209) for simultaneous yaw-pitch (external) thrust vectoring. Furthermore, a number of multi-axis (internal) thrust vectoring studies, e.g., the (now cancelled) X-29A program employing the GE-ADEN and the GE-gimballed nozzle (see Lecture VI), have been reported (220). A minor US program is also being conducted now in this laboratory to evaluate the pros and cons of simultaneous yaw-pitch-roll (internal) thrust vectoring (179). This program includes laboratory tests and flight testing of vectored RPVs equipped with various two-dimensional nozzles, ranging from 2 to 46.7 NAR. It also includes a comparison of axisymmetric thrust-vectoring nozzles with $2D$ nozzles.

These design differences may be highly critical in the final assessment of fighter aircraft combat-ability and effectiveness in the future. Hence, it is imperative, and

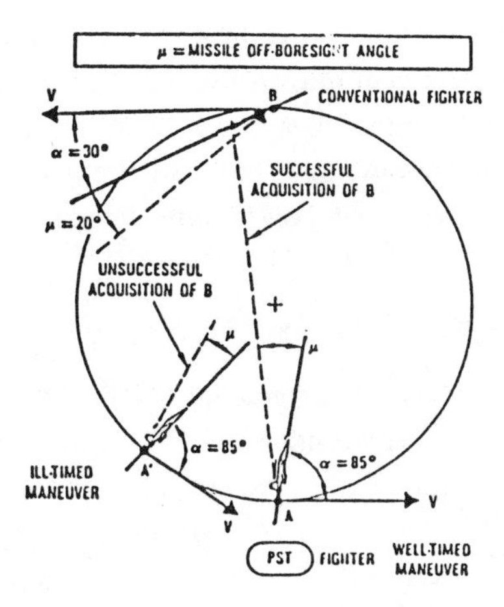

Fig. 3. **Vectored Fighters Exploit the Nose Pointing Capability Achieved Through PST or PSM (Pure Sideslip Maneuvering) (Tamrat, Northrop, 185).**

This figure provides a simple example for the offensive exploitation of a PST-nose-pointing capability of a vectored fighter. Here the conventional (B) and vectored (A) fighters are turning in a circle with neither gaining any advantage. Then, at position A, an execution of, say, 90 degrees, post-stall maneuver by the vectored fighter will make the conventional fighter vulnerable to missile/gun attack (in this example, a successful acquisition of B in the missile off-boresight angle).

Combat applications of pure sideslip maneuvers (PSM) are numerous. They include air-to-ground maneuvers (Fig. 15), escape maneuvers, and some highly promising alternatives to PST maneuvers. The last category may be demonstrated by modification to the PST-maneuvering shown in this Figure: Instead of the (energy degrading) pitch-PST maneuver depicted, one may consider a roll followed by RaNPAS yaw/pure sideslip maneuver, which may be prolonged and/or repeated without significant speed/energy degradation. The use of this option may become significant in close-combat, target-rich environments. Thus, PSM/RaNPAS maneuvers are less dangerous than the PST/RaNPAS ones.

Combat agility requirements, when maneuvering well beyond the maximum lift angle of attack, are treated differently from those for the conventional angle-of-attack regime, since at such extreme angles of attack the fighter depends mainly on its thrust vectoring to maintain both lift and flight-control power (185).

The very selection of the proper position, or the execution timing, is key to winning in PST maneuvers. For instance, a PST maneuver executed from postion A' will fail to gain a missile/gun pointing advantage. Thus, the vectored fighter, having missed the proper position/timing to shoot, will become "a sitting duck", for it has lost speed during the PST maneuver, and, hence, the potential to change flight path to evade a counterattack (185). It should be stressed, however, that this figure, as well as Figs. 4 and 5, represent only "unit operations" which **may** be conducted within a complex time-space-targets of actual air-combat.

timely, to define and compare the expected advantages of adding (axi or *2D*) simultaneous yaw-pitch-roll thrust vectoring to the current F-15 (S/MTD) – ATF propulsion methodology.

Secondly, one may note that a thrust-vectored version of the Su-27 is now being developed. Thirdly, one must notice that the Soviet simulations of vectored propulsion/flight control, have been reported by a propulsion institute, and not by a flight-dynamics institute, as is the tradition in the West. The reason behind this is, probably, the realizaiton that vectored aircraft agility improvements require novel integrated propulsion/flight-control programs. This philosophy is also reflected by a new Russian book on thrust vectoring (226). This book combines aerogasdynamics, flight/propulsion characteristics and powerplant design philosophy in a single text.

1.3 The Search for Vectored-Thrust/Super-Agility "Metrics"

The very definition of agility is still being debated (cf. parag. 3.5 and 5, and Figs. 9, 19 and 20 below, Lecture II, and references 184, 195, and 196).

Anticipating the introduction of vectored aircraft technology, McAtee, of GD (196), has, in 1987, defined agility as composed of two complementary concepts: Maneuverability and controllability. PST maneuverability is then called "supermaneuverability", while PST controllability is named "supercontrollability" (see below). The last two concepts, combined, may also be termed "super-agility" (see 3.5 below).

2. HOW TO ORDER THE DEFINITIONS OF NEW INTERCONNECTED CONCEPTS

Following the definitions of "pure" and "partial" vectored aircraft, we shall stress the higher combat effectiveness of pure, over partially-vectored aircraft. Within this initial framework, we shall also assess the pros and cons of the X-31 program, as well as that of the F-15-S/MTD and other advanced programs.

During this process we shall distinguish between the advantages obtainable by upgrading programs of existing fighter aircraft (e.g., vectored F-15, F-18 and vectored F-16), and the radically-different task of designing high-performance, Rapid-Nose-Pointing-and-Shooting (RaNPAS), Post-Stall-Technology (PST), low-signatures, pure vectored fighter aircraft [with simultaneous yaw-pitch-roll vectoring capabilities].

2.1 PST-Vectored RPV as a Research Tool

We shall then introduce a new research tool: The mini, PST, RaNPAS, Pure Vectored RPV (PVR). It has been demonstrated, on the basis of flight tests conducted by this laboratory, that such low-cost, R&D&T tools, can replace some of the more expensive roles of wind-tunnel and complex computer simulations of actual, thrust-vectored-aircraft-maneuverability/controllability capabilities.

2.2 Cold Propulsion

Finally, we shall assess the pros and cons of "cold propulsion", e.g., of the effectiveness of STOL, low-IR-signatures, inlet-distortion-free, cold-jet, PST-PVRs. Such PVRs may be based on, say, multi-stage, multi-axis, ducted fans, ducted prop-fans, or sim-

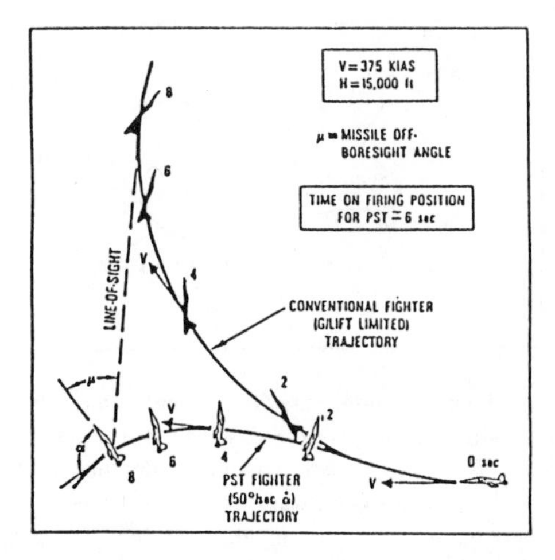

Fig. 4. **The Effect of Angle-of-Attack Rate Of Change On Close-Combat Effectiveness.**

The combat scenario shown here is based on a point-mass trajectory model, as employed by Tamrat of Northrop (185). At t = 0 sec., the vectored fighter pitches up, at an AoA rate of 50 deg./sec., to rapidly reach a final AoA = 90 deg. The conventional adversary, which, for the sake of comparison, is assumed to start at the same initial conditions, pulls up to the aircraft load-factor/lift-limit (apparently the only option available), and climbs to positions 2 to 8. In such a theoretical comparison, according to Tamrat's results, the vectored fighter may have the adversary within its missile/gun envelope for 6 sec., beginning at 2 sec. This "time within firing envelope" (TWIFE), may be considered as a nominal yard-stick for PST combat advantage. (See, however, the remarks in the following figures and in the text.) Alternatively, one may employ PSM/RaNPAS maneuvers.

The **Pougachev's Cobra** maneuver demonstrated by the Su-27 (without TV) may be considered as a unit operation in this category. TV is now being added to the Su-27 (see also Fig. II-7).

ple, off-the-shelf, centrifugal compressors feeding sub- or supersonic cold-jet nozzles with cold, compressed-air, for STOL, low-IR-signature, vectored propuslion. While limited to specific applications, such [cold] vectored aircraft may be applicable as STOL vehicles, or used as, say, low-signature, vectored, PST, cruise-missiles, equipped with simple, (non-PST), semi-recessed, s-shaped, high-aspect-ratio inlets.

3. THE FUNDAMENTAL DEFINITIONS

3.1 Pure Vectored Aircraft

Vectored aircraft may be divided into those that are "pure" or "partial".

In pure thrust vectoring (Fig. 1) the flight-control forces generated by the conventional, aerodynamically-affected, external control surfaces of the aircraft, are replaced by the stronger, internal, thrust forces of the jet engine(s). These forces may be simultaneously, or separately, directed in all directions, i.e., in the yaw, pitch, roll, thrust-reversal, and forward thrust coordinates of the aircraft.

Since engine forces (for post-stall-tailored inlets), are less dependent on the external-

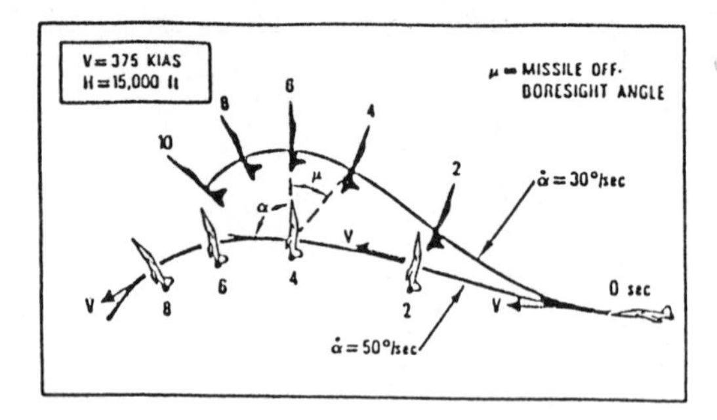

Fig. 5. **The TWIFE Yardstick and the AoA Rate Of Change (Tamrat, Northrop, 185).**

Using similar modelling as those employed in Fig. 4, Tamrat estimates the relative combat advantage of one vectored fighter over another with AoA rate difference of 20 deg./sec. The more agile fighter has a TWIFE of about 2 sec. This value was increased to 10 sec., when the superior fighter was engaged against an adversary having AoA rate capability of only 10 deg./sec. Again, PSM/RaNPAS maneuvers may be considered instead.

flow, than the forces generated by the aerodynamic control surfaces, the flight-control forces of Pure Vectored Aircraft (PVA), remain effective even beyond the maximum-lift Angle-of-Attack (AoA), i.e., PVA are fully controllable even in the domain of Post-Stall Technology (PST), as defined in Fig. 2. Thus, vectored flight provides the highest payoffs at the weakest domains of aerodynamically-controlled aircraft (i.e., at high-alpha–beta values, at very low (or zero) speeds, high altitude, high-rate spins, very-short runways, and during all PST, Rapid Nose-Pointing-and-Shooting (RaNPAS) maneuvers).

Consequently, no rudders, ailerons, flaps, elevators and flaperons are required in pure vectored flight, and even the vertical stabilizers may become redundant. Thus, by employig FBW, FBL and Integrated Flight/Propulsion Control (IFPC) systems, vectored aircraft may resemble "flying wings", or variously-shaped "body lift objects" which can provide Pure Sideslip Maneuvers (PSM), or even PSM/RaNPAS maneuvers (cf. Figs. 1 and 3 and Lecture II).

In pure-vectored-takeoff-methodology one first turns the jets upward, and rotates (nose liftup) the aircraft at a much lower speed (i.e., using a much shorter runway), than with conventional-technology aircraft. Then, under the automatic guidance of IFPC, the jets are turned down, adding direct lift and supercirculation to proper wing sections (cf. Fig. II-5). Approach and landing methodologies of PVA (Fig. II-6) also afford the use of very short runways. Thus, STOL, V/STOL, STOVL and VTOL performance become the "natural" domains of PVA (cf. Appendix B).

3.2 Thrust Vectored Pure Sideslip Maneuvers (PSM)

Pure sideslip maneuvers are defined as sidewise translations with unchanged nose pointing and without tilting the aircraft sideways (i.e. no banking). When combined

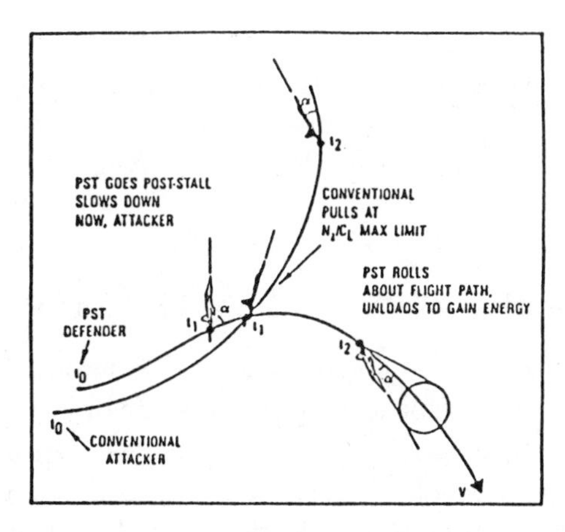

Fig. 6. **A Simplified Defensive, PST, Sub-Step Maneuver (after Tamrat of Northrop, 1988, Ref. 185).**

In future aerial combat, according to Tamrat, pointing the nose/weapon of the aircraft at the adversary first will be required to win, since pointing first means having the first opportunity to shoot (cf., e.g., Figs. 4 and 5). Thus, according to Tamrat, the point-and-shoot capability requirement, results from the introduciton of the all-aspect missile, which, if pointed at the adversary, can be launched from any aspect, including the head-on encounter.

At slower speeds, according to Tamrat, nose pointing capability of future fighters is increased by utilizing their capability to fly at large post-stall angles of attack (up to 90 degrees). In turn, post-stall maneuvering capability is the SOLE RESULT of the new, potential availability of powerplant and controls technologies for thrust vectoring. Consequently, Tamrat concludes, thrust vectoring is the basic requirement for all future fighter aircraft.

An important use of various vectored-PST maneuvers is during the defensive portion of air combat. This schematic figure shows an over-simplified situation in which the vectored fighter is initially above and in front of the conventional adversary. At the initial position the vectored fighter starts a vertical pitch-up vectoring to reach AoA up to 90 degrees. This PST-DECELERATION forces the adversary to overshoot.

When this is done, the adversary will be in front and high vs. the vectored fighter (role exchange). After delivering its weapon, the vectored fighter may roll about its flight path, reduce AoA, and accelerate to regain energy. For a variant to this over-simplified sub-step maneuver see Fig. 3 and Lecture II, including a PSM/RaNPAS maneuver.

with RaNPAS, it would be called PSM/RaNPAS-maneuvers (cf. Fig. 3). Such important maneuvers (see below) become feasible with the introduction of PVA technology. There are a number of basic propulsion/flight-control rules and some technical solutions to this kind of pure vectored flight (cf. parag. II-3 for detail and definitions of preliminary-design criteria, control rules and associated dimensionless numbers).

The simplest pure sideslip flight (say, to the left hand side of the aircraft), may be carried out by deflecting the right-engine jet to the yaw angle marked in Fig. 1 (until it coincides with the Center of Pressure in the Y-direction – CP_Y), while simultaneously reducing the thrust of the left engine to the same level as that left over for the forward thrust of the right engine [cf. Fig. 1]. This combined operation (which must be contolled by a proper IFPC system programming), keeps the nose-pointing unchanged, while pushing the aircraft to the left without banking. In fact, pure sideslips are feasible with and without canard-configured PVA. Most important, rapid nose-pointing may be combined with PSM to provide endless PSM/RaNPAS maneuvers.

3.2.1 Combat Applications of PSM or PSM/PST-RaNPAS

Combat applications of pure sideslip maneuvers are numerous. They include air-to-ground maneuvers (Fig. 15), escape maneuvers, confusing-the-adversary maneuvers, and some highly promising alternatives to pitch-only, PST maneuvers. The last category may be demonstrated by Fig. 3: Instead of the (energy-degrading) pitch-PST maneuver depicted, one may consider a 90-degrees-roll, and roll-stop followed by RaNPAS yaw/pure sideslip maneuver, which may be prolonged and/or repeated for other targets but without significant speed/energy degredation. (Pure, tailess, vectored aircraft, of the type shown in Fig. 1, do not incorporate vertical stabilizer(s). Moreover, their total sideslip drag is relatively low. Hence, during pure, 90-degrees RaNPAS-sideslip, PSM-maneuvers, the instantaneous rate of energy degradation should be substantially lower than that encountered during pitch-only, RaNPAS-PST-

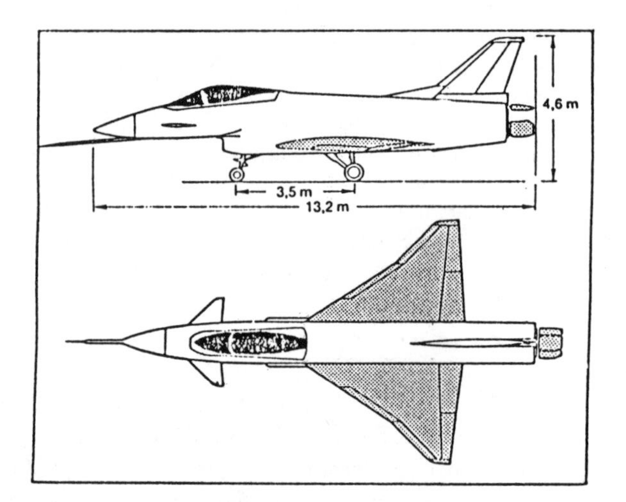

Fig. 7. **The PST, Thrust-Vectored X-31 Design Concepts.**

This figure shows the yaw-pitch-thrust-pedals (beyond-nozzle-exit, free-jet deflectors), of the experimental plane X-31. Expected to be flight-tested in 1990, this plane is to fly at up to 70 deg. AoA. Designed for quick nose-pointing-and-shooting capabilities, such a propulsion-controlled aircraft may significantly increase close-combat effectiveness. Consequently, by using such design concepts, and especially the *Herbst PST concepts* (188), one may help to increase the probability to win, or the kill ratio of PST fighter aircraft, etc. (cf. Table 1).

While the X-31 is highly-promising and advanced, such a thrust-vectoring design methodology may suffer from certain disadvantages (see Lecture VI for more detail).

maneuvers.) Thus, the use of this option may become significant in close-combat, target-rich environments, in which minimum degradation of energy becomes imperative. A number of new RaNPAS-PST-PSM-maneuvers can be worked out on the basis of this methodology. In fact, such maneuvers are being investigated now by this laboratory, using vectored RPVs (Lecture IV).

However, the expected stability-IFPC problems associated with such potential maneuvers should not be underestimated (179).

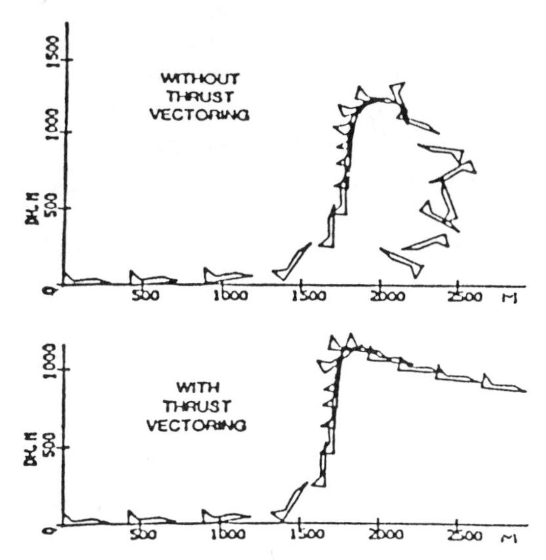

Fig. 8. **The Introduction of Vectored Fighter Aircraft Causes Such Classical Concepts as 'Spin-Danger' and 'Performance-Stall-Limit' to Become Obsolete, Thereby Demonstrating the Need for Curriculum Redefinitions in Modern Pilot and Engineering Education.**

This schematic drawing, taken from a recent report of the Central Institute of Aviation Motors in Moscow, USSR, (180), demonstrates a simple performance payoff of vectored fighter aircraft as predicted by their numerical-mathematical analysis of the flight dynamics of Post-Stall Technology (PST), simultaneous-yaw-pitch, Vectored Aircraft.

Finally, one may also stress here the expected combat potentials of this methodology for partially-vectored fighter aircraft. This topic is taken-up next.

3.3 Partially-Vectored Aircraft

Rudders, elevons, ailerons, flaps and elevators are still used in Partial Jetborne Flight (PJF). Thus, the elimination of one or more modes of the pure-vectored elements, such as the elimination of yaw-vectoring and supercirculation in the Harrier, or in the STOL F-15 demonstrator (cf. Lecture V), categorizes the aircraft as PJF. In fact, all vectored aircraft employing external thrust-vectoring devices should also be classified as partially-vectored aircraft which are only capable of PJF, i.e., maneuverability and controllability limits of these aircraft are still affected, to a degree, by the external-flow regime. Thus, this category includes a recent F-14 flight with a single, post-exist, yaw vane, as well as the X-31, and one of the proposed designs for vectored F-16 and F-18 (see Lecture VI). Beyond a limiting AoA, the interaction of the external flow with engine exhaust-jets may become prohibitive for establishing a reliable IFPC during RaNPAS-PST maneuvers (Lecture VI).

Aside, and in addition to these inherent problems, one may stress the fact (which was recently verified by flight-testing of our pure and partially-vectored RPVs), that flight/propulsion control during PJF is much more complicated and cumbersome than that feasible with pure vectored vehicles.

To start with, there are too many variables, most of which become redundant. Furthermore, any reliable IFPC system for PJF suffers from lack of reliable DATABASE for the very definitions of the relevant variables needed, their proper range, limits, and coupling effects to each other during actual flight testing. (To verify this conclusion we have established a PJF program, which compares the performance of pure vectored RPVs with that of partially-vectored ones; e.g., prototype No. 7 of this laboratory is a flying, computerized, 9-feet-long, variable-canard-configured, vectored-F-15 RPV. It is also a STOL-PST demonstrator.)

3.4 Thrust-Vectored, PST-Maneuvers

Figures 3 to 6 demonstrate some of the most-simplified, PST, or PSM, close-combat maneuvers which may be carried-out by vectored fighters (in comparison with those obtainable with conventional fighter aircraft). The conclusions extracted from these simulations are depicted and detailed therein. Other, more fundamental conclusions, are discussed next.

3.5 The Definitions of Agility, Supermaneuverability and Supercontrollability

According to McAtee (196), the quality of agility is the combination of three (measurable) tasks/abilities (Figs. 19 and 20):

1) –The ability to "outpoint" the opponent (pointing at him before he points at you). This advantage must be such that the opponent does not have the opportunity to launch his weapon before he is destroyed. Otherwise, with current launch-and-leave weapons, mutual destruction would result. **It is, therefore, the ability to point the nose/weapon at the enemy quickly, while, simultaneously, computing and locking, so as to minimize the total length of delay times associated with secure locking and obtaining the shortest/optimal missile flight path/time from aircraft release point-attitude to the moving target. This requires aircraft superagility to be well-integrated with missile's high-"g"/speed agility and initial vectoring conditions (cf. II-7.4). It also dictates the development of TV-missiles.** This aircraft ability is measured in terms of Turn Rate/Bleed Rate, as demonstrated in Figs. 19 and 20.

2) – The ability to continue maneuvering at high-turn rates over prolonged periods to

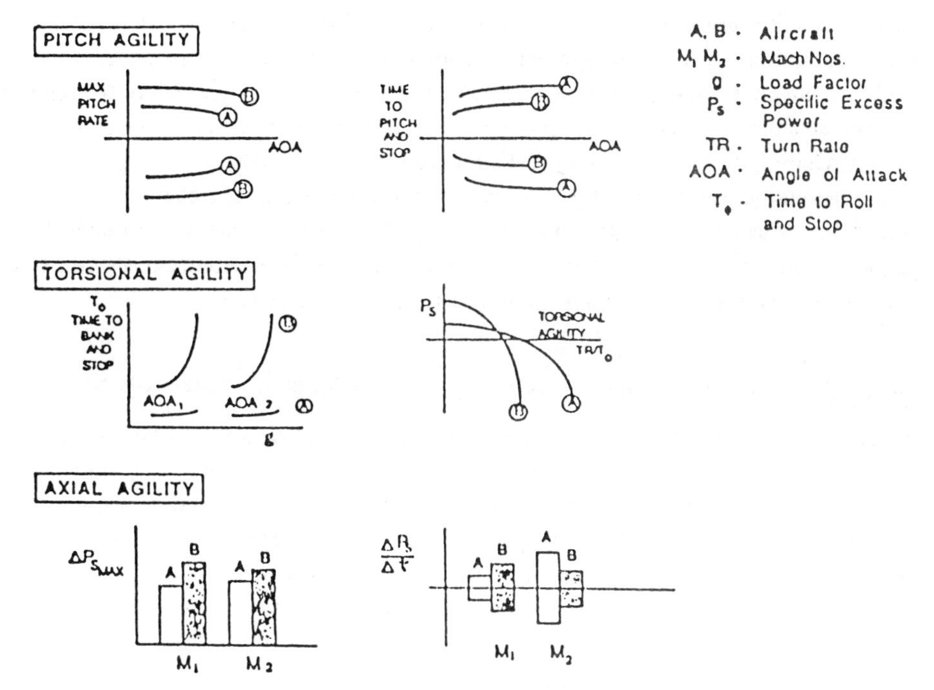

Fig. 9. **Debated Agility Measures.**

The current definitions of agility (cf. parag. 3.5), and of transient measures of merit are still being debated. This figure shows some of which have been proposed. These measures may be considered as the time-rate-of-change of the classical energy measures.

Unlike such simple analyses, Hodgkinson, et al. (184), have used complex battle simulation programs to reproduce many of the decisions and actions of air-to-air combat, and to compare aircraft with various levels of performance and weaponry. Some of their results are shown in the next figure. See also Figs. 19 and 20.

retain the potential for performing defensive maneuvers, or make multiple kills when appropriate. I.e., to defend against attacks from other aircraft, or to accomplish multiple kills if the opportunity exists, an "agile" aircraft must be able to continue maneuvering at high-turn rates over prolonged periods. This ability is measured in terms of Residual Turn Rate, as shown by the middle drawing in Fig. 20.

3) – The ability to accelerate rapidly straight ahead so as to leave a flight at will, to regain maneuvering speed when necessary, or to pursue a departing target when appropriate. This includes the ability to disengage, or escape from a battle without being destroyed in the process, as well as the acceleration necessary to "chase down" an enemy that is trying to escape. This ability is measured by acceleration/speed plots as those demonstrated in Figs. 19 and 20.

Using both his rich combat and (F-16) design experiences, as well as advanced analytical methods, McAtee concludes that these three measurable tasks/abilities are crucial for success in modern close-in combat. Thus, McAtee states, the critical design features for modern fighters are those that enable the pilot to command very high maximum turn rates over prolonged periods, and to perform a 1-g acceleration.

However, good maneuverability must be integrated with effective controllability, i.e., the ability to change states rapidly (control power), and the ability to capture and hold a desired state with precision (handling qualities) (196).

Traditionally, controllability was thought to be degraded at either of two conditions: High Mach number, or high AoA (cf. Fig. 20). However, the introduction of PST and vectored aircraft technology requires reassessment of the second condition. It also requires the introduction of new standards and MIL specs (196).

3.6 Yaw-Pitch-Roll Thrust Vectoring for Breaking the "Stall Barrier".

Pitch and yaw control requirements increase with AoA. For a given roll rate, as AoA increases, the requirements for pitch and yaw forces/moments (for non-thrust-vectoring aircraft), increase exponentially. At the same time, with conventional aerodynamic controls, the forces/moments available decrease as airspeed decreases. Thus, beyond a given limit, conventional control technology becomes obsolete. This technology limit is reached when the size and weight of the aerodynamic control surfaces needed to provide sufficient forces/moments bocome prohibitive. However, the introduction of PST and vectored aircraft technology (together denoted by McAtee as the new domain of "supercontrollability"), requires reassessment of all maneuverability and controllability concepts and requirements.

Thus, according to McAtee, new point-and-shoot weapons have reduced engagement times drastically, leaving aircraft with poor maneuverability and controllability at the mercy of those that can use their agility to kill quickly during close-in combat (cf. Fig. 2, Lecture II).

Vectored PST maneuvers, such as those shown in Figs. 3 to 6 and 16, or those discussed in Lectures II and V, may thus be defined as supermaneuvers.

There are a few dozen candidate supermaneuvers, half of which may demonstrate a real combat promise (195). In Lecture II we provide a few examples for combat payoffs during the proper use, at the proper position/timing, of yaw-pitch-roll thrust vectoring during "angles" and "energy" tactics (see below). These tactics employ supermaneuvers well beyond the current flight envelopes of conventional fighter aircraft.

3.7 A Basic Thurst-Vectored, PST-RaNPAS, Turn-Back Maneuver

Fig. 16 schematically demonstrates a basic, PST, yaw-pitch-roll thrust vectoring for a rapid, short-penetration radius, turn-back maneuver (E.g., an escape maneuver from, say, a SAM envelope, or a RaNPAS, turn-back maneuver in head-on encounters). We shall return below to this basic maneuver. In the meanwhile it should be stressed here that in-flight thrust reversal is much less effective than the nose-up, PST-maneuver shown (cf. Figs. 21).

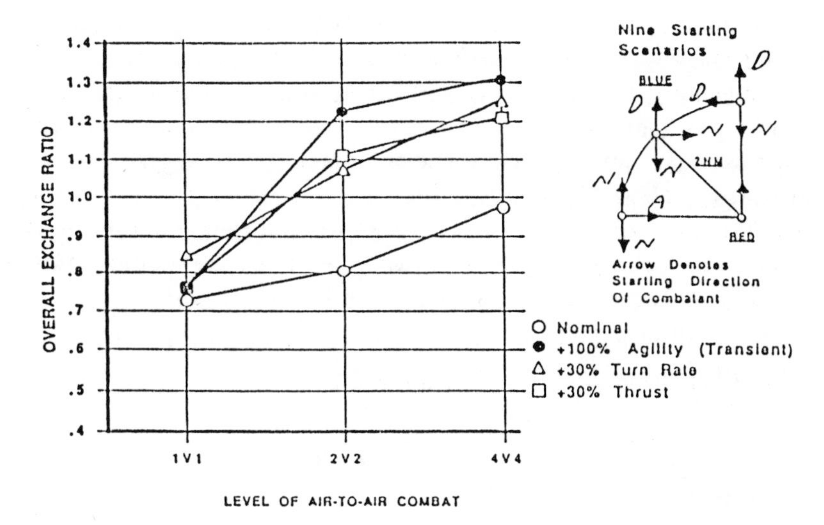

Fig. 10. **In a Target-Rich Environment, a 100% Increase In Transient Agility (in pitch, roll, and axial), Is Similar to the Operational Payoff of a 30% Increase in Conventional Agility (i.e., Enhancement in Turn Rate or Thrust).** Moreover, the operational payoff of a given enhancement, is more pronounced in a target-rich environment. Furthermore, of the three transient agilities shown in Fig. 9, torsional (roll) agility has the largest impact on combat effectiveness.

These conclusions were recently reached by Hodgkinson, Skow, Ettinger, Lynch, Laboy, Chody and Cord (184), using complex battle simulation programs. The simulated results obtained by these authors, also indicate that the exchange ratio between highly capable, matched aircraft, tends to increase as the target environment becomes richer, and, that intense, target-rich WVR (Within Visual Range) combat, is an inevitable feature of tomorrow's war.

Definitions: This figure shows combat effectiveness (i.e., overall exchange ratio, cf. the definition below), as a function of scenario size and level of 'blue' aircraft enhancement. "Nominal" aircraft in the scenarios are F-16-like in performance.

The 'red' aircraft always had nominal characteristics. The 'blue' aircraft either had nominal characteristics, or enhancement in transient and conventional agility. Scenario size was varied from 1v1, to 2v2, to 4v4, to progressively enrich the target environment.

The Exchange Ratio is defined as:

$$\frac{\text{Number of 'red' aircraft killed}}{\text{Number of 'blue' aircraft killed}}$$

Overall Exchange Ratio (OER) is defined as the aggregate from nine scenarios, each with a different starting condition (see the arrows indicated on the upper-right portion of this drawing), so as to vary the 'blue' positions from defensive to neutral to offensive. As such, OER measures combat effectiveness of a 'blue' aircraft enhancement over a range of tactical positions from defensive to offensive.

3.8 A "Nominal", Three Steps, Vectored-PST Maneuver

In assessing various vectored maneuvers we have arrived at a common denomination; PST-RaNPAS maneuvers nominally involve three-steps:

The first one is to acquire the PST AoA value (say, of 90 degrees), by employing a quick, maximum AoA rate of change, momentarily disregarding the energy level (see also below the discussion of "angle" vs. "energy" tactics). The application of this maneuver is highly risky as well as limited, for it is feasible only at the proper position/timing (cf. Fig. 3, and Lecture II). However, a less risky maneuver first involves a 90-degrees-roll, then roll-stop followed by 90-degrees RaNPAS/pure sideslip maneuver (see parags, 3.2 and 3.2.1).

During the second step, the vectored fighter may roll about the velocity vector to disengage, and/or yaw-vector its nose for starting the second part of a 180-degrees turn-back maneuver (cf. Lecture II).

The final step of such 'nominal' maneuvers involves unloading and acceleration to regain the initial speed/energy level.

3.9 Low or High Load Factors During Post-Stall Maneuvers by Vectored Aircraft: Intuiton vs. Physics

The lift coefficient as well as the effectiveness of all aerodynamic control surfaces are drastically reduced during vectored, rapid-nose-pointing, high AoA/Post-Stall maneuvers (Fig. 2). On the other hand, the thrust-generated forces and moments of a properly-designed, internal, thrust-vectoring system, remain almost unaffected by the

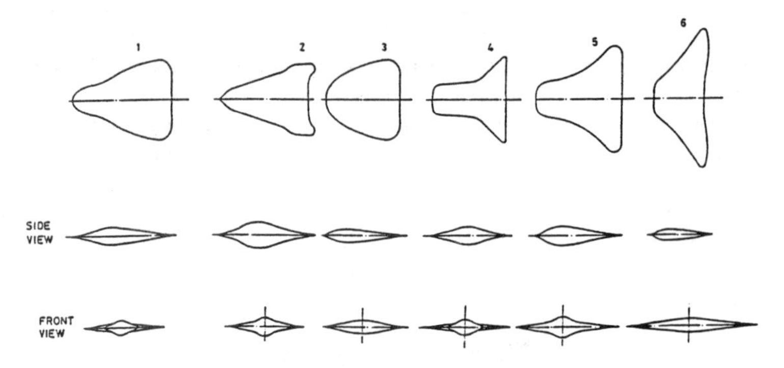

Fig. 11. **A Number of Preliminary (approximately '1-feet') Models Tested in a Subsonic Wind-Tunnel as Potential Candidates for the Final Vectored RPV Flight-Test Program.** See also Fig. IV-3.

The finally-selected configurations for the design, construction and flight-testing of the earliest vectored RPVs, were slightly influenced by the wind-tunnel results obtained for these models. However, the first (approximately 7×4 ft.) flying prototypes were canard configured, and also had different aerodynamic characteristics, proportions, swept-wing angles, etc. These modifications were primarily due to the need to incorporate optimal yaw-pitch-roll thrust vectoring systems, into the prototype final airframe design.

The optimal vectored propulsion system was first evaluated experimentally on a Mabore II, ~400 kg thrust turbojet engine, using the altitude test chamber shown in Fig. IV-6. Test conditions included pure yaw, pure pitch, pure roll, and simultaneous yaw-pitch and yaw-roll thrust vectoring. Maximum tested vectoring angles were plus-minus 20 degrees, in both the pitching and the yawing planes. The flying prototypes were simple, geometrically-scaled-down copies of the aforementioned optimal propulsion system.

external-flow conditions, i.e., thrust vectoring is most effective at the weakest domain of aerodynamically controlled aircraft.

Thus, as the angles of attack are increased beyond the maximum value, the load factor on a PST-maneuvering vectored aircraft, may not be as classically computed. In general, the load factor on a PST-maneuvering vectored aircraft depends on the specific design of the thrust vectoring system, on the time-varying directions and values of the vectored-jets deflected, on engine throttle, on the turn rate/radius, body-wing-AoA/sideslip-angle, speed, altitude, the direction of the gravitational vector, canard/elevators/flaperons deflections/loads, the time-variations in the proper drag components, etc. Moreover, if the aircraft slows down just prior to a vectored-controlled, turn-maneuver (with or without thrust-reversal), the load factor is reduced during the turn performance (cf. Figs. 3 to 6).

Since the lift coefficient falls down at high-alpha values, a properly-designed propulsion/flight control system should maintain the proper load-factor/acceleration-force according to mission and the pilot's demands, using thrust vectoring forces and moments to replace the loss in lift force and the loss in moments generated by the aerodynamic control surfaces. Furthermore, as the altitude is increased, the thrust, and, hence, the vectoring moments and forces (and, thus, the total load factors), are reduced, when other parameters remain unchanged.

Still further, one must distinguish between the different maximum *"g"*-components which a pilot can sustain for a given duration (in the positive or negative pitch plane, in the yaw plane and during head-on braking). One must also differentiate between thrust-yaw, thrust reversal, and thrust-pitch forces for yaw, pitch, thrust-reversal, or simultaneous yaw-pitch, yaw-pitch-roll, or yaw-pitch-roll/thrust-reversal maneuvers (cf. Lectures I and II).

Consequently, for vectored aircraft performing PST-maneuvers, the instantaneous, and the "time-averaged" load factors may be designed to be even lower, and shorter, than those intuitively assumed initially for similar, but unvectored, rapid-nose-pointing-and-shooting (RaNPAS) maneuvers.

Practically it means that thrust-vectored, post-stall maneuvers, can be safely employed to increase survivability and killing ratios without surpassing human and structural limitations.

It should also be stressed that high angles of attack can only be used for low-speed flight increments, for instance, during a short-time, rapid turning for pointing-and-shooting, or for a proper escaping maneuver from, say, a surface-to-air missile envelope (cf. Figs. II-2 and II-3).

Nevertheless, such maneuvers do not require so high angles of attack when proper thrust vectorization, say, during pure sideslips in the yaw plane, is used (cf. Figs II-1a and 1b and parag. II-3.2). Alternatively, higher speed maneuvers become possible with vectorization at relatively moderate AoA combined with rapid yaw vectoring at the proper timing (cf. Figs. II-2 and II-3). Restricting the vectorization method, as, for instance, has been designed for the F-15 S/MTD, may, however, increase the load factor at low supersonic flight conditions (cf. Fig. V-9).

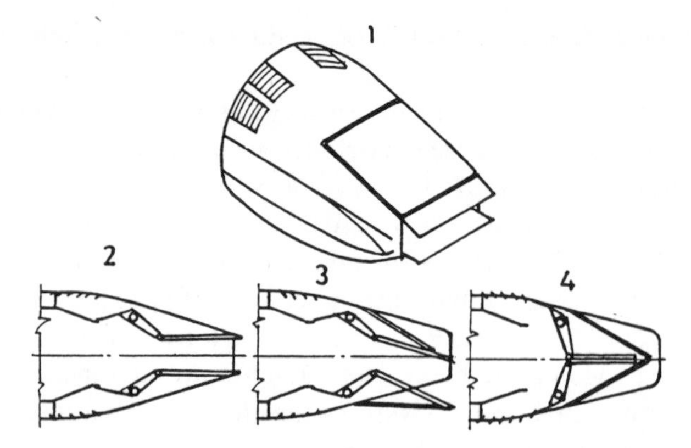

Fig. 12. **Two-dimensional/Converging-Diverging (2–D/C–D) nozzles may supply the same, or somewhat higher thrust levels than those of conventional (circular) nozzles. Moreover, the thrust level may further increase during subsonic thrust vectoring (cf. Fig. 13 below and Figs. 17 and 18 in Lecture I).**

1 — *GE/PW* design for low aspect ratio, pitch-only, pitch/thrust-reversal nozzle for an upgraded F-16 partial-vectoring program (Lecture I).(Axi-TV nozzles may also be available).

2 – The *GE/PW* nozzle during unvectored flight.

3 – The *GE/PW* nozzle during down-pitch vectoring.

4 – The *GE/PW* nozzle during full thrust reversal. The venetian-type vanes are oriented approximately 45 degrees forward. Note: During the approach phase (cf. Fig. 14), the venetian-type vanes are oriented about 135 degrees to the back, the throat remains partially open, the engine throttle is fully open, and the diverging flaps are vectored down. This type of thrust vectoring reduces the approach speed, and, following touchdown, also the landing distance (for the engine spool-up time required in conventional thrust reversing, has been saved). However, the cost, weight and complexity of this kind of thrust reversal may be prohibitive (cf. Lecture I for detail).

4. "ENERGY" VS. "ANGLES" FIGHTS

Fighter pilots of the F-4 have been among the first flying-qualities tacticians who distinguish between the well-known combat payoffs of transient agility during "energy" and "angles" fights (184).

In an "angles" fight, positional advantage is gained regardless of energy expense.

In an "energy" fight, the objective is to gain an energy advantage, and, then, at an appropriate time, to trade that energy advantage for position.

According to Hodgkinson, Skow, Ettinger, Lynch, Laboy and Chody of Eidetics International and Cord of AFWAL/FIGC, Wright Patterson AFB, USAF (184), an "angles" pilot tactician maneuvers his aircraft so that a snapshot opportunity presents itself before he himself has spent his energy. Consequently, a typical "angles" maneuver is to use altitude to sustain the time the aircraft spends close to the corner speed, so that the "angles" tactician can maintain a small turn radius and a high turn rate, thus achieving a snapshot.

On the other hand, a typical "energy" maneuver is a rolling scissors in which relative energy is sustained while a number of firing and disengagement opportunities present themselves.

Following these notes the aforementioned authors introduce a number of important criteria and definitions for transient, torsional, axial and pitch agility and for combat payoffs of good flying qualities.

5. PITCH, TORSIONAL and AXIAL AGILITY CONCEPTS

A typical maneuver using transient agility might use high pitch rates beyond maximum lift, a missile launch, and a rapid unload and acceleration. Thus, the time scales of engagements like these have been considerably compressed. Consequently, measures of agility or transient dynamics, which recognize an aircraft's ability to "SUPERMANEUVER" beyond current flight envelopes for very short times, are required.

Pitch agility, for instance, may be defined as time to pitch up to a designed AoA and time to pitch down. There has, traditionally, been no explicit requirement for time-to-pitch in the flying qualities requirements documents (184).

Torsional agility refers to the capability of an aircraft to change the plane of its maneuver. Though this chiefly involves a rollng maneuver, the necessity to roll more nearly about the wind axis at elevated AoA, or to perform a loaded roll, has led to proposals to include in the definition of torsional agility times-to-bank and stop, and turn rate divided by time to bank and stop. The latter expression is an attempt to augment a traditional agility measure with a time function so that it would have the appearance of a second derivative term.

Axial agility is the capability of an aircraft to change rapidly its thrust and drag. Here the main contributors are engine thrust transient response and braking deployment time. However, current flying qualities specifications contain no requirement on engine thrust responses (184). This brings us to an important difference between the time lag in thrust reversing and that of yaw or yaw-pitch thrust vectoring.

6. TIME-LAGS IN THRUST VECTORING

The current engine specification – MIL-E-5007 – now contains only a requirement for engine response to reach 95% of commanded thrust in five seconds. Since the aforementioned definitions of agility strongly depend on this requirement, especially for vectored aircraft, it is imperative to re-examine this specification as well as the near-term, technology limits in this field.

To start with one must stress the following points:

1) – To consider only the nominal time-lag for idle-to-maximum power transient may be highly misleading, for such engine time-lags increase with altitude, inlet distortion, nozzle position, engine life, etc.

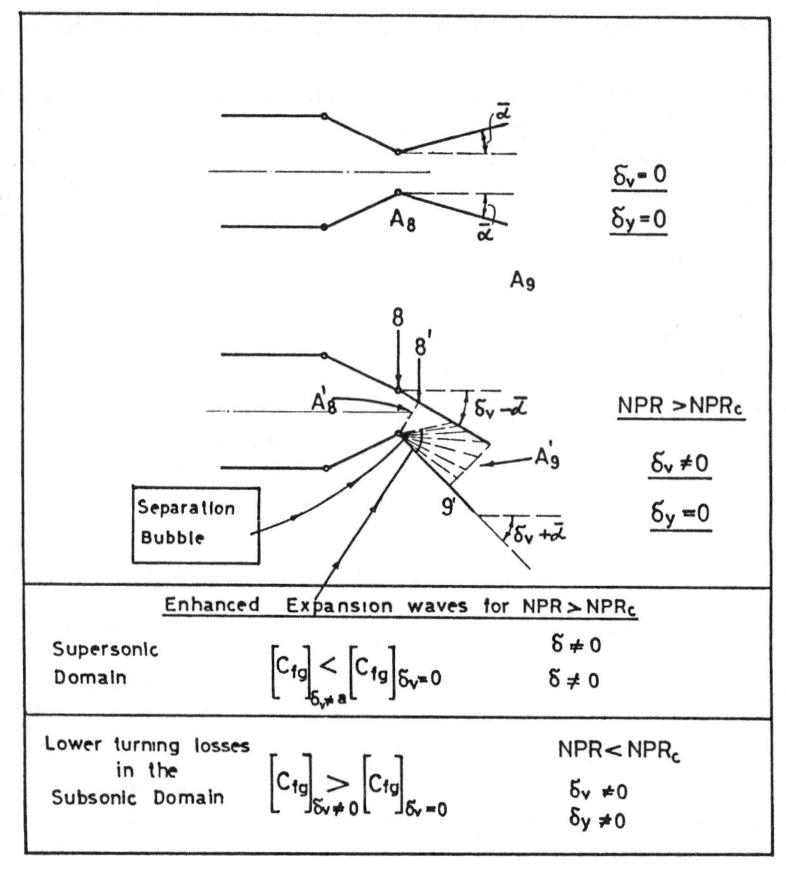

Fig. 13. **Thrust Vectoring May Increase the Subsonic Thrust Coefficient Beyond That of the Unvectored Values.**

During thrust vectoring, at a given value of NPR, one must keep the values of A_8' and A_9' as a function of $(\cos \delta v) \times (\cos \delta y)$ (cf. Lecture I, eqs. 11 to 13).

For NPR > NPR(critical), i.e., in the supersonic domain of the nozzle flow field, the expansion bubble just downstream of the lower throat corner (in the specific case depicted), lowers the value of the "effective" NPR. Consequently, the thrust coefficient during supersonic vectoring may be lower than that for unvectored operation.

Nevertheless, for NPR < NPR(critical), the vectored thrust coefficient may be higher than that for unvectored operation. This may result from the higher payoffs of the "straight" flow passing the upper corner, than the (subsonic) losses associated with the lower corner. Indeed, some of the experimental results reported in Lecture I, as well as some of those obtained by this laboratory, substantiate these effects.

It should also be noted that during pitch vectorization the proper throat cross-sectional area moves from 8 to 8′, and the proper exit area, from 9 to 9′. Similar transformations (in the plane perpendicular to the drawing page) take place during yaw vectoring (Lecture I). These effects require proper variations/adjustments in all flap positions during pitch, yaw, or simultaneous yaw-pitch vectoring (Lecture I, cf. the proper control rules for yaw-pitch thrust vectoring as given by eqs. 1 to 14).

Conclusion: Pitch-only, or Yaw-Pitch Vectoring Nozzles May Supply the Airframer With Approximately the Same, and in Some Cases Even Higher Thrust Levels Than Those Available From Conventional (Circular, Unvectored) Engine Exhaust Nozzles.

2) – In design optimization an airframe manufacturer can trade off engine life for improved axial, pitch and torsional agility capabilities.
3) – Yaw, pitch, or simultaneous yaw-pitch-roll thrust vectoring responses are faster and more reliable than those for thrust reversing. This conclusion must be added to, and combined with another conclusion discussed in this volume, namely, that the airframe designer can extract significantly higher combat-agility payoffs from yaw-pitch vectoring, than from, say, pitch/thrust-reversing.

7. THE OVERALL COMBAT PICTURE

Unfortunately, the aforementioned conclusions may form only a slice through the overall picture of many other practical factors such as: scenario size (e.g., number of friendly aircraft vs. number of enemy aircraft, their initial positions, attitude, velocities, left-over weapons, etc.), low observability until WVR (Within Visual Range), the technology levels of helmet-sight-aiming and EW systems on both sides, combat-management methodology/systems used, the varying technology limits of all-aspect missiles, the need to evade a missile launched previously by a now-destroyed opponent, etc.

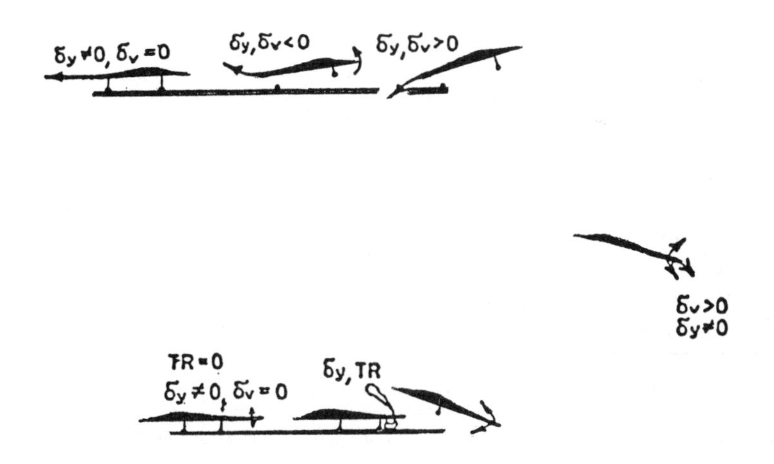

Fig. 14. **Takeoff, Approach and Landing of Vectored Aircraft.**
Thrust vectoring during takeoff may be performed by first turning the jets upward, and, following nose lift-up (rotation), turning the jets down using IFPC-FBW-yaw-vectoring system eliminates the conventional need for vertical stabilizers. PVA lift is enhanced by the "direct lift" of the engines, as well as by circulation (cf. Lecture V).
(below) Approach and landing. For detail see Fig. 12.
Good ground handling requires up to 45 degrees yaw thrust vectoring.

8. THRUST VECTORING AND POST-STALL CAPABILITY IN AIR COMBAT

Philippe Costes, of ONERA, has recently published the results of numerical simulations of one-to-one, close-range, air-gunnery combat maneuvers, from two initial conditions: High and low altitudes (182). His results show that thrust vectoring considerably improves the aircraft maximum turn rate. Stating that such supermaneuverability is a keypoint to the design of advanced combat aircraft, Costes concludes that thrust vectoring is especially effective in close-range combat, below the corner speed, and that its use, during post-stall maneuvers, allows a much quicker nose pointing, and, consequently, first-shootig opportunity. For instance, the probability of victory by vectored aircraft at high-altitude air combat is 0.78 vs. 0.22 for similar, but unvectored aircraft. This gives the vectored aircraft a Kill-Ratio of 3.55 (cf. Table 1). During close-range dogfight, the probability of victory by vectored aircraft becomes 0.89, while that of the conventional combat aircraft is only 0.11. This gives the vectored aircraft a Kill-Ratio of 8.1, under these conditons.

Table 1. Kill-Ratios: Vectored Vs. Conventional Aircraft
(After Costes, [182])

Type of aircraft	Altitude *(m)* Speed *(M)*	Probability of Victory	Kill Ratio $Pv(i)/Pv(j)$
Model A: Conventional fighter aircraft	108000 $M=0.9$	$Pv(A)=0.522$ $Pv(A)=0.477$	1.09
Model D: Vectored fighter aircraft (with PST capability)	10800 $M=0.9$	$Pv(D)=0.78$ $Pv(A)=0.22$	3.55
Model D: Vectored fighter aircraft (with PST capability at close-range dogfight)	1500 $M=0.5$	$Pv(D)=0.89$ $Pv(A)=0.11$	8.1

It should be noted that these simulations are limited to zero sideslip flight conditions. Moreover, these models do not incorporate yaw-vectoring simulation capability. Hence, these results should be viewed as the minimal baseline results for pitch-only, partially-vectored aircraft.

9. MINIMUM TIME TURNS USING VECTORED THRUST

Schneider and Watt, of the US Air Force Institute of Technology at the Wright–Patterson AFB, have recently published an interesting study (218) about optimal con-

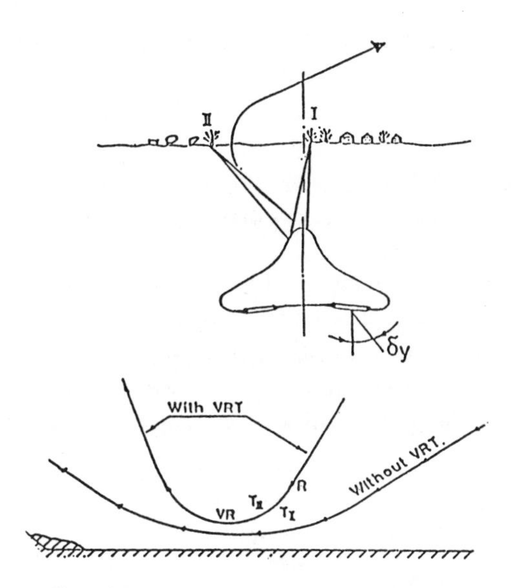

Fig. 15. **The Simultaneous Use of Yaw Vectoring** *-T-* (or "Targeting") with Pitch/Thrust-Reversal *-(VR)-* (With and Without PST, or PSM), May revolutionize Air-to-Ground and Air-to-Air Multiple-Target Performance. It should also be stressed that *R* is a non-essential option (cf. Fig. 3 and Lecture II).

trols and trajectories which minimize the time to turn for high performance aircraft with thrust vectoring capability.

No constrains were placed on the angles through which the thrust was vectored in order to determine how much range of thrust vectoring would be exploited if it were available. The determined controls and trajectories are then compared by the authors against other methods of turning in minimum time, to determine the effects and advantages of thrust vectoring.

The results obtained by Schneider and Watt indicate that the use of vectored thrust to supplement the aircraft's lift, by directing the thrust into the turn, can substantially reduce turning times and increase in-flight maneuverability. **The greater the velocity at which the turn is initiated, the larger the vectoring capability becomes, and the greater the reduction in turning time.**

10. VECTORED AIRCRAFT AND LOW OBSERVABILITY

To what extent should low observability affect the design of vectored propulsion/airframe, and vice versa? Should low detectability be the over-riding principle in the design of advanced, vectored propulsion system? Obviously, wing-integrated, high-aspect-ratio, yaw-pitch-roll, vectoring nozzles/propulsion systems, help reduce RCS/IR/Optical signatures, especially in the absence of vertical stabilizers (cf. Fig. 1).

Designing engine inlets and nozzles to begin and to end, respectively, on the upper wing skin, can reduce detectability from below.

Reduction of RCS by geometrical control of the wing/body shape may follow two methodologies:

1) – Aircraft external geometry is a combination of many plane surfaces.

2) – Aircraft external geometry is dominated by constantly-changing radii of curvature.

Thus, a design feature of the stealth F-117A fighter aircraft, unlike that of the stealth bomber B-2, is the absence of curves. Composed of many plane surfaces (cf. Fig. VI-1), this plane drastically reduces the number of main-lobe reflections that a radar will be able to seek, and track, during a given time period. Since the aircraft vibrates as it flies, the probability of an enemy radar to pick-up an echo from one of these plane surfaces,

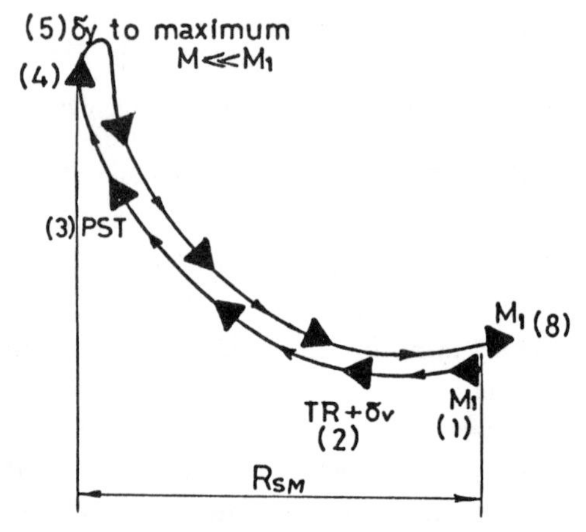

Fig. 16. **A Schematic Representation of a Rapid, 180-Degrees PST/PSM, Turn-Back Maneuver, Using Simultaneous Yaw-Pitch-Roll Thrust Vectoring, Which Significantly Reduces Turn-Time Without Loss of Energy (head-on encounters, or an escape supermaneuver from a SAM envelope).**

States:

(1) – Maneuver initiation at speed M_1.

(2) – Vectored, PST, nose-liftup against the gravitational vector (with or without the option of thrust reversal).

(3) – Aircraft is fully-controlled by vectored yaw-pitch jet-moments/forces in the post-stall domain (Fig. 2).

(4) – Yaw vectoring, at low-speed, rapidly turns the aircraft nose down for the beginning of a rapid acceleration downward (with the addition of the acceleration due to gravity). The timing of this yaw (or pitch) vectoring is determined by the corresponding maximum load-component values allowed for pilot and aircraft. The entire maneuver is carried out at full engine throttle. **Note:** The vectored fighter aircraft is momentarily at a "sitting duck" situation. However, the conventional fighter still has its nose pointing away from the vectored fighter. It should be stressed, from the very beginning, that this state is a highly risky one in target-rich environments. Consequently, it may be considered only for 1v1, close-combat encounters, or for a short radius/rapid escape maneuver from, say, a SAM envelope.

(5) — Back to the initial speed/energy level, while saving about 4 seconds in comparison with a conventional fighter aircraft (see Fig. 17).

on two or more successive sweeps, are very low. Consequently, the probability of detection, or the "averaged" RCS of this stealth aircraft is very low.

On the other hand, the B-2 shape (cf. Fig. VI-1) is constantly curved, so that it scatters the electromagnetic waves over different spatial directions. Thus, in principle, each point on the B-2 surface reflects a low-intensity electromagnetic radiation into a different spatial direction. It should be stressed, however, that RCS depends on the illuminating radar-wave length, and on the detector vs. aircraft-attitude angle, as well as on numerous other factors (see below).

Adding to this design philosophy the elimination of vertical stabilizers, straight-line wing-trailing-edges, straight-cut inlet lips, etc., and the absence of rudders, or 90-degrees-corners [cf., e.g., the B-2 all-wing shape], as well as the inclusion of burried engines, weapons and payloads, semi-burried, upper-wing, *s*-shaped engine inlets (which also provide engine protection against bird ingestion), and fully-recessed, *2D* engine nozzles (cf. e.g., Fig. VI.1), further reduces RCS/IR/optical signatures. In fact, the *2D* nozzle's stealth benefits are greater than has been widely supposed. The *2D* nozzle also affords the use of radar-absorbing structures that cannot be incorporated when the engine has a circular, axisymmetric nozzle.

Then, one may construct the aircraft primarily of RAM-composites, and especially of carbon-based composites (cf. e.g., the RCC properties as given in Appendix B, Fig. 1), while eliminating material and surface discontinuities, improperly-designed antennas, etc., and including special surface smoothness characteristics and skin treatments, as well as highly-integrated active-passive-*EW* systems, adaptive management of power, the use of internal, very-low-sidelobe antennas, constant variations in frequency and waveform, and such internal, saw-type, wing-leading-edge-structures as in the SR-71.

Thus, by employing such an over-riding design philosophy, sometimes in contradiction with classical aerodynamic principles, one may further reduce detectibility.

C^3I systems of stealthy aircraft also bring more integration of active and passive *EW* systems across a broad frequency spectrum as well as of radio-frequency, electro-optical, infrared and optical sensor systems. Battle management, flight/propulsion control and intelligence gathering must also be well-integrated, using non-radiating systems and a Common Integrating Processor – CIP.

10.1 RCS/IR Databases for Vectored Aircraft

Three different methods are being used to evaluate the RCS and IR signatures of a flying platform:

1) – Actual measurements from various viewing angles – a highly costly and time-consuming method, especially for a flying RPV, or for a prototype. (Examples of the temperature profiles inside hot jets emerging from high-aspect-ratio, two-dimensional, engine exhaust nozzles, are given in Figs III-8 and 9. Note also, that even the mixed, relatively low-temperature jets leaving the (ATB) B-2 propulsion system,

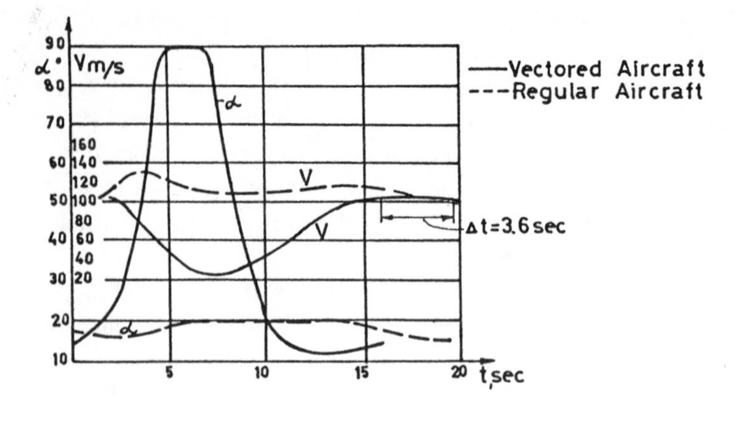

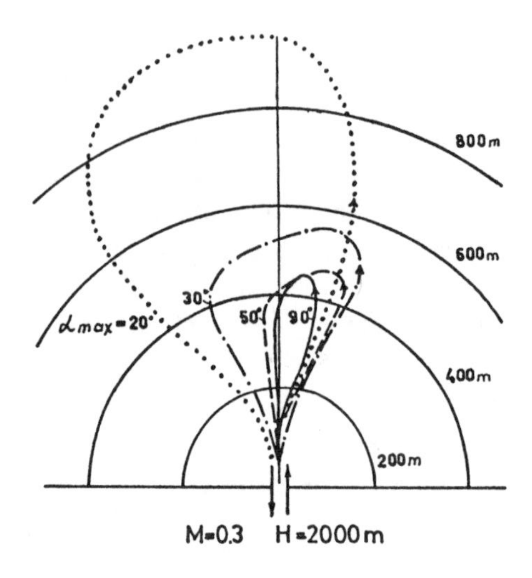

Fig. 17. **The Herbst PST Variations of AoA and Speed of the Vectored Fighter Aircraft Shown in Fig. 16, in Comparison With Those for a Conventional Fighter (upper drawing).**

About 3.6 seconds have been saved. During the second leg of this supermaneuver, the vectored fighter aircraft can acquire the conventional fighter in the missile off-boresight angle, while the reverse aquisition, by the conventional fighter, is impossible.

(lower drawing) – The penetration radii for various AoA maneuvers as those depicted in Fig. 16 (after Herbst, 154).

are still detectable (but from a much-shorter range), by heat seeking sensors.)

2) – The measurements of static, scaled-down models – a method which may introduce errors, but is much less expensive, or time-consuming. When scaled-down models are used for static or dynamic RCS measurements, the frequency of the illuminating source must be increased to correspond to the scaled-down-ratio of the model.

3) – Computational modeling – a method suffering from similar problems as those associated with scaled-down modeling, plus the difficult task of exact modelling such

parameters as curved, or multi-flat shapes, internal RAM/structural systems or engines/inlets/nozzles/jets, as well as material and surface discontinuities, engine inlet lips (cf. Chapter VI), engine fan/compressor/turbine blades, weapons, payloads, vertical stabilizers, radoms, pilot canopies, aircraft deflected flaps, elevators, rudders, etc., including those contributed by passive and active antennas and various *EW* systems.

Obviously, there is no point to reducing aircraft RCS if it is allowed to radiate (say, by using improperly-designed, active, *EW* systems), or if it carries high-RCS weapons and antennas.

10.2 Stealth RPVs

Currently, there are many low-flying, long-loitering, unvectored, *EW*-RPVs. Obviously, whatever can be well-done by a propeller-driven RPV, or by (unvectored) jet-

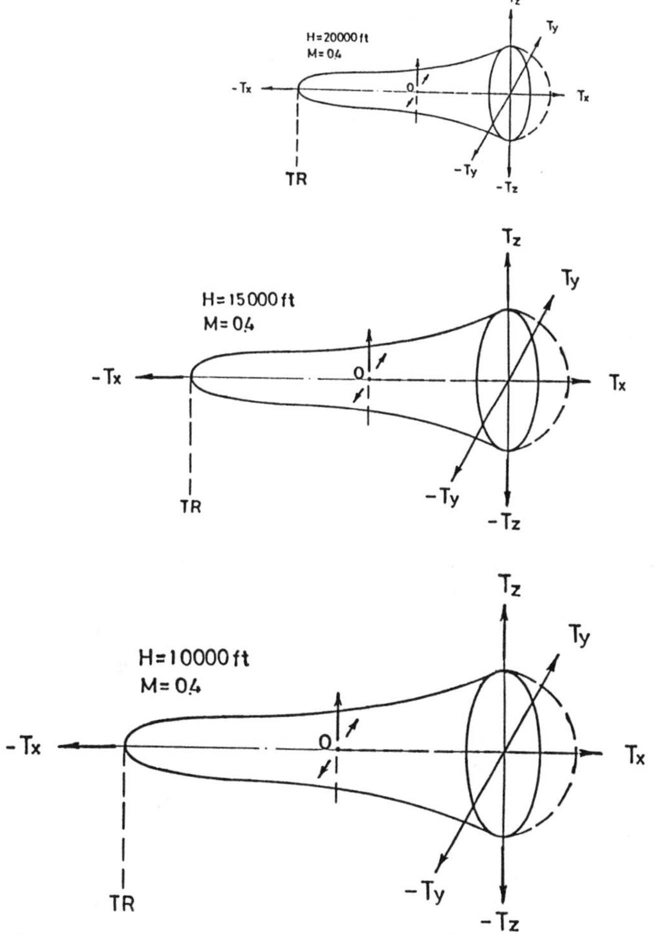

Fig. 18. **Expanded Performance Envelopes for Vectored Fighter Aircraft.**

propelled RPV, does not require the extra capability of a vectored RPV. This leaves out only a specific family of missions which require, at least in the final mission-targeting phase, a highly-maneuverable, vectored RPV, including the R&D-type of vectored RPVs discussed in Lecture IV. Another basic requirement may be VTOL/stealth capability. The last topic is intended for Volume II.

11. THE TIME GAP BETWEEN NEW PROPULSION SYSTEMS AND NEW C³I SYSTEMS

There is a time-difference between the pace of evolution, and maturity, of propulsion systems, and that of integrated C^3I systems. While the former changes about every ten to twelve years, it takes the latter only five to six years to mature into a new technology. This means that the designers of advanced airframe/propulsion systems can complete the integration process of their system with advanced C^3I systems, only during the second half of the "evolution leg" of each new generation of advanced platforms/propulsion systems. This constitutes a practical reason for the need to address the design problems of IFPC and low-signature systems only later, in a second volume, or in a second edition to this volume. It also means that flight testing of vectored RPVs may accelerate the build-up process of IFPC data bases.

12. WELL-INTEGRATED LABORATORY/FLIGHT-TESTING METHODOLOGIES

The evolution of vectored-propulsion/vectored-aircraft requires the development of a well-integrated laboratory and vectored-RPV flight-testing methodology for data extraction. Such a methodology is currently being developed by this laboratory. It constitutes three, highly-integrated, feed-back phases:

1) – New ideas, or design modifications emerging from flight testing, are tested on a small (about 1 kg/sec airmass flow rate), highly-instrumented, thrust-vectoring-nozzle test rig (cf. Fig. IV-7), or on a PST-inlet test rig of a similar size.

2) – Pre-optimized thrust-vectoring nozzles, or inlets, emerging from phase one are scaled-up and installed on a full-scale jet engine. Thrust-vectoring engine tests are then carried out in a highly-instrumented (altitude) engine test rig. This rig provides pre-optimization metrics during unvectored baseline tests, and during pitch, yaw, yaw-pitch-roll thrust vectoring tests. All tests must be conducted with a proper engine inlet; e.g., a standard bellmouth, or a Stealth/PST-tailored inlet. Thus, a complete Stealth/PST-vectoring propulsion system must include thrust-vectoring nozzle well-integrated with the engine/wing/inlet/IFPC systems, and tested under various combinations of NPR, altitude, air-speed, AoA, sideslip, and vectoring angles (cf. Fig. IV-6).

3) – Pre-optimized nozzles emerging from phase two are scaled-down back to the

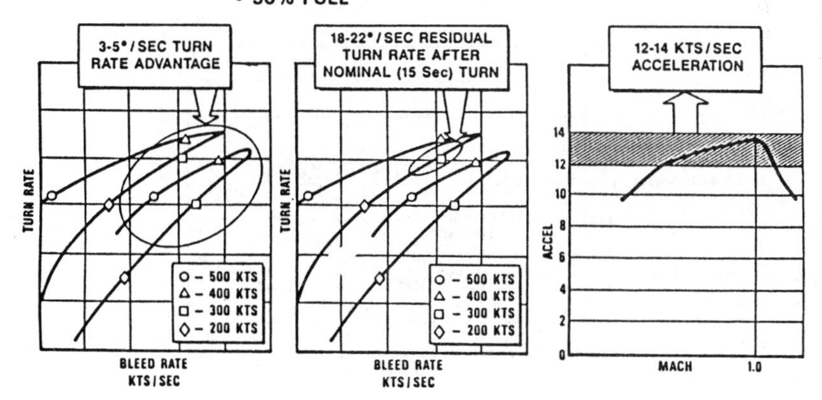

Fig. 19. **The McAtee Dynamic-Speed-Turn (DST) plots are Useful to Evaluate Fighter Aircraft Maneuverability.**

The combat tasks of generating higher turn rates (left figure), retaining residual maneuverability (middle figure), and accelerating rapidly (figure at right), are relevant to the definition of agility requirements (cf. parag. 3.5). However, good maneuverability must be integrated with effective controllability, i.e., the ability to change states rapidly (control power), and the ability to capture and hold a desired state with precision (handling qualities) (196).

Traditionally, controllability was thought to be degraded at either of two conditions: High Mach number, or high AoA (cf. Fig. 20). However, the introduction of PST, and vectored-aircraft technology, requires reassessment of the second condition. It also requires the introduction of **new standards** and MIL specs for vectored engines, nozzles, inlets, agility metrics, and IFPC.

phase-one size, and installed on the proper powerplant of a vectored-RPV-prototype of a vectored aircraft equipped with PST-inlets.

This phase includes flight-testing of thrust-vectored RPVs [which, in turn, have been designed "around" the newly-developed, yaw-pitch-roll, propulsion systems — cf. Figs. 1 and 11]. Alternatively, they include flight testing of, say, 1/7-th scaled-down, highly-instrumented, radio-controlled, vectored RPV-models of existing fighter aircraft [e.g., a vectored F-15, or a vectored F-16, which are both being considered by the USAF for various vectored-propulsion upgradig programs].

Using specially-developed, onboard-computers for vectored-propulsion/airframe data extraction/recording, the aforementioned metrics are extracted during well-defined thrust-vectored maneuvers, including PST-maneuvers. Performance-metrics of the unvectored aircraft models are employed as Baseline-1 (for a comparative study of the payoffs obtainable with various vectored propulsion designs).

Secondly, pitch-only metrics are established as Baseline-2. Thus, using, for instance, vectored F-15 and vectored F-16 prototypes, Baseline-2 may constitute the metrics of the current *GE/PW*-2D-CD "pitch-only" nozzle technology. (Such nozzle designs are now installed on the F-15 S/MTD, cf. Lecture V.)

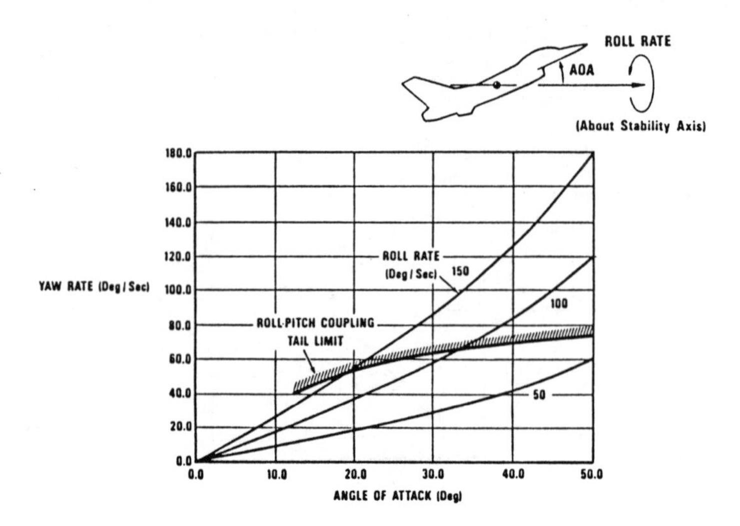

Fig. 20. **Pitch and Yaw Control Requirements Increase with AoA.**

For a given roll rate, as AoA increases, the requirements for pitch and yaw forces/moments (for non-thrust-vectoring aircraft), increase exponentially. At the same time, with conventional aerodynamic controls, the forces/moments available decrease as airspeed decreases. Thus, beyond a given limit, conventional-control technology becomes obsolete. This technology limit is reached, when the size and weight of the aerodynamic control surfaces needed to provide sufficient forces/moments, become prohibitive. (After McAtee of GD, 196). However, the introduction of PST and vectored aircraft technology (together denoted by McAtee as the new domain of "supercontrollability"), requires reassessment of all maneuverability and controllability concepts and requirements.

Thus, according to McAtee, new point-and-shoot weapons have reduced engagement times drastically, leaving aircraft with poor maneuverability and controllability, at the mercy of those that can use their agility to point-and-shoot quickly during close-in combat (cf. Fig. 2, Lecture II).

Finally, simultaneous yaw-pitch-roll metrics are established, and compared with Baseline-2 metrics, or with ETV metrics. Employing our newly-developed data extraction techniques for thrust vectoring, the next optimization step sometimes includes a return to phase one, and repetition of the whole cycle (see also the figure depicted in the Preface, and Lecture IV).

13. NEW EDUCATIONAL SHIFTS

The USAF Test Pilot School has recently revised its curriculum to place more emphasis on highly-integrated military aircraft entering the near-term active inventory. Consequently, the classical flying-quality courses that had dominated the traditional education of pilots, engineers and navigators, have now been significantly compressed. Newly designed courses which integrate the various aspects of system engineering with flying qualities and Command, Control, Communication and Intelligence (C^3I) systems, have been introduced instead. This change is probably the

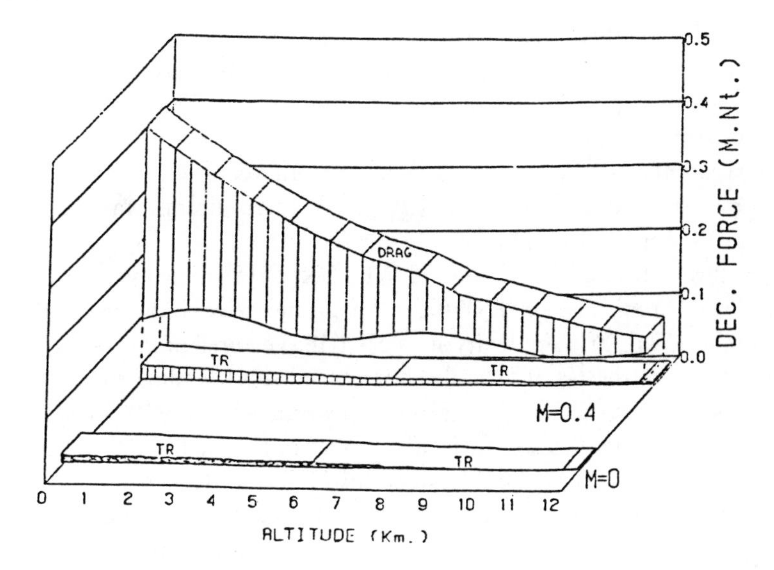

Fig. 21. **Flat Plate Drag Vs. Reversed Thrust Variations with Altitude and Speed: PST breaking is more effective than*TR*.**

first important milestone in the educational shifts that are expected to take place in all academic institutions whose curriculum deals with advanced aviation.

14. CONCLUDING REMARKS

– The availability of PST vectored fighters, helmet-sight-aiming systems, all-aspect missiles and the new generation of *EW* systems, require reassessment of the optimal balance between aircraft agility and effectiveness, and the agility and efectiveness of missile/helmet-sight-aiming systems (§ 3.5).

– Whatever is the aforementioned balance, high-performance fighter aircraft will gradually be based on improved thrust vectored propulsion/maneuverability/controllability (Lectures I, IV and V).

– New point-and-shoot weapons have reduced engagement times drastically, leaving aircraft with poor maneuverability and controllability at the mercy of those that can use their agility to point-and-shoot quickly during close-in combat (§ 3.5).

– Since future fighter aircraft would be thrust-vectored, and since thrust-vectoring engines would be used for enhanced maneuverability and controllability, as well as for brute-force propulsion, one must first define new propulsion concepts and new measurable "metrics", which would be employed in a realistic comparison of vectored aircraft maneuverability-controllability with that of conventional fighter aircraft (Lectures I, V).

– The evolution of this highly-integrated propulsion technology depends on the gradual replacement of partially thrust vectored aircraft by pure vectored aircraft.

– Controllability and supercontrollability of pure vectored aircraft is much safer and easier than that of partially vectored aircraft.

– Maneuverability and supermaneuverability of pure vectored aircraft is considerably superior to those obtainable with partially vectored aircraft and conventional fighter aircraft (Lectures I to VI).

– In-flight thrust reversal is not required in PST/RaNPAS maneuvers (Fig. 21). It is much less effective than nose-up, PST/thrust-vectoring deceleration, and is, therefore, adding unrequired weight, complexity and cost to the vectored propulsion system and to the aircraft. Nevertheless, it provides unmatched short landing runs.

– Pure sideslip thrust vectoring maneuvers constitute RaNPAS-PST/PSM methodology which does not dissipate as much energy as a similar, pitch-only, PST maneuver (§ 3.2). Further investigations should assess this option in detail.

– Aerodynamic Flight Control is to be gradually replaced by Thrust-Vectoring Flight Control. Pitch and yaw control requirements increase with AoA. For a given roll rate, as AoA increases, the requirements for pitch and yaw forces/moments (for non-thrust-vectoring aircraft), increase exponentially. At the same time, with conventional aerodynamic controls, the forces/moments available decrease as airspeed decreases. Thus, beyond a given limit, conventional control technology becomes obsolete. This technology limit is reached when the size and weight of the aerodynamic control surfaces needed to provide sufficient forces/moments become prohibitive.

– The introduction of PST/PSM-vectored-aircraft technology requires reassessment of all previous maneuverability and controllability concepts and powerplant definitions/requirements (§ 3.2).

– In line with the new concepts of RaNPAS-PST-PSM-Vectored-Aircraft, the aeroengine and airframe communities must search now for propulsion/airframe systems with the following concepts in mind:

(i) – A common set of engine-related agility-performance "metrics", or measurable thrust-vectored maneuverability and controllability parameters (Lectures I–III).

Such metrics may be presented as three-dimensional depictions of powerplant steady-state or dynamic responses, somewhat similar to those proposed recently to depict aircraft agility. These depictions may include specific throttle/pitch/yaw/roll transient capabilities for any practical combination of thrust-vectoring flight condition, i.e., one metrics map for each performance set of throttle, yaw-pitch-roll settings, NPR, NAR, Y, D, T/W, AoA, sideslip rate, inlet geometry, speed, altitude, maneuver mode, IFPC mode, etc.

(ii) – The cargo and civil aircraft industries may exploit some of the military methodologies of vectored propulsion and controllability, for instance, by introducing low-drag, cost-effective, STOL, pure-vectoring design methodologies and flight training. Advanced TV systems can also increase flight safety.

(iii) – In the military domain, the aforementioned definitions, examples and conclusions form only a slice through the overall picture of many other relevant factors such as: scenario size (e.g., number of friendly aircraft vs. number of enemy aircraft, their initial positions, attitude, velocities, left-over weapons, etc.), low observability until WVR (Within Visual Range), the technology levels of *EW* systems and of helmet-sight-aiming-systems on both sides, combat-management methodology/systems used, the varying technology limits of all-aspect missiles, the methodology to evade a missile launched previously by a now-destroyed opponent, fighter-strike missions, etc. (§ 3.5).

(iv) – One must also stress the importance of upgrading existing fighter aircraft having (at least) $T/W > 0.6$. High-aspect-ratio, split-type, yaw-pitch-roll, vectoring nozzles, have the potential of replacing the horizontal and vertical tails of existing fighter aircraft. Thus, by adding FBW, or FBL, to such propulsion systems, one may discard the vertical stabilizer, and the elevators of, say, the F-15 and F-16 aircraft, and replace them with "tailless", efficient, low-cost, yaw-pitch-roll vectoring nozzles (Lectures IV–VI).

(v) – Integrating the propulsion, control, avionics, weapons, and airframe elements, to maximize performance, must become the central goal of well-integrated R&D&T strategies, and of new tactics for fighter pilots. Altogether, these new concepts amount to a revolution in the educational process of engineers, and of pilots (§ 13).

(vi) – New integrated testing/design methodologies, involving cost-effective, integrated jet-propulsion-laboratory/vectored-RPVs-flight testing, may help advance this field (§ 12).

(vii) – Automatic, "vectoring-PST-inlets" must be developed and laboratory/flight tested for each PST aircraft, namely, for upgraded F-16, F-15 and F-18, for PVAs, and for post-ATF vectored aircraft.

(viii) – The *2D* nozzle's stealth benefits are greater than has been widely supposed. It also affords the use of radar-absorbing structures that cannot be incorporated when the engine has a circular, axisymmetric nozzle. Thus, the STOL and super-maneuverability benefits attainable by yaw-pitch *2D–ITV* nozzles may be viewed only as additional advantages to the provision for stealth that yaw-pitch, *2D–ITV* nozzle's blended integration into the airframe provides.

The "Why?" and "How?" of this forthcoming change are described in the main text.

LECTURE I

FUNDAMENTAL CONCEPTS REVISITED AND REDEFINED: PRELIMINARY NOTES

"With the help of physical theories we try to find our way through the maze of observed facts."

Leopold Infeld and Albert Einstein

"Revolutions are ambiguous things. Their success is generally proportionate to their power of adaptation and to the reabsorption within them of what they rebelled against."

George Santayana

I.1 New Missions and Modified Design Trends

Traditionally the engine has been considered to have little influence in the flight-control design. Consequently, aircraft dynamicists tended to develop sophisticated airframe dynamic models in conjunction with only a *rudimentary* model for the engine.

This was the so-called *"Big-Airframe, little-engine"* approach.

On the other hand, engine manufacturers had traditionally used the opposite approach, almost ignoring the best *integration methods* that might be required by future *airframe designers.*

The recent introduciton of new, vectored-airframe missions and performance has drastically changed these attitudes (see below).

Almost suddenly it was realized that there is no unified approach, nor integrated design tools and criteria to handle the new problems properly.

Simple *additions* of the two technologies, in some linear simulation methods, simply giving *equal favor* to both, have been quickly found to be *inadequate*, or even *misleading.*

A new, really integrated methodology, had to be developed quickly, and, apparently, *from no verifiable base of low-risk technology.* This, in turn, was to revolutionize the mode of thinking of *pilots* and of *mission decision-makers*, as well as the entire approach to *aerospace engineering education and practice.*

In this lecture we first examine the origin of early *conceptual design limitations.* We then review a new class of (patented) vectoring components and their inherent new type of *unit operations* and potential *new missions.* While the new vectoring components are quite simple in conceptual design (some would even say *"trivial"*), their recent introduction to the aviation industry has generated *a substantial revolution.* This revolution has its origin in the earlier introduction of the well-known *Harrier-type technologies.* However, the new revolution in thinking, research, design and testing, has already generated *its own momentum*, with its own terminology of fundamental concepts, design methodology, performance and mission definitions.

Consequently, we begin this lecture by revisiting and redefining the *fundamental concepts* associated with the new vectoring nozzle components and their associated *unit operations*. In the next Lectures we shall proceed to definitions and classifications of *cold propulsion, pure vectored aircraft, internal and external thrust-vectoring systems, supercirculation, and robot aircraft.*

During this course we shall also examine the main "60 variables" associated with vectored propulsion, and, especially, with *integrated flight-propulsion modes of design, control and performance payoffs.*

Special attention will be given to the recent introduction of *a new research tool: The vectored RPV.*

In turn, the newly designed vectored RPVs, or in their global engineering meaning, *'robot aircraft'* (including Advanced Cruise Missiles (ACM) and drones), have already generated a whole new class of *special missions*; missions *unattainable* by conventional propeller-powered RPVs, or by current-technology, jet-powered robot aircraft.

This particular revolution in the design, performance and mission definitions of robot aircraft will be examined in Lecture IV.

We turn now to the first subject, namely, *the inherent prejudice toward innovative designs in vectored propulsion and vectored aircraft.*

I.2 Early Conceptual Design Limitations

Military aircraft exhaust nozzles have evolved from fixed, subsonic, convergent types, to supersonic, variable-area, convergent-divergent systems. Traditional thinking, combined with some nacelle/fuselage structural limitations of specific aircraft, have, so far, dictated the restrictions of these nozzles to a *circular* cross-section and to a simple, unvectored, brute force 'forward'.

This situation has been changed in recent years, mainly because new *survivability* improvements, over current-technology aircraft, have been stressed in a number of categories:

- *Inflight thrust vectoring.*
- *Drastic reductions of takeoff and landing distances by thrust vectoring and by thrust reversal.*
- *Enhanced maneuverability through simultaneous roll-yaw-pitch thrust vectoring.*
- *Significant reductions of minimum flight speeds by the combined unit operations of vectoring* (see below).
- *Low observability* (i.e., low RCS, optical, noise, and IR signatures associated, in part, with this new technology).

Engine manufacturers had first tried to meet these requirements by simplified *additions* of existing, *proven components*. The result was *large increases in weight and complexity*, and/or substantial *reductions in performance.*

Such early designs resulted in reduction in expected performance envelopes and increases in weight and complexity. Even nowadays some designers simply add such proven components as *wedges*, reverser *clamshells, gimbal rings* and *plugs*, while retaining the old-fashioned control rules of current-technology exhaust nozzles and powerplants.

I.3 Internal Thrust Vectoring (ITV)

It was in 1976, that PWA obtained a patent (No. 3,973,731) on a 2-D/C-D thrust-vectoring nozzle shown in Fig. 1. *This design contains only four major moving parts. Yet it provides the best overall performance when yaw vectoring is not required.* However, as was stressed in the Introduction, simultaneous yaw-pitch-roll thrust vectoring is a basic requirement of all advanced, PST/PSM-RaNPAS, vectored aircraft.

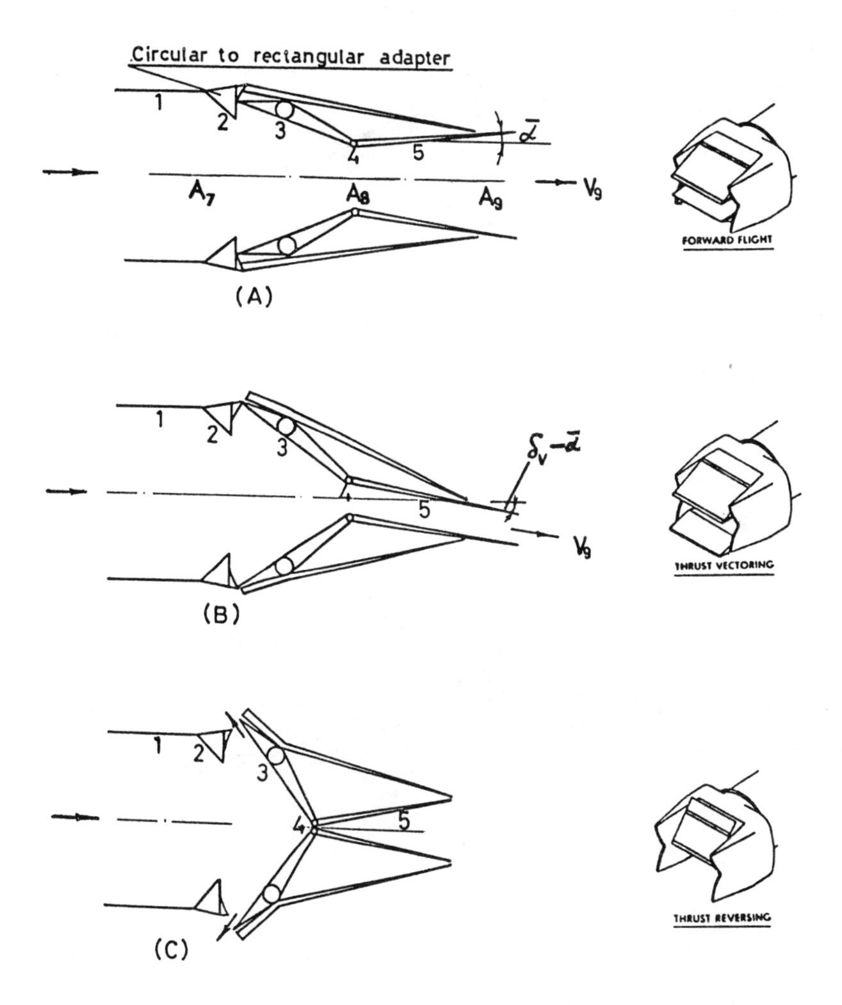

Fig. 1. **The vectoring nozzle shown here is termed two-dimensional, converging-diverging, or 2D-CD.** It allows pitch vectoring and thrust reversing *without* thrust *yawing.* It may also be referred to as *TV/TR nozzle.* (See also Fig. 3)

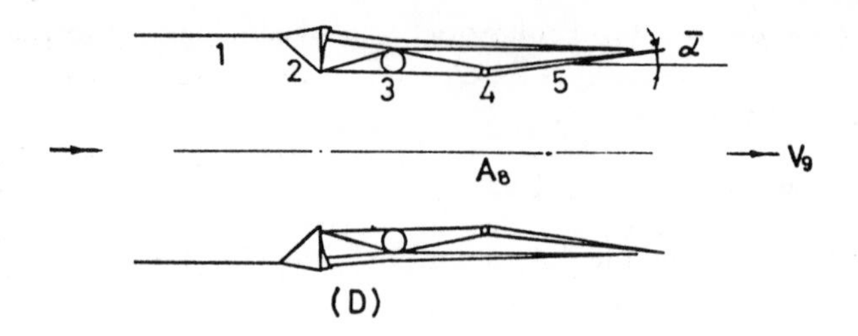

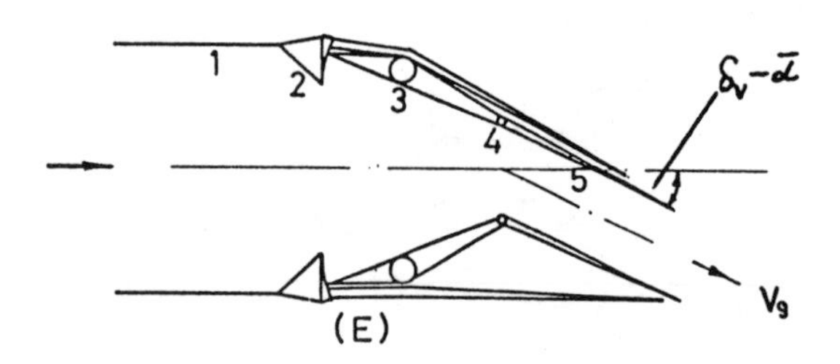

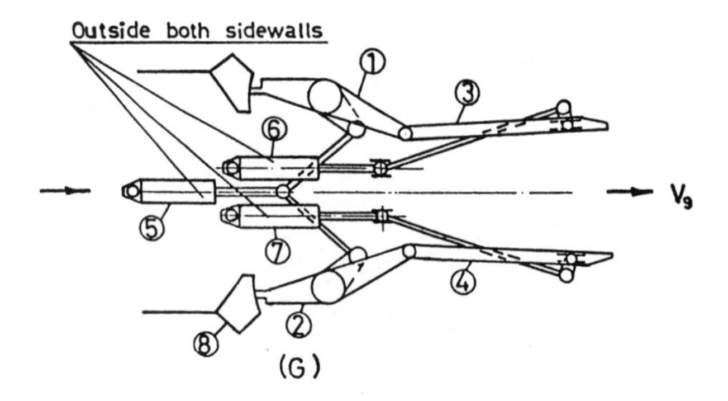

Pitch plane primary thrust vectoring of ±20 degrees is available at all engine power settings, including maximum *A/B*.

Independent controls of nozzle throat area, A_8, (see Fig. 11 for definitions) and nozzle exit area, A_9, are also provided, resulting in optimum area ratio matching for maximum internal performance (cf. Fig. 11 and eqs. 5 and 6).

Closure of A_8 by the convergent nozzle flaps, and simultaneous porting of the exhaust through venetian-type, vectoring/reversing vanes, is available with engine throttle settings up to, and including "Intermediate Power (or "full dry power").

Actuation of nozzle surfaces is accomplished with 6 actuators (3 inside each sidewall as shown in Fig. 8). Two additional ones are required for the yaw vanes control in our new yaw-pitch designs.

Average actuation rate is 30 degrees per second. Nozzle reversing is possible throughout the dry engine power range (excluding *A/B* thrust reversal). However, this design has been modified for the F-15 S/MDT (cf. Fig. 3).

The porting of the exhaust flow through rotating vanes (Figs. 4 and 7) allows a wide range of force control versatility in the nozzle pitch plane, including *axial force modulation to achieve longitudinal moment, acceleration and velocity control, especially during landing approach phases (Fig. II-6).*

Reverse thrust is achieved while maintaining the engine throttle at the "intermediate" (or "dry") power setting, or less. Up to 67% of the "forward dry" gross thrust is available for rapid (less than 1 sec) deceleration on the ground and in flight.

Legend

1) Circular engine duct.
2) Circular-to-rectangular duct (*this section may become A/B*, as for instance, in Figs. 13 and 14).
3) Central hinges of converging flaps.
4) Nozzle throat hinges.
5) Diverging flaps (vectorized only in the pitching mode).

Subfigures:

A) $\delta_v=0$ at full military power setting.
B) $\delta_v\neq 0$ at full military power setting.
C) Full thrust reversal deployment.
D) Full *A/B* at $\delta_v=0$.
E) Full *A/B* with pitching vectoring (throat area should be increased.
G) Sidewall, nozzle control actuators ($\delta_v=0$).

Notes:

1. Sidewalls may be truncated as in Figs. 15 and 16.
2. Throat area during vectoring is defined in Fig. 11, and its control adjustments by eqs. 5 and 6.
3. For yaw thrust vectoring see Figs. II-1 and II-2.
4. Cooling methodology is illustrated in Fig. 8.

Legend for Fig. G:

5) Converging nozzle flaps actuator (symmetric mode of linear displacement during VR flight control with IFPC).
6) Uper diverging flap actuator controls ($\delta_v-\alpha$) pitch vectoring angles in VR-IFPC modes of control.
7) Lower diverging flap actuator controls ($\delta_v+\alpha$) pitch vectoring angles in VR-IFPC modes of control.
8) *The design of this section may vary considerably.* First and foremost it includes circular-to-rectangular ducts, *streamlined flow-dividers/structural struts*, etc. [Figs. 13 and 14]. In some designs it includes *Venetian-type vanes which are fully controlled during approach-landing* (cf. Fig. 7) *and in air-to-air and air-to-surface maneuvers.*

The General Electric Co., PWA, and other research bodies, have since improved this patent by various means, including the use of *Venetian-type vanes* during thrust reversal (Fig. 7). Other improvements, including an Israeli Patent Application from 1987 (No. 80532), have introduced *simultaneous yaw-pitch vectoring vanes/stators inside the converging, or converging-diverging types of exhaust nozzles.* Combined with numerous other improvements, to be described later, the introduction of simultaneous yaw-pitch-roll vectoring, or even yaw-pitch-reversal vectoring, has opened the way to the design of pure vectored aircraft *whose dependence on aerodynamic flight control surfaces is minimal. Consequently, aircraft capable of super-agility at post-stall/PSM conditions may now become a real possibility.*

The aforementioned inventions and improvements are *readily adjustable to fit different engine-cycles, or aircraft installation requirements.* Their performance payoffs begin at around o.6 thrust-to-weight ratio. Hence, many older fighters may become candidates for becoming vectored aircraft.

I.4 External Thrust Vectoring (ETV)

Figs. 4 and 5 show a comparison of yaw-pitch angle envelopes for three-vane and four-vane External Thrust Vectoring (ETV) configurations. A comparison of ITV with ETV is to be discussed in Lecture VI.

I.5 Subsonic-Supersonic Concepts of Engine Exhaust Nozzles Revisited and Redefined

Fig. 1 is a schematic view of an engine exhaust nozzle during unvectored and vectored operation. A_8 represents the throat area, A_9 the exit nozzle area, while $\tilde{\alpha}$ represents the divergence (half-angle) deflection required to maintain optimal $A_9/A_8 = A_e/A_t = \varepsilon$ area ratios for achieving maximum thrust efficiency, C_{fg}, during supersonic engine operation, i.e., for Nozzle Pressure Ratios (NPR) above a critical NPR_c value.

Tables are available in standard textbooks, and in engine simulation programs, for the calculation of ε as a function of NPR. However, as will be shown below, C_{fg} is not the only coefficient to be considered for the evaluation of converging-diverging, or converging, exhaust nozzles. For that purpose one needs at least four engine nozzle coefficients: C_{fg}, C_{D8}, C_A, and C_v (their definitions and physical meaning will be given at a later stage of this course).

What should be stressed at this early stage of the course are the following preliminary notes.

- Pitch vectored, variable, two-dimensional (*2D*) engine exhaust nozzles, may be formed with only *four flat flaps* (supersonic) [only two are required for converging-type subsonic nozzles]. Cf. Figs. 1, 4, and 7 to 14.
- On the other hand, current-technology (axisymmetric) nozzles are formed, and controlled, by *a large number of movable metal flaps*, which, being rotated together,

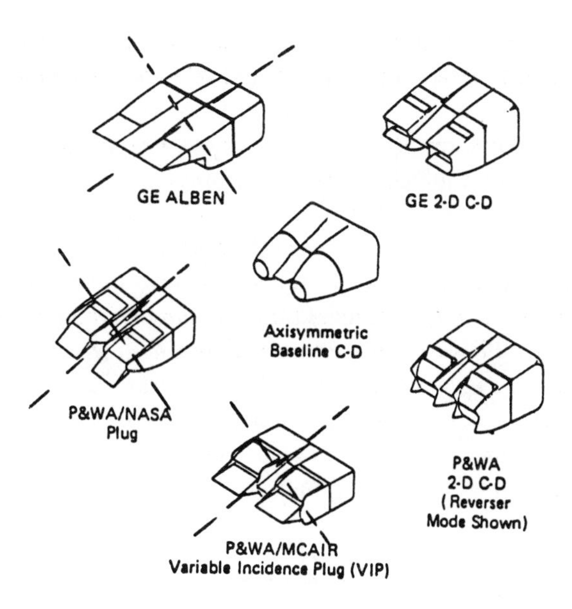

Fig. 1a: **A number of early-design, low aspect-ratio, vectoring nozzles proposed by GE and PWA**. (Cf. Appendix A for more details. The ones crossed were found to be less useful.)

form an *almost circular* (axisymmetric) nozzle. However, the large number of flaps generate *gas leakge* and *cooling (thrust) losses, cf. Fig. 2.* Moreover, they generate velocity and thermal boundary layers, which deviate from simple *2D*, or axisymmetric forms. Consequently, vectored nozzles may demonstrate *equal, or even higher thrust efficiencies than those obtainable today with current-technology nozzles, as shown in Fig. 3.*

- During pure pitch vectoring ($\delta_v \neq 0$, $\delta_y = 0$), the geometry of the four *2D/CD* flaps may appear as in the scheme shown in Fig. 11.
- During supersonic operation (i.e., when $NPR > NPR_c$, where the subscript *c* refers to the critical NPR which separates between subsonic and supersonic operations), the *expansion waves* aft the throat corner are enhanced during vectoring. The characteristics of these expansion waves depend on the δ_v deflection value and they may be enhanced by a *'separation bubble'* formed *downsteam* of the hinge corner, well inside the *supersonic stream.* A few compression lines, fore and aft the 'bubble', may complicate the actual mechanism.
- The throat area variation is thus given by

$$\frac{A'_8}{A_8} = \cos\delta_v \tag{1}$$

or, during *simultaneous pitch-yaw vectoring,* by

$$\frac{A'_8}{A_8} = \cos\delta_v \cdot \cos\delta_y \tag{2}$$

where A'_8 is the effective throat cross-sectional area defined by point 8′ in Fig. 13, Introduction. However, eqs. (1) and (2) *neglect the effects of the engine*, as described below.

- To maintain a predetermined $A'_e/A'_t (= A'_9/A'_8 = \varepsilon_v)$ ratio, the effective nozzle exit area A'_9 should also be subject to the condition

$$\frac{A'_9}{A_9} = \cos\delta_v \cdot \cos\delta_y \tag{3}$$

- However, to maintain *the same* mass flow rate *through the engine*, for δ_v^-, δ_y^- *deflections at a given NPR value, the effective throat area* A'_8 must be subject to the condition

$$A'_8 = A_8, \quad \text{or} \quad (\dot{M})_{\delta_v, \delta_y} = \dot{M}_{(\delta_v = \delta_y = 0)} \tag{4}$$

Consequently, the flaps in the throat area must be "opened" by a factor of

$$\frac{A_8^{\delta_v \delta_y}}{A_8} = \frac{1}{\cos\delta_v \cdot \cos\delta_y} \qquad \text{(5)} \quad \textbf{Control Rule No 1}$$

- Similarly, to maintain the required A_e/A_t value, the flaps in the nozzle exit area should be "opened" by a factor of

$$\frac{A_9^{\delta_v \delta_y}}{A_9} = \frac{1}{\cos\delta_v \cdot \cos\delta_y} \qquad \text{(6)} \quad \textbf{Control Rule No 2}$$

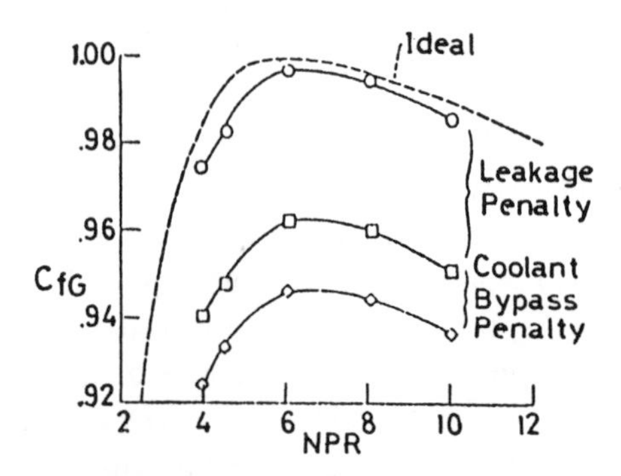

Fig. 2. **Typical variations of exhaust nozzle thrust efficiency, C_{fg}, with Nozzle Pressure Ratio (NPR) for a constant** A_e/A_t ($\equiv A_9/A_8$) value. The Straight and Cullom results shown are for the *circular* (axisymmetric) nozzle of the J-85 engine [48]. *Leakage and cooling (thrust) losses are mainly attributed to current-technology, many-flaps nozzles.* These losses are expected to be minimized in the new, vectoring, *2D–CD* nozzles manufactured with or without RCC materials (see text and Figs. C-1 to C-6).

Thus, the thrust coefficient of vectoring nozzles may be higher than that for conventional (circular) nozzles, even for current-technology (cooled) flaps, i.e., the leakage penalty shown would be substantially reduced in properly designed *2D-CD thrust vectoring nozzles.* This expectation has been recently verified experimentally by our jet propulsion laboratory tests with a number of new thrust vectoring nozzles.

- For *"down" pure pitch vectoring* the *'upper'* divergent flap is tilted according to

$$\textbf{Tilted angle} = \delta_v - \bar{\alpha}, \tag{7}$$

where δ_v is counted positive for nose down pitching moments. This causes the sonic line to move from point 8 *to point* 8'. Thus, point 8 remains in the *subsonic* domain during *supersonic vectoring operation (of the engine). Consequently, whatever is the value of* $(\delta_v - \bar{\alpha})$, expansion or compression waves would *not* be generated at point 8 on the "upper" hinge of the throat, while *only* enhanced expansion waves would be generated aft the "lower" hinge, as shown in Figure 1.
- The net result is, therefore, asymmetric operation during which the enhanced expansion waves cause the expansion ratio from throat to exit to be under lower NPR values, than those computed by standard textbooks, i.e., in the *supersonic domain*

$$\text{NPR}_{\text{vectoring}} < \text{NPR}_{\text{nonvectoring}} \tag{8}$$

- The final result should, therefore, give the designer somewhat *lower thrust efficiency values for* $NPR > NPR_c$, i.e., for most of the supersonic domain.
- However, for a properly designed nozzle, the *opposite conclusion* may be expected for *subsonic operations!* To start with, one may notice that the flow at point 8 would be such as to improve the vectoring of the emerging jet, with minimal, or no throat-corner losses on the "upper" hinge, while those on the "lower" hinge would not generate expansion-waves losses, but would be limited to known, well-calibrated subsonic flows around a single corner. However, for *subsonic* operation the divergent (half) angle $\bar{\alpha}$ should normally be close to *zero. Hence, the boundary-layer separation tendency at the lower hinge would be less damaging than the payoffs supplied by the "upper" hinge-flap geometry.* We therefore conclude this section by five preliminary control rules for primary and secondary effects in vectoring, i.e.,

$$[C_{fg}]_{\text{vectoring}} < [C_{fg}]_{\substack{\delta v=0 \\ \delta y=0}} \quad \text{for} \quad \text{NPR} > \text{NPR}_c \ \text{(supersonic)} \tag{9}$$

$$[C_{fg}]_{\text{vectoring}} \geq [C_{fg}]_{\substack{\delta v=0 \\ \delta y=0}} \quad \text{for} \quad \text{NPR} < \text{NPR}_c \ \text{(subsonic)} \tag{10}$$

For primary thrust vectoring we may, *tentatively*, write

$$T_x = C_{fg} \cdot T_i \cdot \cos\delta_v \cdot \cos\delta_y \qquad \textit{C.R. No. 3} \tag{11}$$

$$T_v = C_{fg} \cdot T_i \cdot \sin\delta_v \cdot \cos\delta_y \qquad \textit{C.R. No. 4} \tag{12}$$

$$T_y = C_{fg} \cdot T_i \cdot \cos\delta_v \cdot \sin\delta_y \qquad \textit{C.R. No. 5} \tag{13}$$

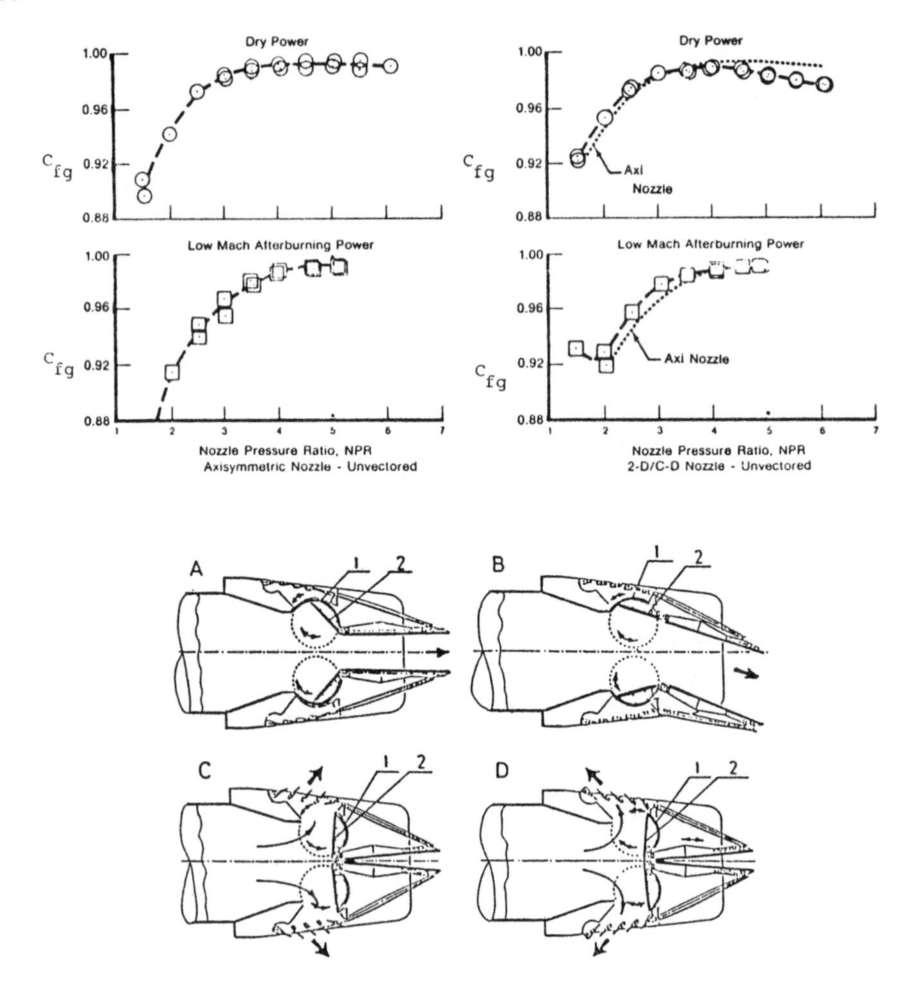

Fig. 3. **Vectoring 2*D-CD* nozzles (at the unvectored position) may produce somewhat higher thrust than conventional circular (axisymmetric) exhaust nozzles, up to $\mathbf{NPR_c}$.** Rotatable converging flap (2) and cylindrical seal-section (1) allow TV/TR modes *A, B. C, D.* Vectoring effects are shown in Fig. 6.

where T_i is the *ideal thrust without vectoring, given by* (⋆)

$$T_i = \dot{M}_{\text{actual}} \left[RT_T \frac{2\gamma}{\gamma - 1} \left(1 - \left\{ \frac{P_a}{P_T} \right\}^{\gamma-1/\gamma} \right) \right]^{1/2} \tag{14}$$

(⋆) For R&D projects at the Jet Propulsion Laboratory we employ enthalpy changes of a real gas (air+fuel-reaction products) to estimate T_i (without relying on the perfect-gas equations), i.e., "ideal thrust" does *not mean "ideal gas". Cf.* § I-6 below.

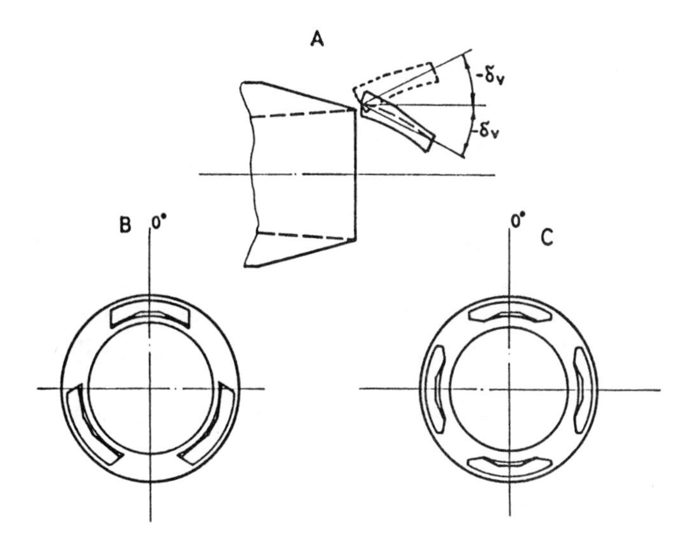

Fig. 4. **External Thrust Vectoring (EVT) as advanced by Berrier and Mason of NASA (208, 209).** A – Sideview. B – Backview, 3 pedals. C – Backview, 4 pedals.

The pedals may be rotated *asymmetrically* during *approach* or ground-handling to trim *pitching moments*, or to eliminate undesirable ground effects.

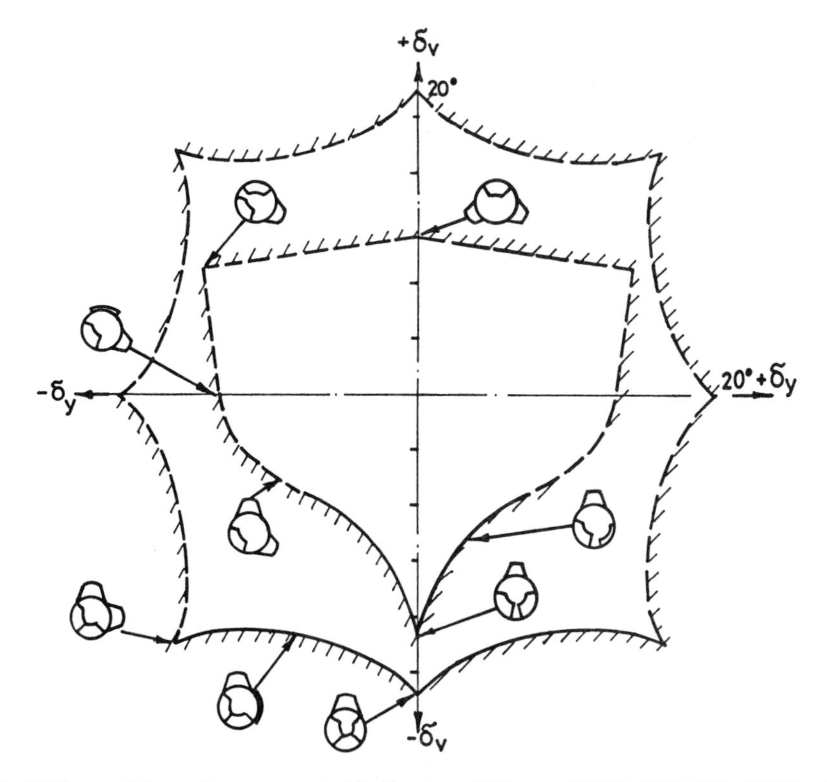

Fig. 5. **ETV yaw-pitch envelopes as reported by Berrier and Mason of NASA (208, 209).** (Cf. Fig. 4).

An *experimental confirmation* of eqs. 9 and 10 is shown in Fig. 6.

The designer of vectored propulsion systems, whether limited to pitch-only vectoring, or not, may, therefore, expect *higher thrust efficiency values up to about NPR=4*, i.e., in the subsonic and lower supersonic domains.

Apparently, the highest efficiency is obtainable around $\delta_v = 10$ degrees and NPR=3 to 4. *However, for initial design purposes one may consider the range $\delta_v \leq 20$ degrees.*

No experimental results for the effects of δ_y-vectoring on thrust efficiency have yet been reported in the literature. One reason for this is the fact that only a few designs of internal vectored propulsion systems with simultaneous yaw and pitch vectoring exist today, and their geometrical and experimental characteristics are still not available for publication.

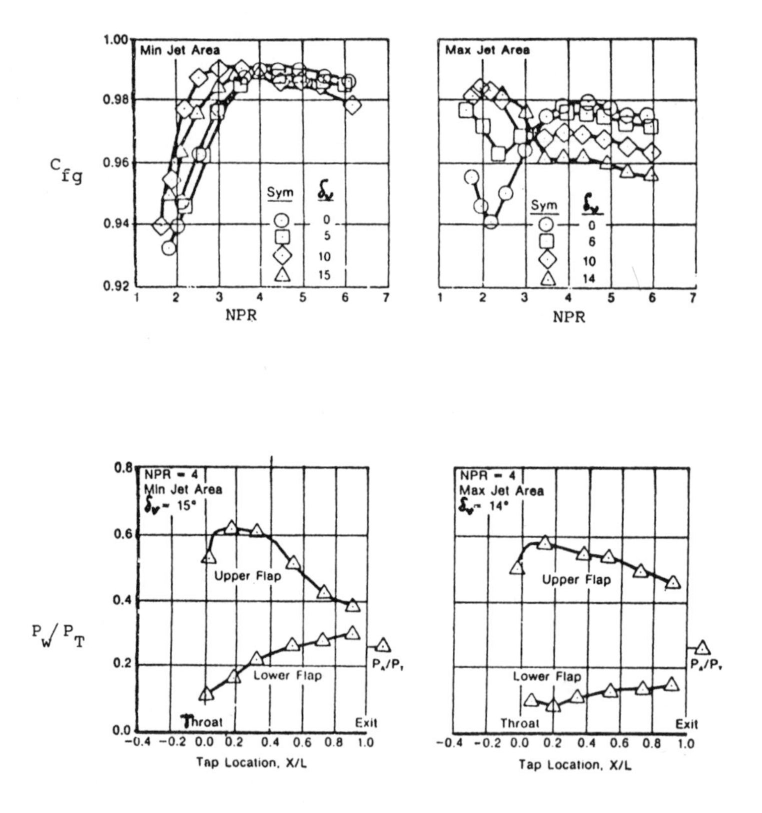

Fig. 6. **As demonstrated by Straight and Cullom of NASA (48), the thrust efficiency of vectored exhaust nozzles may increase during deflections, up to NPR=4 (i.e., mainly in the "subsonic" domain).** Maximum efficiency is obtainable with $\delta_v = 10$ degrees. The experimental data shown are for a *2D-CD* vectoring nozzle installed on a General Electric J-85 engine (cf. Table III-1 and Figs. III-4, 5, 6, 7 for additional details). The lower figures show the variations of wall static pressure on the "upper" and "lower" diverging flaps at NPR=4, $\delta_v = 15$ degrees. *These highly interesting results were reported in 1983 by Straight and Cullom of the NASA-Lewis Research Center [48].* Cf. Fig. C-7.

I.6 How To Compare The Performance of Different Thrust-Vectoring Nozzles

The central element in Internal Thrust Vectoring (ITV) is the nozzle design and its integration with the wing and the flight control system.

Unfortunately, no systematic study, or design, has yet been reported in the literature on a practical solution for simultaneous yaw-pitch-roll ITV design, control system and integration potentials.

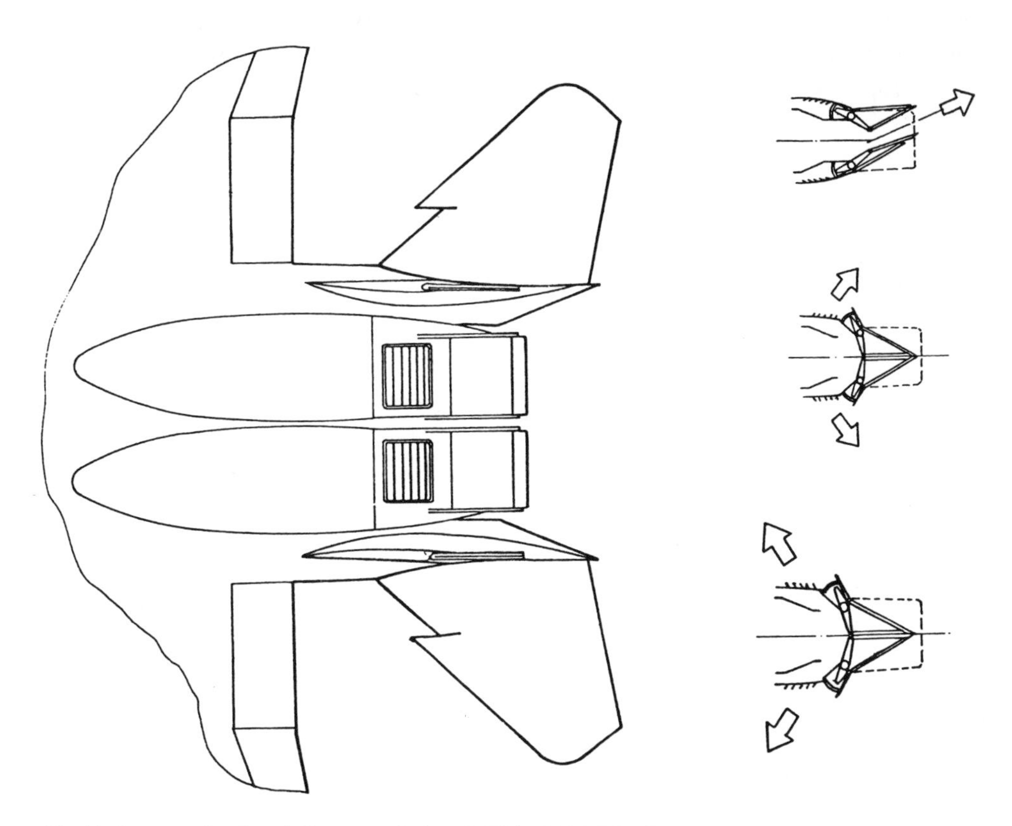

Fig. 7. **An example of partially-vectored aircraft.** This F-15 STOL fighter demonstrator has been flight tested in 1989 (13, 18, 62, 70, 71, 111, 128, 234).

Instead of the simplified *TR* design shown in Fig. 8, this design is based on *venetian-type of thrust reversal vanes.* These are *closed* during *pitch vectoring (upper right), open in the downstream direction during landing final approach (center), and opened in the upstream direction after touch-down (lower right).* However, thrust reversal is *less* important than thrust vectoring and may not be required (cf. Fig. 21, Int.).

For various nozzle details see Figs. 13 and 14. Note that *up-pitch vectoring* is also useful in *reducing* runway distance in takeoff, namely, in *rotating* (nose liftup) the aircraft at *much lower speed than with conventional aircraft.*

Then, using IFPC, the jets are "vectored down" for getting the maximum $(\Delta C_L)_v + (\Delta C_L)_{SC}$ (cf. Lecture III, Fig. III-18).

Furthermore, during experimental evaluations of the nozzle coefficients for such yaw-pitch-roll systems, in comparison with pitch-only thrust-vectoring nozzles, one needs an exact thermodynamic basis to perform the comparison properly. However, in establishing such a basis one often encounters the following fundamental problem with respect to the thrust coefficient C_{fg}:

IDEAL THRUST does NOT mean a permission to use ideal gas equations in isentropic adiabatic expansion.

Instead, one must compute isentropic expansion of a REAL gas, namely, of air plus the proper concentrations of the reaction products of burning a given jet fuel. Thus, one must employ the proper ENTHALPY tables to determine the value of the ideal thrust at each test point. This procedure, in itself, may cause two additional problems:

(i) Proper enthalpy tables may not be available for each chemical composition. Hence, one may have to compute them by using well-tabulated enthalpy values of the chemical components which must be known to the experimentalist as comprising the fuel.

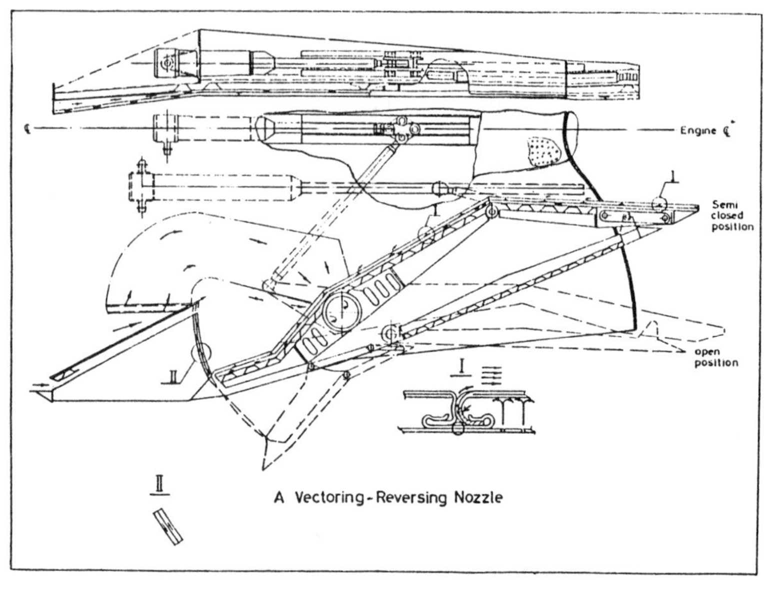

Fig. 8. **An example of an early vectoring-reversing nozzle (no yaw vectoring) (49).** Note the *impingement-film-cooling methods* employed to cool the *C-D* flaps. These cooling flows generate *two boundary layers*, one inside the flap wall *(impingement cooling)* and one outside it *(film cooling). For heat transfer rates in subsonic and supersonic flow regimes see Eqs. III-1, 2 and Figs* III-1 to 6. Note also the location of the actuators *in the sidewalls* (upper part of the drawing). Sidewall *truncation effects* on nozzle efficiency are shown in Figs. 15 and 16. The design of the circular-to-rectangualr transition duct may be done by following the *superellipse equations* (Figs. III-17).

However, for *high NAR*, like the ones shown in Figs. 13 and 14, this simplified design approach may be *misleading. Also misleading is the calculation of AR effects on nozzle weight in isolation* [cf., Fig. III-14].

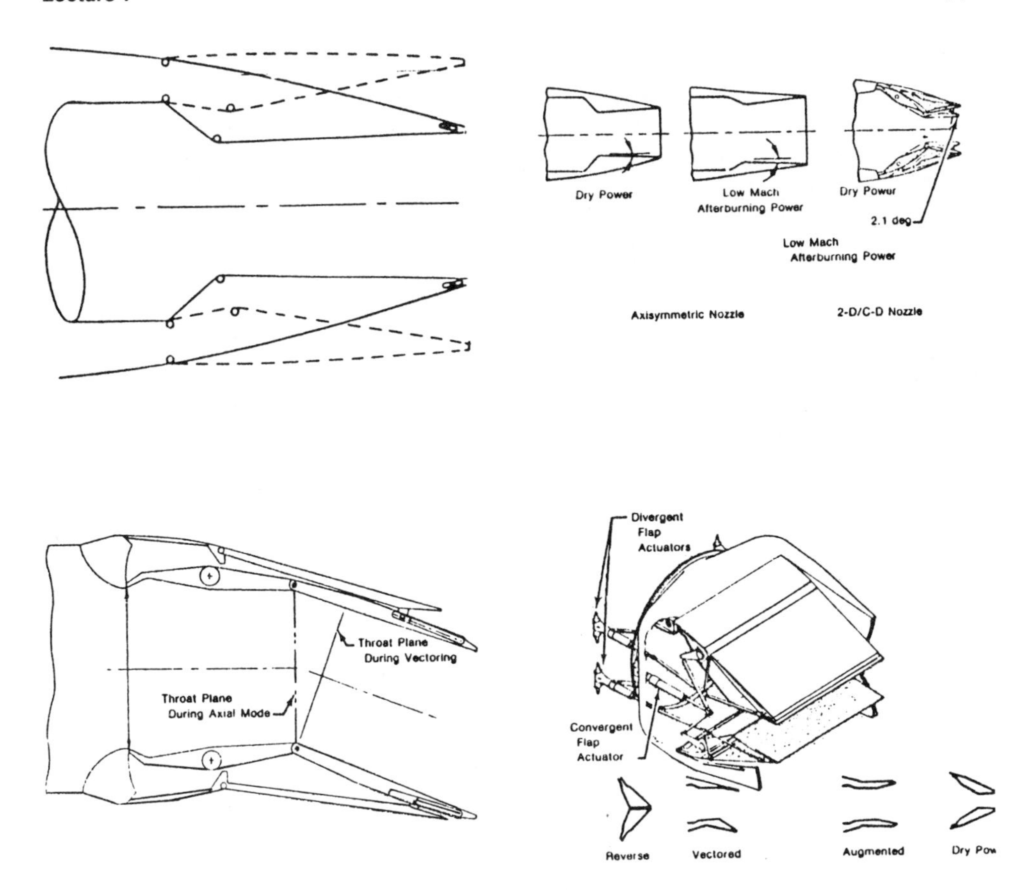

Fig. 9 (upper left). **Schematic view of current-technology *axisymmetric* nozzles.**

Fig. 10 (right). **Axisymmetric and 2-*D/C-D* nozzle geometry comparison (No yaw thrust vectoring) (cf. Figs. 9, 10, 11, 12).**

Fig. 11 (lower left). **In pure pitch vectoring the throat cross-sectional area changes as shown.** *In yaw vectoring a similar change takes place in the perpendicular plane.* (Cf. eqs. 5, 6).

Fig. 12 (lower right). **Typical 2-*D/C-D* actuation system (49).** Low *AR* vectoring nozzle without yaw thrust vectoring (cf. Fig. 8).

(ii) Varying the throttle changes RPM, and, accordingly, the NPR values change. Yet, during this change the chemical composition may change too, due to such factors as combustion efficiency variations and an air-to-fuel ratio change, etc. (cf. Lecture III).

I.7 Subsonic Yaw-Pitch-Roll Thrust Vectoring

The test results shown in Figs. 19 and 20 have been extracted from our early laboratory test series of a subsonic, simultaneous yaw-pitch-roll, internal thrust-vectoring nozzle, with NAR=46.6. Test results have been corrected to standard sea-level temperature

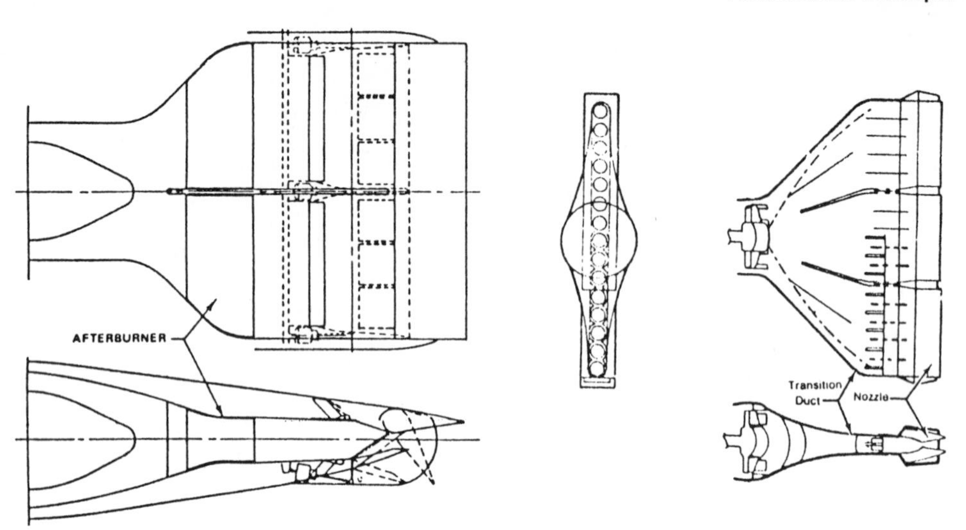

Fig. 13. **G.E. High *AR* SERN (without yaw vectoring) (49).**

Fig. 14 (right). **NAR = 15 nozzle with flow dividers, but without yaw vectoring.**

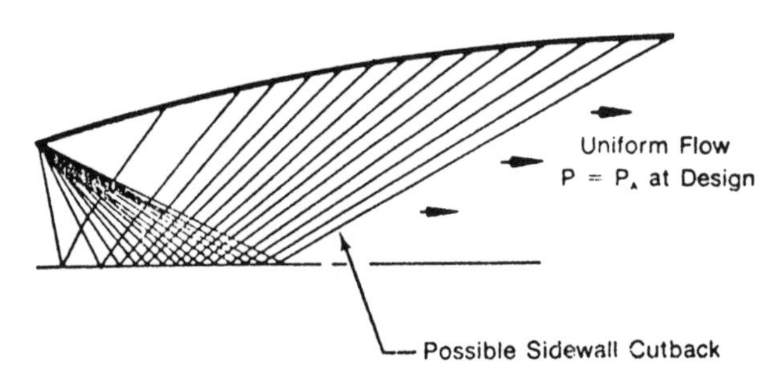

Fig. 15. **Figs. 8 and 12 show that the *sidewalls* of pitch-only, vectored, *2D-CD* nozzles, may be shorter than the diverging flaps.** This is done to reduce nozzle weight (49).

For design cases where sidewall cutbacks are important, one may first examine the Mach lines in this figure and, accordingly, design sidewall cutbacks.

and pressure. The turbojet engine employed delivered up to 350 kg. thrust with a well-calibrated, axisymmetric nozzle. Yaw and pitch angles are the geometric angles of the converging flaps. The fuel used is JET-*A*-1. The engine is equipped with a standard bellmouth inlet. In general, no significant degredation of thrust levels, nor significant SFC increases, were encountered during yaw, pitch, yaw-pitch and yaw-roll thrust vectoring tests.

During yaw-pitch-roll-vectoring flight tests (using vectored RPVs) the nozzle exit area was maintained unchanged by a special control system.

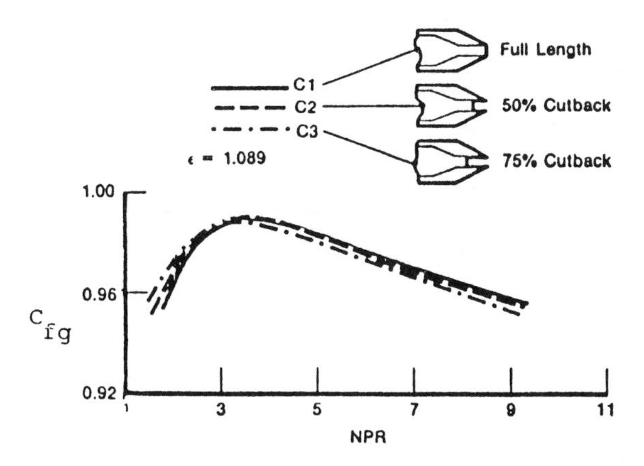

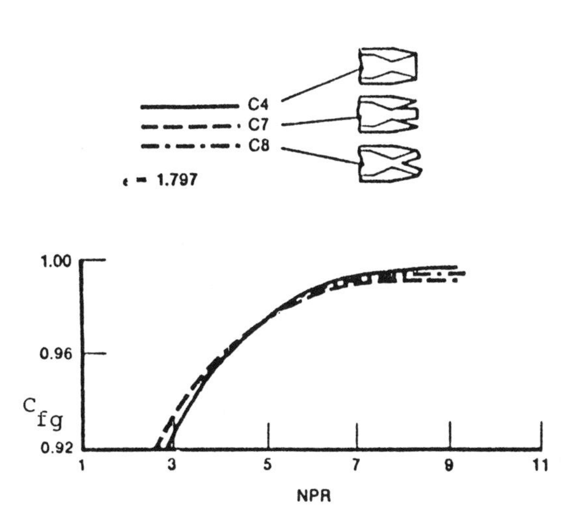

Fig. 16. **Following design consideration shown in Fig. 15, one may examine the slight changes in the thrust coefficient due to 50% and 75% cutbacks in diverging-section sidewalls.**
The data are taken from Ref. 49 for a *2D-CD* vectoring nozzle. *Note a slight increase in C_{fg} values in subsonic operation, with truncated sidewalls, in comparison with untruncated ones.*

Symmetry comparisons between LEFT-RIGHT-YAW vectoring is easily evaluated by using such graphs. Moreover, such graphs may be used to verify the degree of REPEATABILITY of the testing methodology and the test equipment.

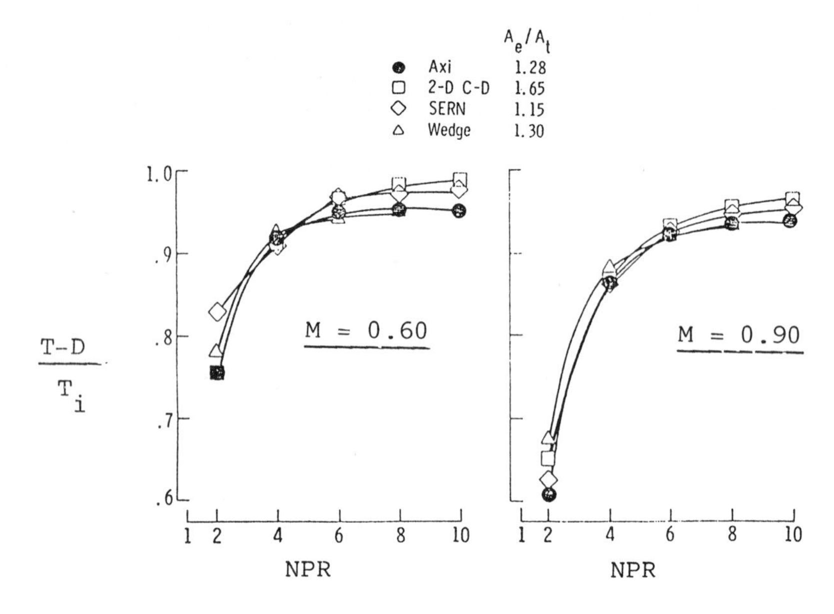

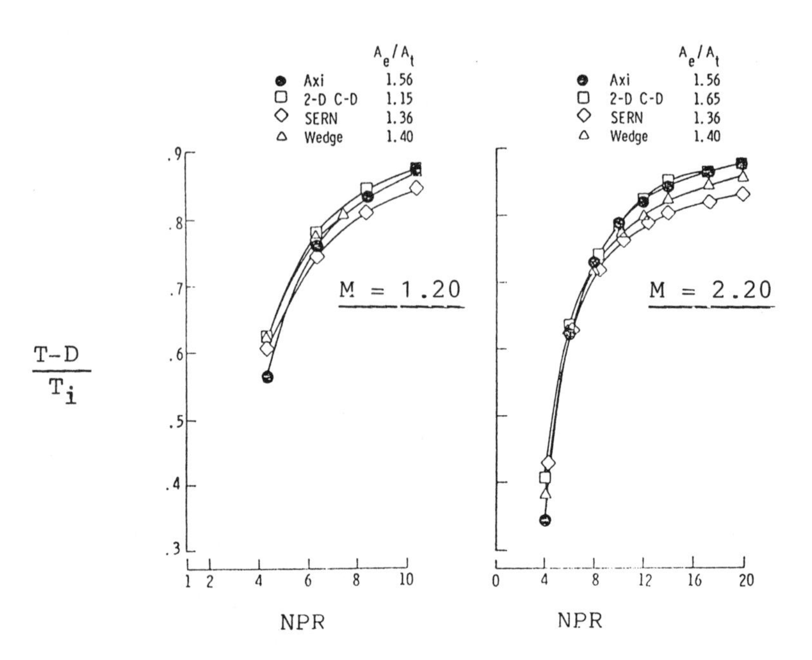

Fig. 17. **As the flight Mach number is increased, net thrust ratio $(T-D)/T_i$** variations demonstrate a slight advantage for the **2D-CD vectoring nozzle over its competitors** (cf. Fig. 1a). This advantage applies through all NPR values [159].

For additional details concerning weight advantages in comparison with other nozzles please turn to the Appendices.

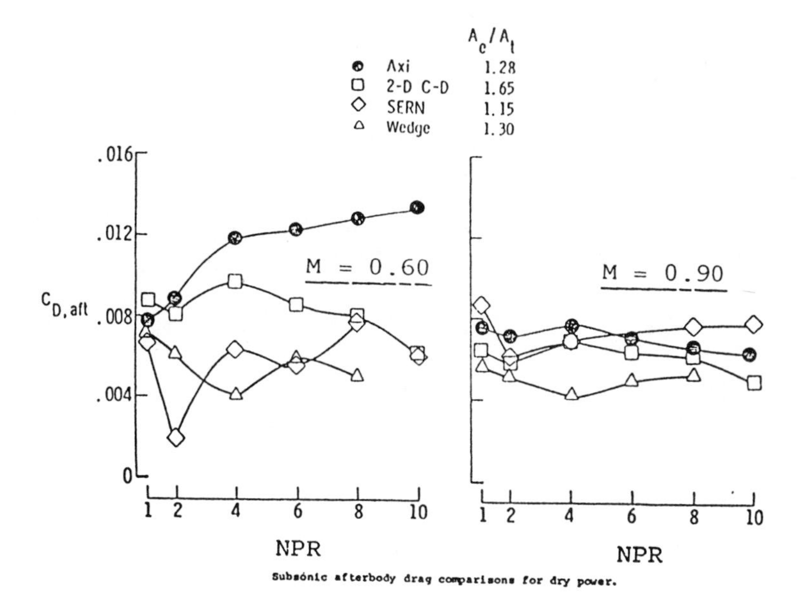

Subsónic afterbody drag comparisons for dry power.

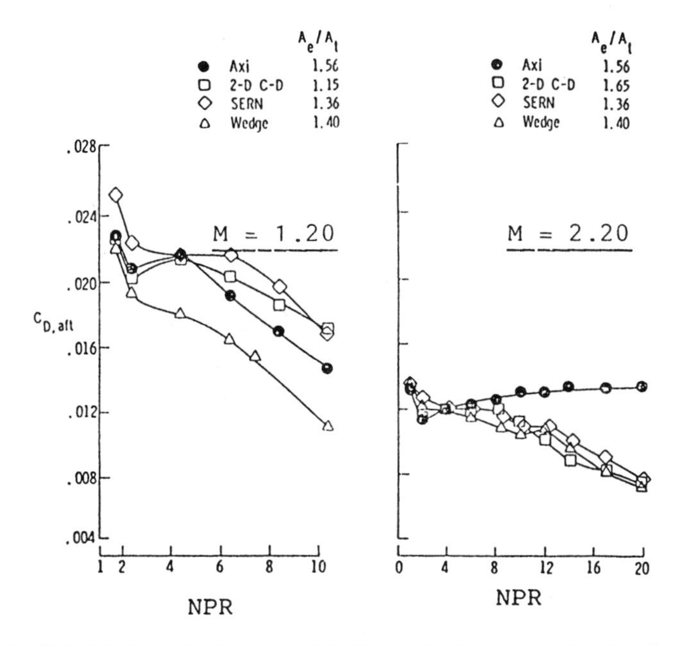

Fig. 18. **As the flight Mach number increases, 2*D-CD* nozzles demonstrate low drag penalty in comparison with current-technology nozzles throughout most NPR conditions tested [159].** However, with some airframes (e.g., the F-16), the axi-TV nozzle generates lower drag than the low *AR* 2*D–CD* nozzle. For further details please turn to the Appendices.

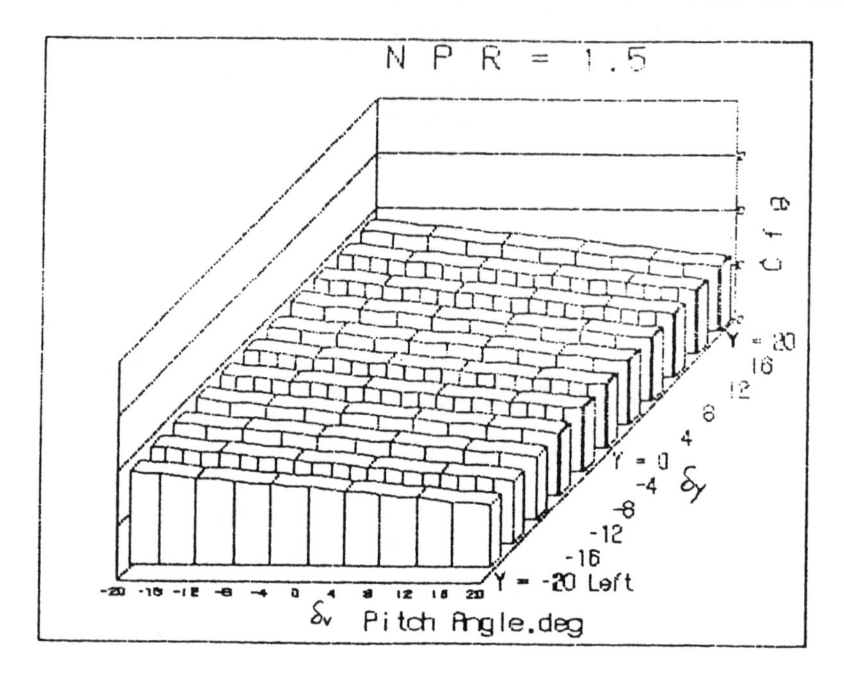

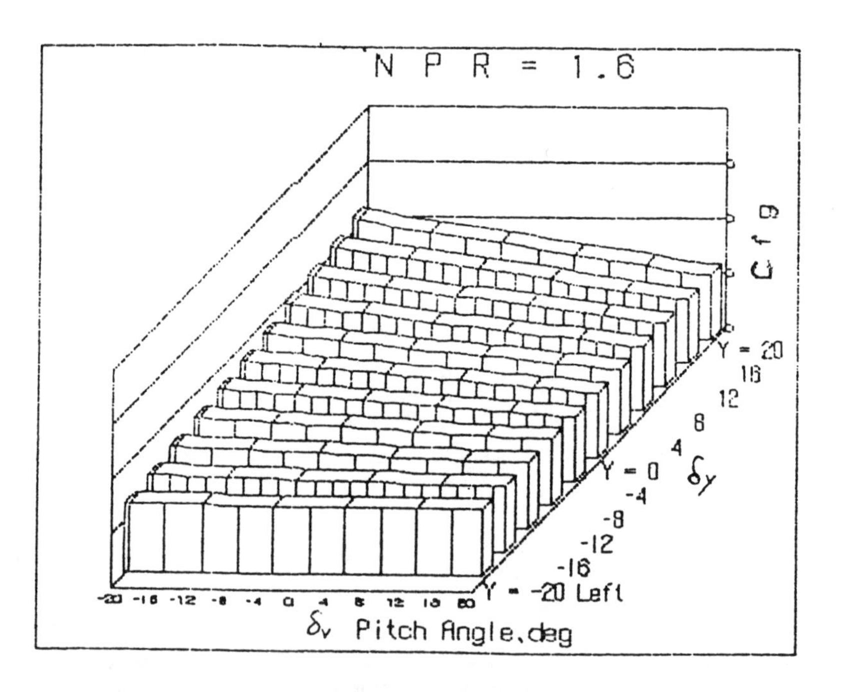

Fig. 19. **An example of our JPL graphical evaluation of thrust efficiencies during simultaneous yaw-pitch thrust vectoring.** $AR = 46.6$.

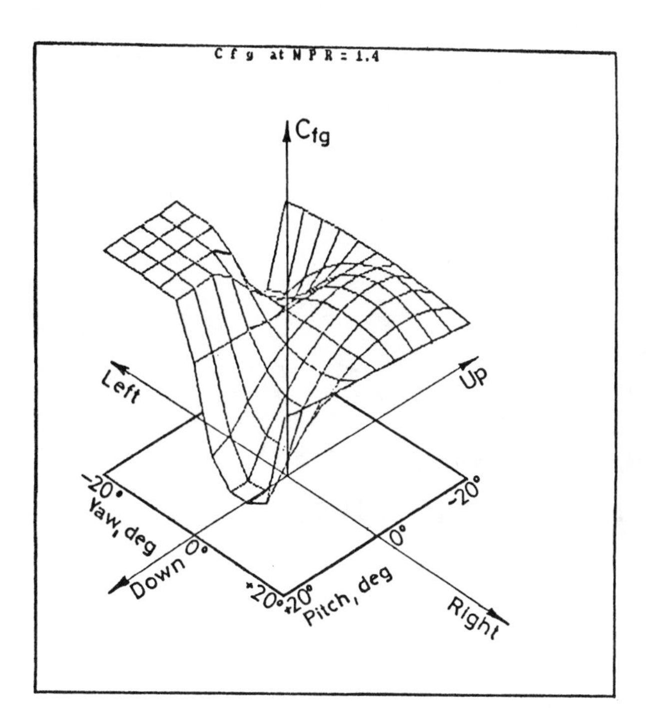

Fig. 20. **How to report and compare different thrust efficiencies during simultaneous yaw-pitch thrust vectoring?** This figure provides a possible graphical procedure. $AR = 46.6$

LECTURE II

VECTORED AIRCRAFT AND SUPERMANEUVERABILITY

Consider what effects that might conceivably have practical bearings you conceive the objects of your conception to have. Then, your conception of those effects is the whole of your conception of the object.

C.S. Peirce (1878)

II.1 Yaw-Pitch Vectoring and Supercirculation vs. The Harrier Technologies

Thrust reversal for vehicle deceleration on the runway, following touch-down, is in common use in transport aircraft. The benefits of the Harrier (AV-8) are also well-known. (Harrier-induced technologies are briefly reviewed in Chapter VI and in Appendix B.)

However, the Harrier technologies do not provide efficient yaw thrust-vectoring, nor the additional payoffs of supercirculation. High drag is also associated with these technologies.

What is new is the continuous, highly integrated attempt to replace all, or most aerodynamic control surfaces, by the simultaneous use of yaw, roll, and pitch thrust vectoring, (with, or without, simultaneous thrust reversing). The actual innovation is to employ simultaneous yaw-pitch-roll vectoring to gain revolutionary enhancement of performance throughout and "outside" the flight envelopes of conventional fighter aircraft, while also reducing drag, radar cross section and infra red signatures. Other benefits include drastic reductions in runway required distances (dry, wet, or icy). In short, vectored aircraft means jetborne flight, or, simply, jetborne aircraft. It also means post-stall maneuverability and controllability (Introduction).

* * *

The concept "vectored flight/propulsion" has added new dimensions to mission definitions and to the design of high-performance fighter aircraft. Thrust vectoring provides *additional forces and moment effectors for optimization and maximization of performance at takeoff, cruise, air-to-air, and/or, air-to-ground missions, and in landing.* It can be used for augmentation of agility, stability, safety, and control at the very weakest domains of aerodynamic-control surfaces. Thus, it can be used for the expansion of flight envelopes, and for flight-control improvement and/or emergency-redundancy systems.

Yaw and pitch vectoring, through multiple vectoring nozzles (for single as well as for multiple-engine aircraft), or through external pedals/flaps, provides *unprecedented roll and turning rates.* The simultaneous use of jet-yaw, pitch-and-thrust reversal fur-

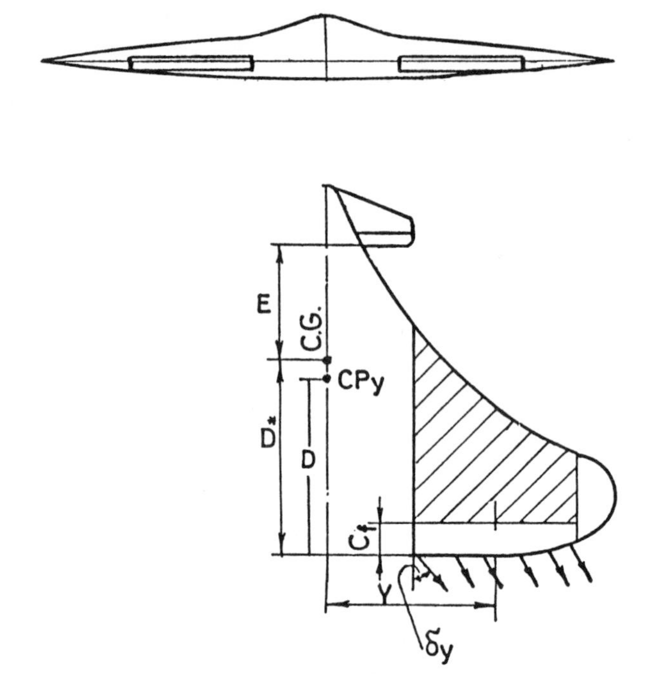

Fig. 1a. **Schematic view of a Pure Vectored Aircraft (PVA), in Pure Jetborne Flight Control.** Note the rolling arm Y, the yaw-directional jet angle δ_y, and the arm D^* for pitching moments of the vertical force $T_v = T \cdot \sin \delta_v \cdot \cos \delta_y$. Yaw force is $T_y = T \cdot \sin \delta_y \cos \delta_v$. Axial thrust is $T_x = T \cdot \cos \delta_v \cdot \cos \delta_y$.

Shaded area represents wing area affected by *supercirculation* (cf. Figs. III-7 and III-8) of the vectored jet in a pitching mode.

1 – Variable canard is differentially controlled. It becomes less effective at PST-maneuvers (see however Fig. 1c).

2 – Wing's sectional area S_j where $(\Delta C_L)_{SC}$ applies (cf. Fig. III-7).

Note the following cases:

(i) "Pure" Sideslips: The yawing thrust-vector of one engine points in the sidewise center-of-pressure CP_Y direction, causing "pure" *sideslips,* while T_x for the 2nd engine is reduced as dictated by eq. 4; (ii) δ_y is the same for both engines causing strong clockwise yawing-moments; (iii) the same as in (ii), but anticlockwise, (iv) *engine-out control:* with $\delta_y \neq 0$ and $\delta_v \neq 0$ with a differentially-controlled canard and C_f-flaps operating as ailerons to maintain stability; (v) other combinations of pitch, yaw and thrust-reversal vectoring in flight, and in landing (cf. Figs. 1b to 1f, and 6; also Fig. III-1).

Performance:

1) *Minimal dependence on aerodynamic-flight-control-surfaces.*
2) *Surpasses stall limits of conventional flight envelopes without loss of control and agility.*
3) *The aircraft is able to point in one direction (to fire its weapons), while 'translating' in another direction (to improve survivability).* Cf. eqs. 8 and 11 and the definitions of "direct lift" and "pure" sideslips (II-3, II-3.1).
4) The aircraft will be able to perform as, for instance, described in Fig. 2, *thereby increasing survivability potentials and simultaneously providing an off-boresight weapon aiming capability.*
5) By rejecting other "direct-lift" type surfaces, tails, stabilizers, flaps, flaperons, elevators, etc., as well as the canards, *RCS and other signatures may be substantially reduced; cf. Figs. 1b to 1f.*
6) *STOL or V/STOL performance are substantially enhanced, thereby becoming the "natural" domain of vectored aircraft.*

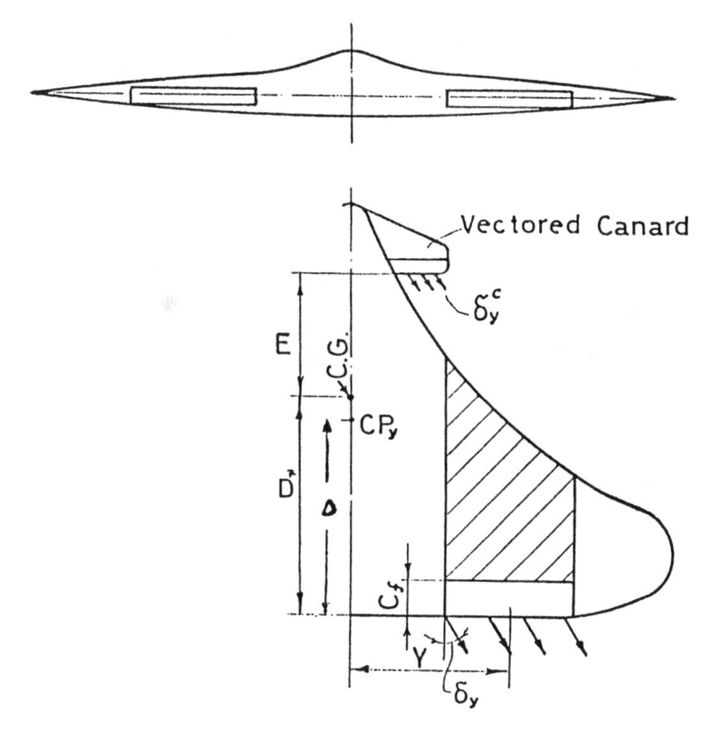

Fig. 1b. **Pure Vectored Aircraft Equipped With Vectored Canards.**

The canards themselves may be fixed or variable. δ_y should normally be less than δ_y^c in pure *sideslip "translations"* (see eq. 9).

Low pressure, *cold fan air* from turbofan engines may be used to control the vectored canards. Note also the distance between CP_y and CG.

Note: In the view of a few designers the optimal location for a canard is on someone else's airplane. Those who subscribe to this assertion are invited to examine the next figures.

ther enhances performance (see, however, our reservations in regard to thrust-reversal methodologies, as enumerated below).

Maximization of benefits to be derived from thrust vectoring requires a new, highly-integrated approach to the design process, including entirely new applications of propulsion, aerodynamics, flight-mechanics, control, materials, and structures technologies. *However, the integrated approach to these disciplines is not available yet, except for partially-vectored aircraft, to be defined next.*

II.2 Pure Jetborne Flight Control

Vectored aircraft may first be divided into those that are "pure" or "partial". Vectored propulsion, in turn, can be divided into internal or external thrust vectoring. In this Lecture we consider the payoffs obtainable from internal thrust vectoring. External thrust vectoring is discussed in Lecture VI.

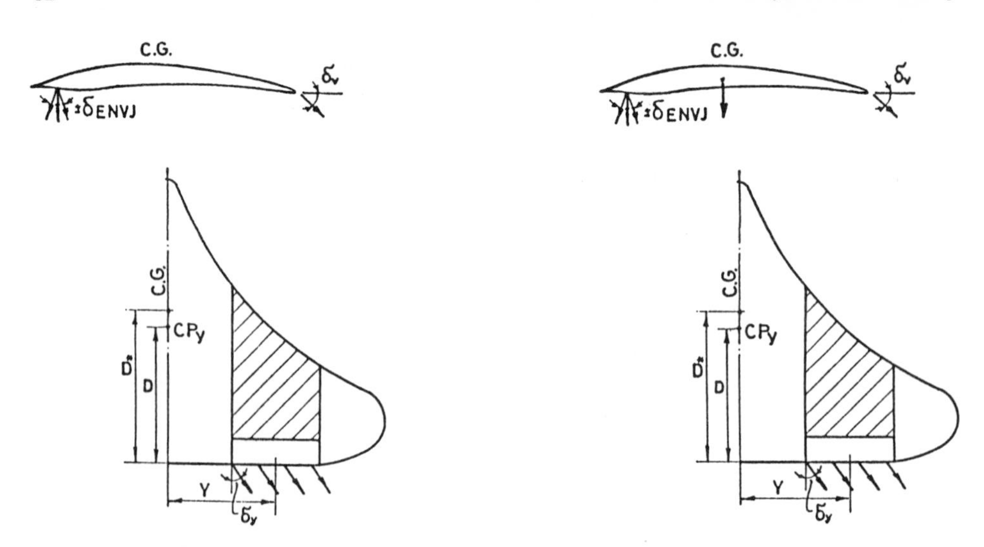

Fig. 1c. (left) **A schematic view of PVA equipped with *Extended Nose Vectoring Jet (ENVJ) for balancing moments and improving PST-Super-agility (without the use of canards)***

Fig. 1d. (right) **A schematic view of a V/STOL ENVJ-PVA.** An example of a central V/STOL nozzle as used by our team is shown in Fig. 1h.

II.2.1 Pure Vectored Aircraft in Pure Jetborne Flight Control

In pure thrust vectoring, the conventional, aerodynamic-control-surfaces of the aircraft, are replaced by the stronger, internal, engine forces. These forces may be simultaneously, or separately, directed in all directions, i.e., during pitch, yaw, roll, thrust reversal and forward thrust–control of the aircraft.

Moreover, pure (internal) thrust vectoring may entail the use of exhaust nozzles-wing structural-integration, so as to obtain optimal *supercirculation* (see below).

Thus, in pure vectored aircraft there is no need for flaperons, rudders, ailerons, elevons and flaps, and even the vertical stabilizers may become redundant. However, to improve maneuverability, a vectoring nose-jet, or differentially-controlled canards, may be required to balance moments and improve rate of turns, dynamic stability, etc. (see below, and Figs. 1a, to 1f).

I.2.2 STOL or V/STOL By Vectored Aircraft

A typical landing procedure is demonstrated in Fig. 6, with nozzle positions as in Fig. I-7. Lower approach speeds, as well as a rapid change of the Venetian-type vanes, allow the use of very short runways. The internal structure of such nozzles is shown in Fig. I-8 with Figs. I–13, 14 showing them with high NAR. Fig. I-6 stresses the high nozzle coefficients of *2D–CD* vectoring nozzles and Fig. I-8 demonstrates elaborate cooling methods for non-RCC nozzles.

Pure vectored flight normally involves the *simultaneous* use of a few effects, e.g.,

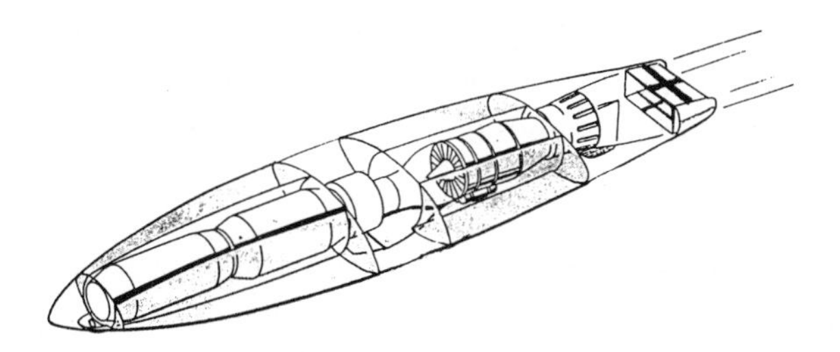

Fig. 1e. **A possible design for an advanced pitch-roll-yaw, thrust-vectored, cruise missile with low signatures.** Note the split-type *2D-CD* thrust-vectoring nozzle which allows simultaneous yaw-pitch-roll thrust-vectored maneuverability or even supermaneuverability. Axi-TV nozzles* cannot provide sufficient roll thrust-vectoring moments, nor supercirculation lift gains and deflection angles greater than 13 to 15°.

thrust reversal combined with pitch and yaw vectoring. *This may be highly useful in close (viffing) combat engagements (Fig. 9), or in low-speed runway approach (Fig. 6).* Combined pitch and yaw vectoring is useful in high-maneuverability engagements, or in landing sites with strong cross-winds (combined with roll moments). Vectored propulsion/flight also allows decoupled flight maneuvers, in which the aircraft translates up, down, left or right, without banking, or changing attitude!

II.2.3 Partial Vectoring or Partial Jetborne Flight (PJF).

Rudders, elevons, ailerons, flaps, canards, and elevators are still used in conjunction with thrust vectoring during Partial Jetborne Flight (PJF). Thus, the elimination of one or more modes of the pure-vectored elements, such as the elimination of yaw-vectoring and supercirculation in the Harrier, or in the STOL F-15 demonstrator, categorizes the aircraft as PJF.

The flight experience of this laboratory, during 1987, employing pure vectored-RPVs, has demonstrated, for the first time, *the ability of a properly-spaced, high-aspect-ratio nozzles**, to act as "elevons" in engine-out flight control, and in landing situations* (cf., *A, C*, Y* and *D* dimensions in Figs. 1 and III-1). Part of this course is based on this experience, and also on the experience gained in testing various vectored engines in the laboratory (see below).

II.3 Dimensionless Numbers For Pure Vectored Aircraft

Force, moment, energy, and Specific Excess Energy (SEP) equations must be derived for the various types of PVAs shown in Figs. 1a to 1g. These equations may become the basis of various *simulation programs* of vectored aircraft performance.

---

* Cf. Fig. 1i.

** NAR=46.

In addition, one may search for dimensionless numbers which allow scale-up and scale-down design considerations. One may, therefore, search for dimensionless numbers which are, under certain conditions, *independent of*:

1) *The thrust level of the engines.*
2) *The number of engines.*
3) *The external aerodynamic forces and moments acting on the aircraft.*
4) *The major aerodynamic design parameters of the aircraft.*

For this purpose one may first define two dimensionless numbers for the type of vectored propulsion shown in Fig. 1a:

$$N_1 = \frac{\text{Vectoring Yawing Moment}}{\text{Vectoring Pitching Moment}} \tag{1}$$

$$N_2 = \frac{\text{Vectoring Yawing Moment}}{\text{Vectoring Pitching Moment}} \tag{2}$$

For the example shown one may further assume that:

a) The exhaust-nozzle-mass-flow-rate in single-engine PVAs is equally divided between the "left" and "right" thrust-vectoring-nozzles, so that, during all vectoring angles,

$$T^R = T^L. \tag{3}$$

b) In multiple engine PVAs eq. 3 holds for both right and left-hand engines (unless pure sideslip flights are conducted – cf. the Introduction and eqs. 6 to 10 below).
c) During vectoring $\delta_y^L = \delta_y^R$ and $\delta_v^L = \delta_v^R$, except in sideslips and rolling.
d) During rolling $\delta_v^L = -\delta_v^R$, or vice versa, while $\delta_y^L = \delta_y^R$.

Let us consider now a steady-state, simultaneous yaw-pitch vectoring. Now, by employing eqs. I-11, I-12 and I-13, and by neglecting the coupling between yaw and roll, as well as the canard effects, one obtains:

$$N_1 = \frac{\cos\delta_v \cdot \sin\delta_y}{\sin\delta_v \cdot \cos\delta_y} \qquad \textbf{(Control Rule No.6)} \tag{4}$$

$$N_2 = \frac{D\cos\delta_v \cdot \sin\delta_y}{Y\sin\delta_v \cdot \cos\delta_y} = \frac{D}{Y}N_1 \qquad \textbf{(Control Rule No.7)} \tag{5}$$

These numbers do not depend on the thrust level of the engine(s), nor on the number of engines.

However, the couplings and the canard effects may not be negligible. Moreover, large β, $\dot{\beta}$, α *and* $\dot{\alpha}$ values may change engine-intake-performance (Appendix F), as well as the left/right-wing-lift values, aerodynamic moments, etc. We shall return to this problem below.

II.3.1 Pure Sideslip Translations

Various decoupled flight maneuvers, which allow the aircraft to "translate" "left", "right", "up" or "down", without banking, or changing attitude, are readily controlled by well-designed vectored vehicles.

Consider, for instance, the following design option for a (steady-state) "left"

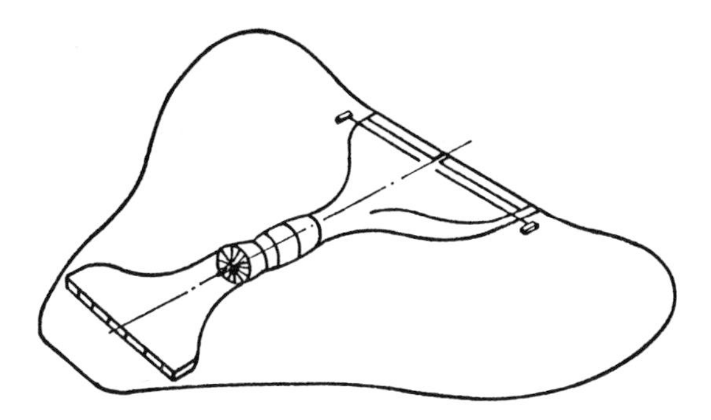

Fig. 1f. **An early concept of a vectored RPV proposed by this laboratory.** The thrust-vectored roll rate of this design was poor, due to the small yaw-arm length Y (Fig. 1a). Hence, its propulsion system was replaced by a split-type roll-yaw-pitch TV nozzle.

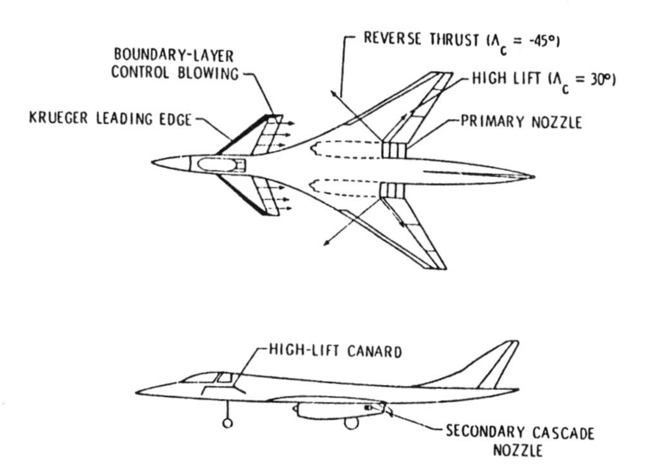

Fig. 1g. **A configuration proposed in 1983 by Paulson and Gatlin [Ref. 155] for an advanced (partially-vectored) fighter**. It shows a specific design for main-vectoring-nozzles-*AR* and location, as well as a boundary-layer-control-*blowing over high-lift canards* [see also Lecture V and the notes in Figures 1a to 1d].

"translation" without changing vehicle's attitude, but with thrust reduction of the left-hand engine, (see, however, § II-3.5).

Consulting Fig. 1a first, one may choose to perform a pure *"left" sideslip* maneuver, by keeping the throttle of the right-hand engine *unchanged, but directing its yawing thrust in the CP_y direction*, i.e.,

$$\delta_y(\text{Right Engine}) = \sin^{-1}[Y/D], \tag{6}$$

where Y and D are defined in Fig. 1a.

Yet, to balance the yawing moments, the left-hand thrust, T^L, (with $\delta_y^L = 0$) must be reduced to

$$T_x^L = T_x^R - T_y^R\, D/Y + F_{cp}\,(D^* - D)/Y \tag{7}$$

i.e., to

$$T_x^L = C_{fg} T_i \left[\cos\delta_v \cos(\sin^{-1}\{Y/D\}) - \frac{D}{Y}\ (\cos\delta_v \sin(\sin^{-1}\{Y/D\}) \right] + F_{cp}\frac{(D^* - D)}{Y}, \tag{8}$$

where F_{cp} is the aerodynamic drag force resulting from the (steady-state) sideslip which operates at CP_y.

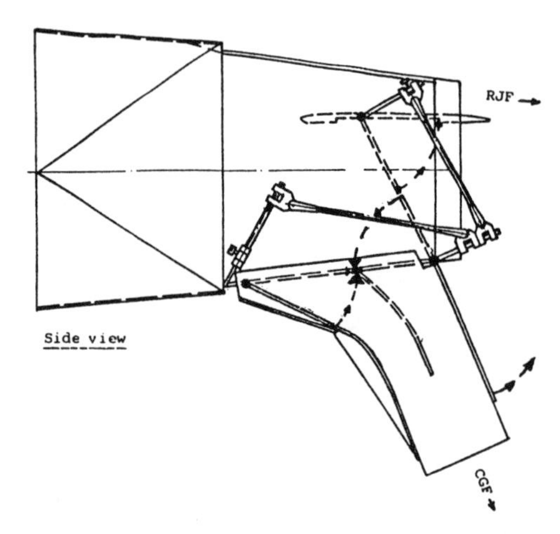

Fig. 1h. **The Turgemann-Friedman-Cohen vectoring nozzle for pure vectored RPVs of the type described in Fig. 1d** (was flight tested in 1987 on our No. 5 V/STOL vectored RPV, with canards instead of ENVJ).

Note that this nozzle divides "rear jet flow" (RJF) from "CG down flow" (CGF), and is normally closed (forming a flat surface under the wing).

The dashed-line arrows indicate the respective flaps rotations during closure of the nozzle in normal cruise flight.

A dimensionless number for such slideslip maneuvers may thus be defined by **Control Rule No.8:**

$$N_3 = \frac{\sum F_y}{\sum F_x} = \frac{T_y^R - D_y}{T_x^L + T_x^R - D_x} = \tag{9}$$

$$= \frac{C_{fg} T_i [\cos \delta_v \cdot \sin \{\sin^{-1} Y/D\}] - D_y}{C_{fg} T_i [\cos \delta_y \cos(\sin^{-1} Y/D)] - \frac{D}{Y}} \Big/$$

$$\cdot [\cos \delta_v \sin(\sin^{-1} Y/D)] + F_{cp} \frac{(D^* - D)}{Y} + C_{fg} T_i \cos \delta_v \cos(\sin^{-1} Y/D) - D_x$$

where D_y and D_x are the *drag components in the y and x-directions*, respectively.

Small controlled variations in δ_v and δ_y may be required in actual sideslip translations (due, for instance, to variations in lift/drag forces on the left and right wings, etc.). Moreover, other combinations of δ_v and δ_y are possible without changing engine thrust of both engines (see § II–3.5). A similar dimensionless number may be derived for *"upslip" translations*,

$$N_4 = \frac{\sum F_z}{\sum F_x}, \tag{10}$$

as well as for the vectored-canard, and ENVJ-PVAs shown in Figures 1b, 1c and 1d.

II-3.2 Pure Side-Slips in Vectored–Canard and ENVJ-PVAs

Vectored-canard and ENVJ-PVAs can perform *sideslip maneuvers by* rotating *all jets in one direction.* In this case one must keep in mind that the rear δ_y*-angles should normally be less* than the canard δ_y^c-angles, due to smaller nose or canard thrust levels and the different moment arms, i.e., under steady-state condition (cf. Fig. 1b);

$$\frac{T_y^c}{T_y} = \frac{D^*}{E} \tag{11}$$

For this type of vectored aircraft *the work-load on the pilot (or on the IFPC system) is much less during pure sideslips.* Consulting Figs. 1b and 1c one may also conclude that, unlike the previous case, no throttle-variations *are required in sideslip "translations"* of this PVA.

The moment dimensionless numbers for this type of PVA become:

$$N_5^c = \frac{\text{Yawing Moment}}{\text{Pitching Moment}} = \frac{T_y \cdot D^* + T_y^c \cdot E}{T_v \cdot D^* + T_v^c \cdot E} \qquad \textbf{(C.R. No. 9)} \tag{12}$$

$$N_6^c = \frac{\text{Yawing Moment}}{\text{Rolling Moment}} = \frac{T_y \cdot D^* + T_y^c \cdot E}{T_v \cdot Y + T_v^c \cdot E} \qquad \textbf{(C.R. No. 10)} \tag{13}$$

Since normally $\delta_v \neq \delta_v^c$ and $\delta_y \neq \delta_y^c$, these numbers *depend* on the thrust, as well as on the distribution of "forward" and "rear" thrust levels.

Left/right-wing-lift-variations, aerodynamic moments and other considerations may now be added, as in conventional aircraft design methodologies.

II-3.3 Vectored Flight-Control During One-Engine Flame-Out

Twin-engine vectored aircraft, like the one shown in Fig. 1, may be flight-controlled in the following mode during *one-engine flame-out*:

The vectoring-nozzle-*flap*, sized as $(C_f) \times (A/2)$, of the powerless engine, is still *used as an aileron*, while the thrust on the other engine is reduced to a predetermined minimum and its yaw-thrust-vectoring set at

$$\delta_y = \sin^{-1}(Y/D) \qquad \textbf{(Control Rule No. 11)} \quad (14)$$

This will cause a controlled, sideslip flight. However, depending on, ENJV, or on the canard configuration and the dihedral and mode of control (e.g., differentially or non-differentially), *the degree of sideslip may be minimized, or eliminated altogether.*

Emergency Landing, Or Restarts With Total Power-Failure

Both single and multiple engine vectored aircraft may be landed safely during *total thrust-failure.** Flight controllability, under these conditions, may, however be regained for restarting in altitude flight.

A number of factors should be considered by the designer of a vectored aircraft for this case:

1) *Both vectored nozzle flaps*, each sized at $(C_f) \times (A/2)$, can still operate *in the conven-*

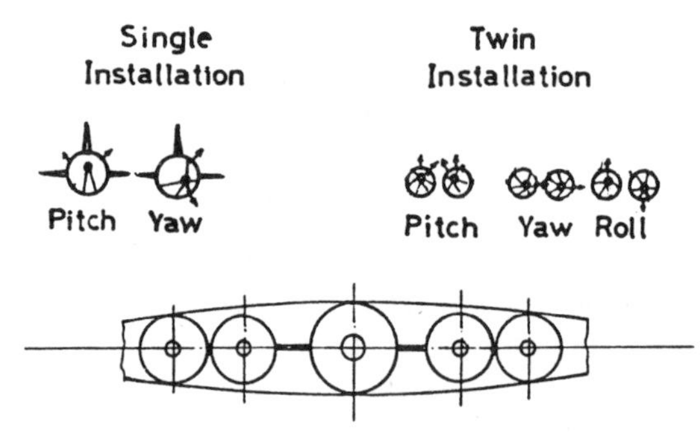

Fig. 1i. **(upper drawing). Yaw-pitch, axisymmetric, thrust vectoring nozzles may provide the optimal mix of low drag and simple modification of nozzle and engine controls.** However, they are limited to low deflection angles and provide no lift enhancement by superciculation. Their signatures may also be larger than those characteristic of high aspect ratio, yaw-roll-pitch *2D–CD* nozzles. (lower drawing) Mini-, Midi-, and Maxi-RPVs may be designed with a central *Vankel engine*, or a *piston engine* driving symmetric pairs of multiple-stage compressors for (cold), high-α, vectored aircraft (front view).

Such vectored RPVs need *not* rely on the performance of PST *inlets* (for these engines *are not sensitive to inlet distortion even at* α *values as high as 90 degrees).* See also Lecture IV and Appendix F.

* This conclusion was verified recently in flying our vectored RPVs.

tional way as moments effectors for flight controllability, supplying pitch and roll control in the usual way of conventional technology flight-control.

2) The canards must be controlled differentially.
3) Canard configuration *with dihedral*, (cf. Fig. 7), or ENVJ are useful in gaining *yaw control* (see also Lecture V).

II-3.4 C_A Variations in Flight Control

An advanced IFPC system for PVA must employ δ_v and δ_y variations for flight control. In turn, these variations would affect engine performance and engine thrust vectoring.

One may, therefore, distinguish between *three vectoring flight-control domains*:

1) *Small, time-modulated,* δ_v-, δ_Y-*variations* affecting only the cruise conditions. Compared with the large values of T, these variations are neglibily small, especially for transient δ_v, δ_y deflections not exceeding a few degrees. Hence, these variations may be considered as *"engine-thrust-loss-variations"* (see Lecture III).

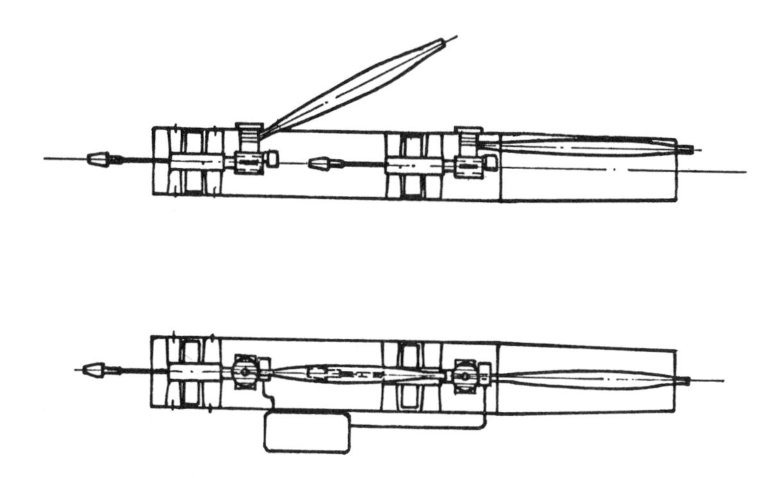

Fig. 1j. **How to By-Bass Inlet Distortion Problems in R&D, PST, Vectored RPVs?** The answer is by Cold Propulsion.

Shown here is the simplest multistage cold propulsion system which overcomes the problem of inlet distortion at high AoA flight conditions. Note that the 2-stroke piston engines do not stall, as does a jet engine equipped with a non-PST inlet at high AoA values. Moreover, high RPM values are always maintained by the two compressors by separating the two spools and the two engines from each other. In one-engine-out situations, the 2nd engine still performs well. Proper side covers allow individual side-wise starting for each engine. Fuel tank is common to all engines.

The 2-stroke engines may be replaced by 4-stroke, or by Vankel-type engines. However, the need to maintain very high RPM values dictate the need for a gearbox. This requirement presents special design problems that are now investigated by this laboratory, using light-weight, aviation-type engines in the range of 30 to 130 HP.

Alternatively, we investigate the use of high-by-pass-ratio turbofans and shrouded propfans, as well as the design and laboratory/flight tests of variable PST inlets.

2) STOL and super-agility δ_y-variations affecting pitch moments.
3) STOL and super-agility δ_y-variations affecting yaw moments.

Consequently, domain-1-type-of-variations may be included in the very definition of C_A, as explained in eqs. III-23-a, b. However, this incorporation is merely a matter of convenience.

II-3.5 Pure Sideslip Without Changing Engines Throttles

Fig. 1k, and eqs. 15 to 20, demonstrate a solution for pure sideslip flight, namely, the elimination of the time-consuming throttle variations of one engine with respect to the other. Thus, one may solve equations (15) to (18) to find the proper δ_y^L, δ_y^R values for obtaining pure sideslips when $T^R = T^L$. The procedure is given below:

$$\sum M_y = (T_x^R - T_x^L)\, Y - (T_y^L + T_y{}^R)\, D^* + D_{cpy}(D^* - D) = 0 \tag{15}$$

$$\sum F_x = T_x^L + T_x^R - D_x - D_{cpx} = 0 \tag{16}$$

$$\sum F_y = T_y^L + T_y^R - D_{cpy} = 0 \tag{17}$$

$$T^R = T^L \quad \text{or} \quad [(T_y^L)^2 + (T_x^L)^2 = (T_y^R)^2 + (T_x^R)^2] \tag{18}$$

Solving (15) to (18),

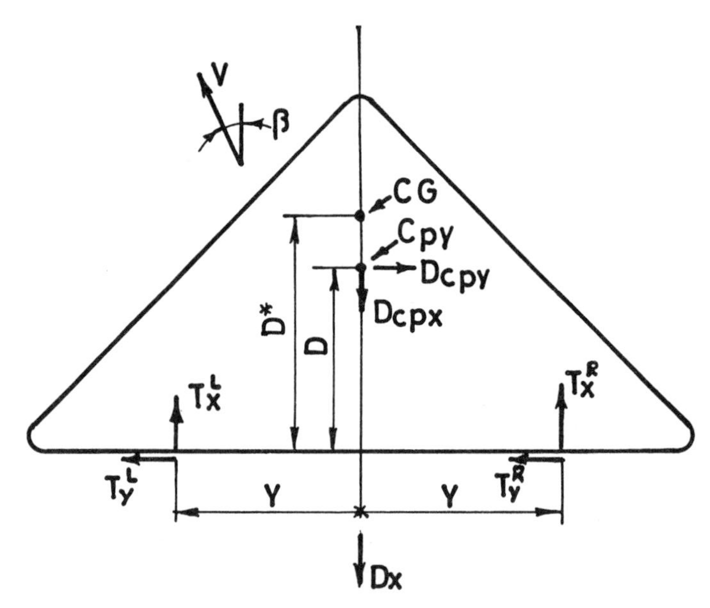

Fig. 1k. **What should be the values of δ_y^L and δ_y^R in pure sideslips involving "no-change" throttle settings?** Proper definitions of relevant vriables for steady-state, pure sideslip-flights involving "no-change" of throttle settings (twin-engine, pure vectored aircraft). For additional definitions see Figs. 1a to 1c and eqs. I-11 to I-13 and II-15 to II-20.

Note that the moments around *c.g.* are counted positive for counterclockwise rotations and *CP* is the sidewise center-of-pressure for the steady-state sideslip motion at angle β. T_x and T_y are the thrust components during roll-yaw-pitch vectoring.

$$\delta_y^L = \text{tg}^{-1}\left(\frac{T_y^L}{T_x^L}\right) \tag{19}$$

$$\delta_y^R = \text{tg}^{-1}\left(\frac{T_y^R}{T_x^R}\right) \tag{20}$$

II-4 Air-To-Air Operations

Air-to-air supermaneuverability operations involving, say (Figs. 2, 3, 4) deceleration with or without thrust reversal, + climbing decelerations at high-α values (up to 90 degrees) + gravitational deceleration, and, then, at *very low speed* (when conventional aerodynamic contol surfaces become ineffective) *to yaw-turn the aircraft nose down, by using full-jet-yaw-vectoring, (engine must stay at full thrust), and accelerating down, and back, to the initial speed and altitude.*

Combined with pure 'sidewise', or 'upwise' translations, such maneuvers result in the

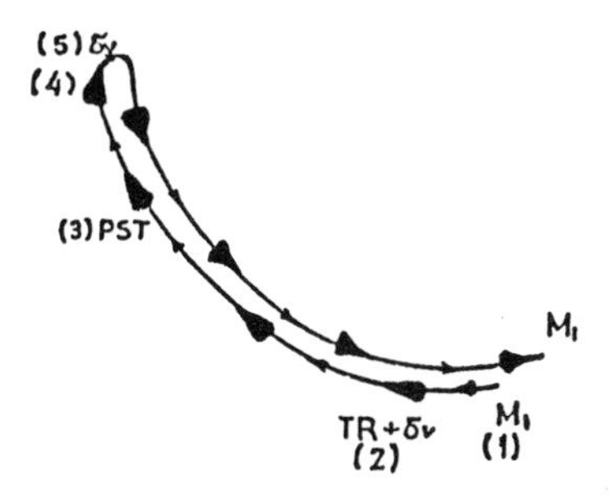

Fig. 2. **A typical PST/PSM by pure vectored flight to gain supermaneuverability in close-combat between a vectored fighter and a conventional fighter (not shown) *in head-on situations*, or in firing missiles *at the edge of an adversary's missiles envelope*, and, simultaneously, turning back to escape from the adversary's missile's envelope.** In any case the load numbers are less than, or equal to those encountered nowadays with conventional fighters (see below).

(1) At $t = 0$, as in the Herbst PST-maneuver, the pilot decides to turn back for a rapid-pointing-and-shooting, or for escaping the adversary's missile's envelope. (2) Increasingly high α values plus *TR*, plus pitching vectoring δ_v may be applied simultaneously. (3) Aircraft enters the post-stall domain. (4) All aerodynamic control surfaces become ineffective. (5) Yaw vectoring is applied for rapid nose-down, δ_y-turn, for the start of diving and acceleration in the gravity vector. Speed is rapidly regained, while the maximum load number allowed has not been surpassed.(8) Aircraft at the initial speed, altitude and energy levels, ready to shoot at a much shorter time and at a much smaller turn-back radius than its adversary. The proper use of a variable canard, or a nose-vectored jet, may enhance this performance. *Engine must stay at full thrust during the entire operation.*

Note also that the use of thrust reversal (point 2) is not essential to gain superiority over the opponent. It is the combined use of pitch and yaw thrust vectoring which matters most! Also note the difference in aims and pilot's methodology between missile avoidance and close-combat situations. *Close-combat* situations, in turn, are subdivided into a number of *sub-categories*, involving, say, 1:1, 1:2, 1:many fighters engaged in various, time-space-attitude-envelopes-weapons-speed situations. Hence, to begin an exact analysis of such situations one must first examine the following figures and the rest of the text.

saving of time for shooting, and, simultaneously, in acquiring superiority-survivability advantages in 1 to 1, 1 to many, and many-to-many close-combat situations (see below).

II-5 Air-to-Ground and V/STOL Operations

Pure vectored propulsion may help not only in gaining superiority in air-to-air combat, or in *escaping missiles, or in aiding low-flying Advanced Cruise Missiles (ACM). It also helps in reducing thermal, radar and optical signatures, and in reducing runway distances, even on icy runways.* Perhaps the most important innovations of vectored propulsion are the new possibilities for its employment, without sacrifice in its performance, in new V/STOL missions. This conclusion applies equally to manned or unmanned aircraft.

Here, one must distinguish between single and multiple engine aircraft, and between manned and Robot Aircraft (RA), such as RPVs, ACM and certain air-breathing, vectored missiles. (The introduction of vectored propulsion into RA technology decreases the conceptual-technological differences between certain advanced missiles and certain vectored RPVs.)*

II-6 Propulsion and Supermaneuverability

A primary factor in vectored aircraft R&D efforts is the design for survivability through supermaneuverability and supercontrollability.

Supermaneuverability includes the capability of a fighter to execute tactical maneuvers with controlled side-slipping, and at angles of attack beyond maximum lift. As an example of supermaneuverability one may re-examine Figs. 2 and 3 for head-on, or escape maneuvers.

The tactical payoffs of such PSM–PST-maneuvers are:

- *A higher survivability against adversary or missile attacks (counterfire denial).*
- *Increased ability to confuse adversary pilots and missiles by using decoupled flight maneuvers that allow the aircraft sharp "translations"; "up", "down", "left" or "right", without banking or a noticeable change of attitude.*
- *Increased first-shot opportunities.*
- *Quicker-into-firing position maneuver.*
- *Maintaining longer firing position.*
- *Quicker and easier to point-and-shoot in multiple-targets situations.*
- *Dictating tactics throughout the entire speed regime.*

---

* Evaluating the feasibility of V/STOL and STOL RA, including our flight testing of single-and-multiple-engine RA, has recently demonstrated the potential payoffs obtainable with 3-jet positions. V/STOL RA, in which one of the "downward-pointing vectored jets" is located at the RA center of gravity, while the other two are being operated as in vectored modes, but with newly balanced moments. Cf. Figs. 10 and 1h and Fig. 15 in the Introduction.

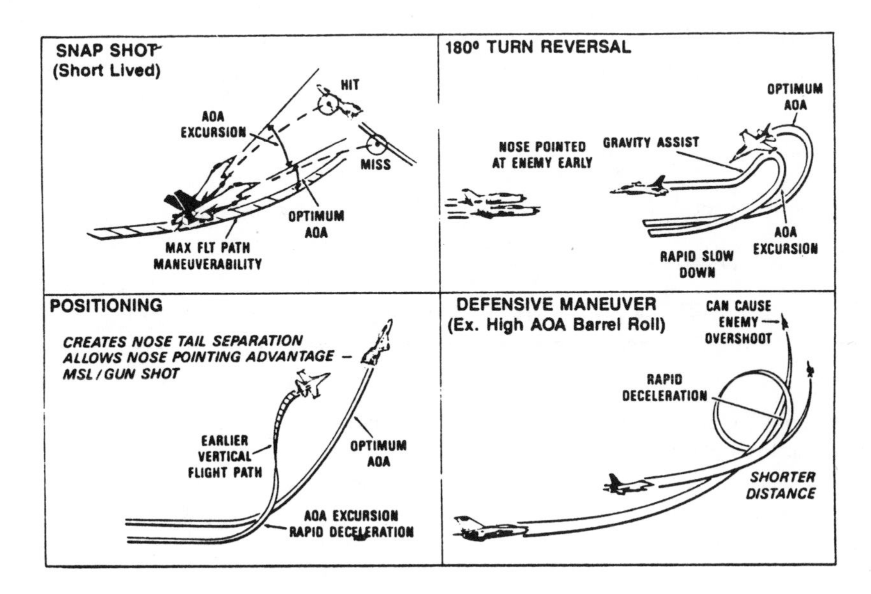

Fig. II-2a. **Herbst PST Excursions, even without thrust vectoring, can be advantageous (196).** According to McAtee of GD, controllability during PST thrust vectoring ("supercontrollability"), presents the industry with the greatest challenge.

- *More missiles deployable.*
- *Lower g-loads on pilots and aircraft structure, and also less time spent at limiting g-loads.*
- *Dictating the course of the engagement.*

An early analysis of supermaneuverability has been reported by Herbst [154]. While limited to zero side slipping, his conclusions are highly instructive for this course. They are based on simplified maneuver elements, and are also aided by manned, close-air-combat simulations. Using such simulations, Herbst defines some maneuver cycles, which are consistent with unconventional air combat maneuvers with all-aspect weapons. *Inter alia*, Herbst recommends a few optional ranges of speed and (simple) vectoring maneuvers, as well as the required T/W ratios and control power to attain super-agility status.

To start with, Herbst doubts the importance of thrust reversing and stresses mainly the great advantages of pitch-yaw-vectoring in Post-Stall maneuvering. He does not, however, deal with PSM.

Under these simulation conditions, a single PST-capable fighter, is able to neutralize two conventional opponents with equal weapons.

Herbst employs computer simulations involving multiple opponents in which the PST-capable fighter gains clear advantages in duel, and in multiple engagements.

His data are based on 3000 simulated engagements on 3 different manned combat simulators operated by operational pilots of three airforces.

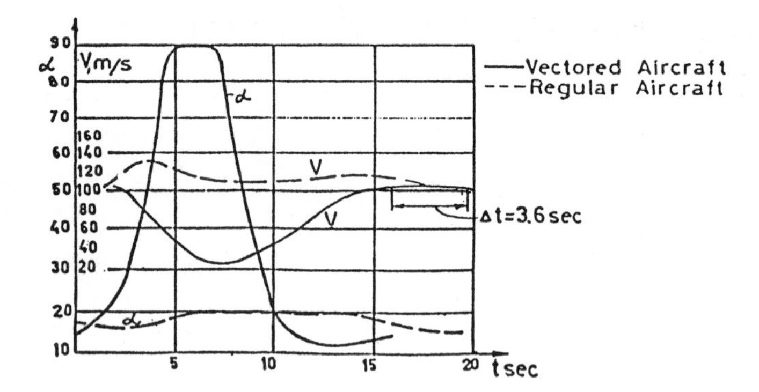

Fig. 3. **Herbst Variations of AoA and aircraft speed for the vectored-supermaneuverability turn-back-maneuver shown in Fig. 2.**

Note the gain of about 4 seconds in comparison with a conventional-technology fighter (dashed line) -Herbst [154]. *This gain may become a key element in enabling vectored fighters to survive and win in battles, both beyond and within visual range of the enemy.* Note also that the lowest speed zone of the vectored fighter is passed through, while the conventional fighter's nose is still pointing in the opposite direction (during a much larger radius of turn). This performance may be further improved by employing high NAR and by the design methods explained in the text.

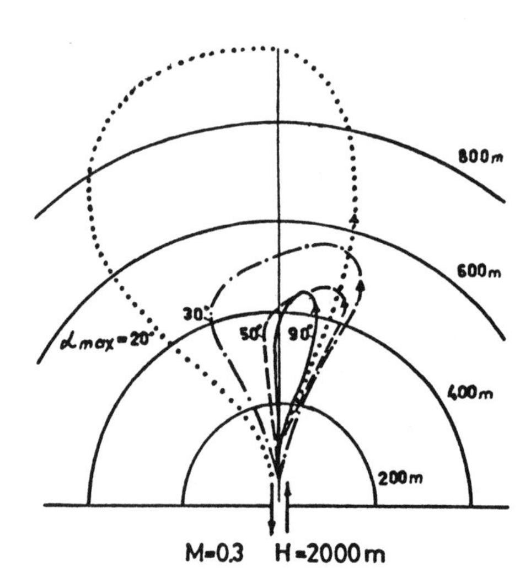

Fig. 4. **Simulated, turning radius values for the vectored-supermaneuverability maneuver shown in Fig. 2.**

This performance may be greatly improved by employing high NAR and supercirculation.

This simulation is limited to zero side-slipping, Herbst [154].

Any of these maneuvers can be designed with load-numbers less than or equal to those encountered nowadays with conventional fighters.

Most firing opportunities occurred just after finishing a PST-maneuver and returning to conventional flight regime.

Pay-offs of such maneuvers are therefore based on a trade-off between *a momentarily loss of energy, versus the gaining of time and positional advantages for a first firing* (cf. Figs. 2, 3, 4). Moreover, with PST capability the fighter can roll around the velocity vector at a constant, very high-angle of attack. Consequently, PST fighters dictate the tactical course of the engagement!

II-7 Maneuver Analysis

Limiting the analysis to zero sideslip conditions during PST-maneuvering, the results shown in Figs. 11, 12 and 14 are calculated for a particular aircraft drag polar, at a particular altitude and for a maximum engine power, using a mass-point analysis [154].

Fig. 11i describes two conditions of a vertical maneuver.

One may note first the zero Specific Excess Power (SEP) line, distinguishing between accelerated and decelerated maneuvers, as well as the structural and maximum lift limits. Thus, any combination of turn rate and Mach number corresponds to a particular radius of turn. These results indicate the following advantages:

- *A turn rate advantage beyond maximum lift AoA at low speeds.*
- *Minimal turn radius advantages at PST–AoA beyond 50 degrees.*
- *Best results are obtained when engine thrust remains at its maximum setting during the entire maneuver.*

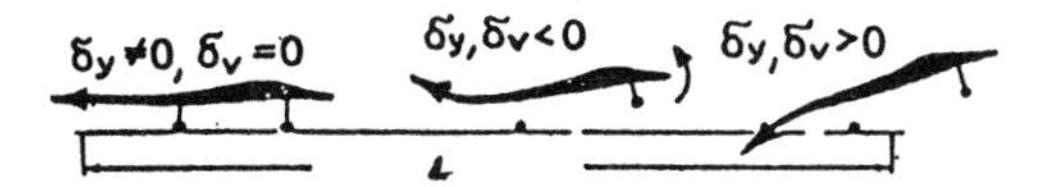

Fig. 5, **Takeoff procedures of pure vectored aircraft.** Note that *the jets are first turned "up", and, following rotation, are turned "down".* Using our vectored RPVs we have tested both vectored and unvectored takeoffs, and found the first minimizing ground roll substantially, especially for high T/W vectored RPVs.

Similar conclusions are obtained from horizontal maneuvers, at a time in which the unvectored aircraft is still outside its firing cone.

Fig. 11e is another representation of the same engagement. It shows how the pilots in the vectored aircraft (left hand side of Fig. 1e) and in the conventional aircraft (right hand side of Fig. 11e) would view their opponents as they look outside their cockpit windows. The proper firing time for the vectored fighter is marked by the large arrow in the middle of the given weapon firing cone.

The vectored fighter performs the low-speed, PST-maneuver far outside the unvectored fighter weapon range. Yet it gets its first firing opportunity in a safe posi-

Fig. 6. **Landing procedure of pure vectored aircraft (STOL Vectored Aircraft).** *Note:* Giving up thrust reversal (*TR*) may increase landing ground roll, but would also save weight, cost and operational complexity. Note also that *pitch vectoring, combined with supercirculation, decreases the approach speed, thereby decreasing landing ground roll, without the need to use thrust reversing.* However, thrust reversal potential deployment on approach – as shown here and as explained in Fig. I-7 – allows the jet engine to be maintained in *full dry power, thus avoiding spool-up delays after touchdown* (as encountered with civilian transport aircraft). Note also that in pure vectored aircraft there is no need to install engines with a fixed pitch deflection, unless cruise drag considerations dictate it. In the following text we assume that δ_v is defined as in Fig. I-1 and engine center-line/pitch deflection may be added or substracted from the proper δ_v values.

The landing phase is an example for the *simultaneous use of vectoring jets in all directions*, i.e., for a partial reversal, a partial forward, yaw, pitch and roll control.

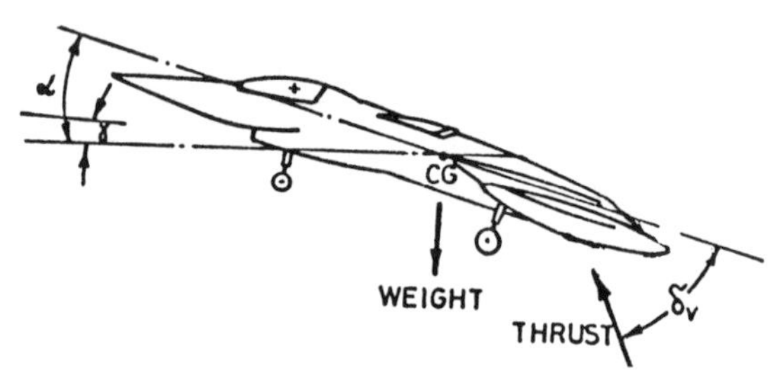

Fig. 6a. **Glide-slope-force-equilibrium on (pitch-only) vectored aircraft.**
Note that:

$$T \cdot \cos(\alpha + \delta_v) = \text{Drag} - \text{Weight} \cdot \sin\gamma$$

$$T \cdot \sin(\alpha + \delta_v) = \text{Weight} \cdot \cos\gamma - \text{Lift}$$

However for future designs, the Vectoring inlet suction force should be separately added to these calculations.

tion relative to the unvectored fighter. The overall result is that the vectored fighter obtains a 'firing period' without facing the possibility of counterfire from the unvectored fighter. In fact, the vectored fighter is not to be found in his opponent's firing cone until the vectored fighter's first missile hits the target. Moreover, the vectored fighter rapidly recovers the energy lost during the PST-maneuver, as shown in Figs. 11a and 12. The area between the solid lines represents the energy of the vectored fighter, while the region marked with dashed lines, illustrates that of the unvectored fighter.

Vulnerability of the Vectored Fighter in PST-Maneuvers

There is no doubt that the vectored fighter is vulnerable during the low-speed/high angle-of-attack phase of the PST-maneuver, especially in multiple engagement situations.

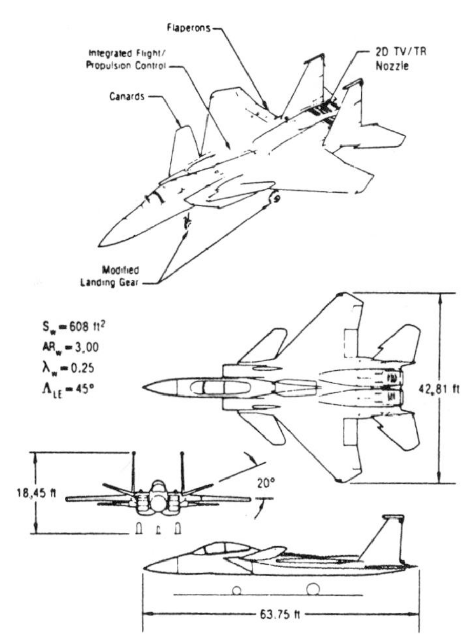

Fig. 7. **The Partially-Vectored F-15 Fighter.**

The new F-15 S/MTD (STOL and Maneuver Technology Demonstrator) is intended to flight-demonstrate *six key technology areas holding promise for ATF-type aircraft, as well as for improved derivatives of current-technology aircraft.* These technology areas include (13, 18, 62, 70, 71, 111, 128, 234):

1) *IFPC systems.*
2) *2D–CD, TV/TR nozzles.*
3) *Enhanced lift systems, including independently-variable-incidence-canards.*
4) *Enhanced pilot/vehicle interface/avionics.*
5) *Rough/soft field landing gear.*
6) *Engine-intake distortion avoidance designs for PST-maneuvers.*

The new fighter is approximately 10% statically unstable with the canards fixed subsonically (see also § V.4). Note; The Soviets are currently testing similar designs on the Su-27-1024 agile interceptor. (The unvectored Su-27 can fly up to 120° AoA during the "Pougachev's Cobra" Maneuver.)

Inlets Suction forces/moments play increasing roles as AoA goes to 90° and beyond. Increasing the throttle may affect attitude. AT AoA = 90° the inlet force is perpendicular to the wing, and the lift force vanishes.

It is, therefore, left to the pilot's judgement, whether or not to perform a PST-maneuver, or to employ other maneuvers, such as various decoupled flight maneuvers in which the vectored fighter translates up, down, left or right, without banking or changing attitude. Such maneuvers may also confuse the opponent and may be combined with modified PSM–PST-maneuvers.

II-7.1 Powerplant, Structural and Controllability Limits

Fig. 13 marks the limits given by engine power, controllability and structural constraints.

There is a low-speed region of possible sustained maneuvers for fighters with $T/W > 1.0$. However, in combat, to retain high SEP (specific excess energy) may be preferred for acceleration, or to gain altitude, etc. See, however, DST in the Introduction.

Nevertheless, as the T/W ratio of high-performance fighters increases, one expects the relative time maneuver in the PST-zone to decrease, thereby to decrease vectored-fighter vulnerability during such PST-maneuvers!

Fig. 14 demonstrates that for $T/W < 0.6$ there is no tactical advantage during PST-maneuvers. Thus, many old fighters may still be upgraded to become vectored aircraft.

II-7.2 Other Requirements for PST-Maneuvers

Herbst recommends the following rules for PST-capability:

1) *Sufficient control power in pitch, yaw and roll at Mach numbers as low as* 0.1 *and with an incidence of up to* 70 *degrees.*
2) High-angle-of-attack compatibility of up to 70 degrees, at Mach numbers as high as 0.6 (4,000m altitude), with respect to aircraft stability and/or engine intake performance (i.e., flame-out limits, etc.).
3) $T/W > 1.2$.
4) *PST-Maneuvers are characterized by high pitch rates and rotation in yaw and roll at the same time.* Coordinated flight with zero sideslip requires rotation around the velocity vector. Since the pilot does not recognize the velocity vector, *the control system has to be designed accordingly.* Thus, a lateral stick input would have to produce more yaw and less roll at increasing AoA. However, Herbst points out

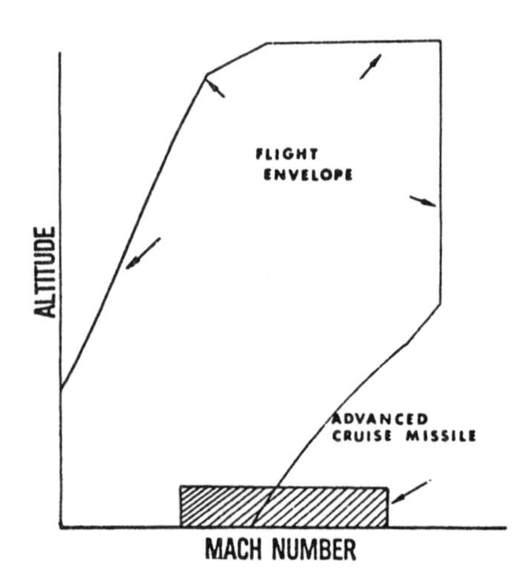

Fig. 8. **Simultaneous yaw-pitch-roll thrust vectoring may also be used to expand the various performance and steady-state flight envelopes of advanced, vectored/stealth cruise missiles.** Advanced Cruise Missiles (ACM) could fly by body lift assisted by thrust vectoring. Such designs may be almost impossible to detect. High-energy boron slurry fuel may be of interest in some missions.

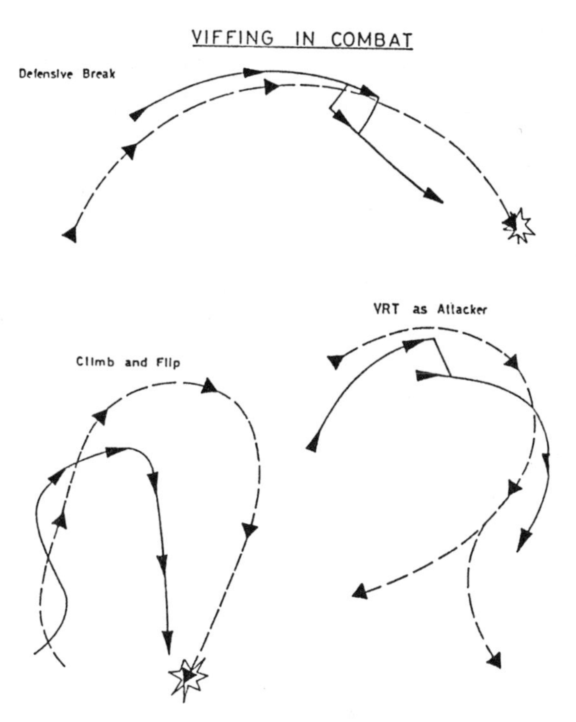

Fig. 9. **Examples of viffing performed by a vectored aircraft in combat with conventional fighter.** During these schematic maneuvers, a vectored fighter may use thrust reversing, or, more efficiently, a PST-braking maneuver (cf. Introduction, Figs. 4 and 5 and Figs. I-2, 3 and 4). Herbst PST may be effectively combined with PSM.

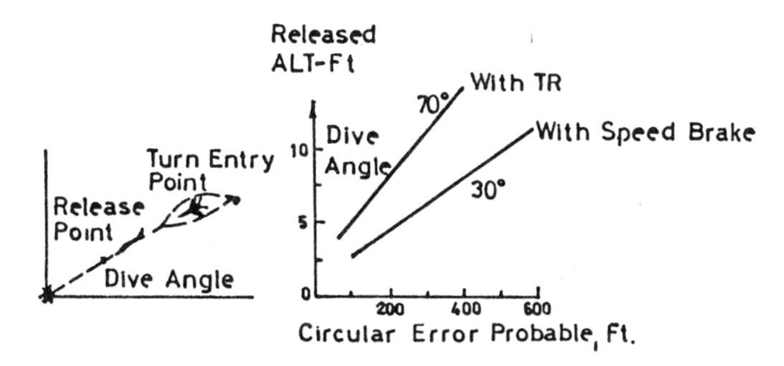

Fig. 10. **The simultaneous use of TR, yaw vectoring and combined pitch vectoring may revolutionize air-to-surface performance of vectored aircraft.** (Dive angle is increased, error decreased. Cf. Fig. 15, Introduction).

that this has caused some confusion among the pilots participating in his simulations (for the pilot tends to use body axis as reference).

Consequently, Herbst developed *a requirement for velocity vector roll acceleration.* It is ploted in Fig. 11f as a function of AoA for a speed of $M = 0.2$ at 6000 m altitude.

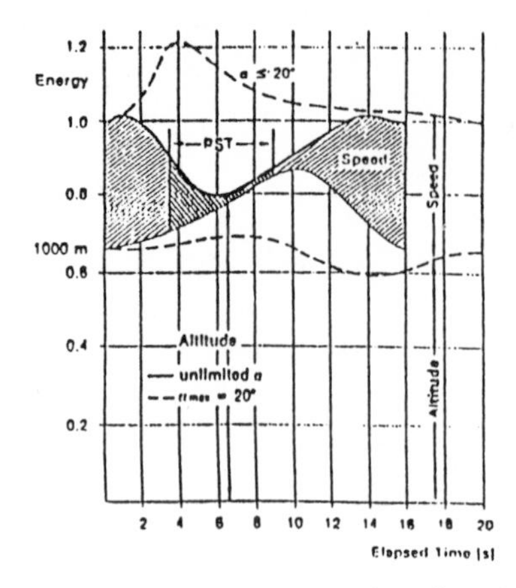

a. Energy management comparison for minimum time maneuvers. Results of trajectory optimizations.

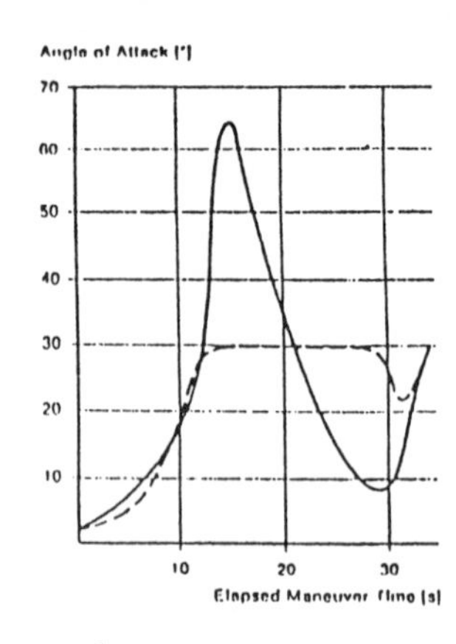

d. Angle-of-Attack during a typical air combat engagement. Result of computer simulations.

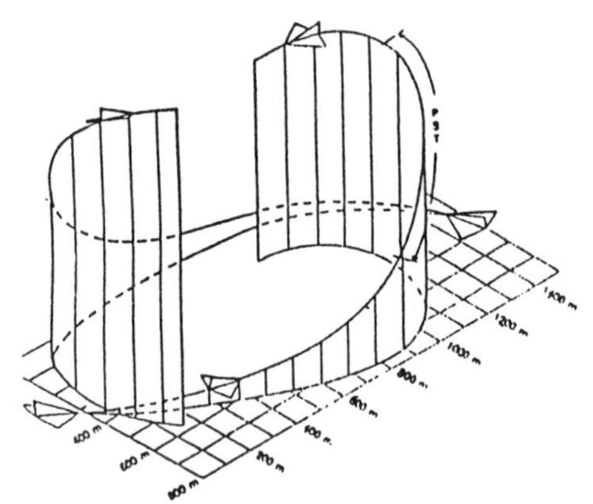

b. Typical air combat engagement of a PST fighter against a conventional α-limited fighter with all aspect weapons. Result of computer simulations.

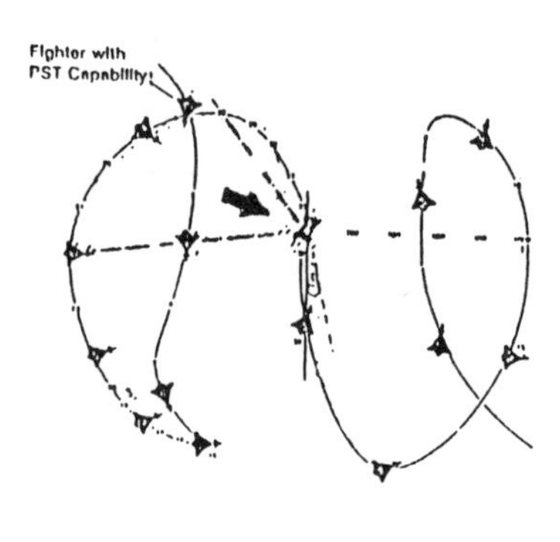

e. Typical air combat engagement. Relative positions of opponents depicting their aspects and firing opportunities. Result of computer simulations.

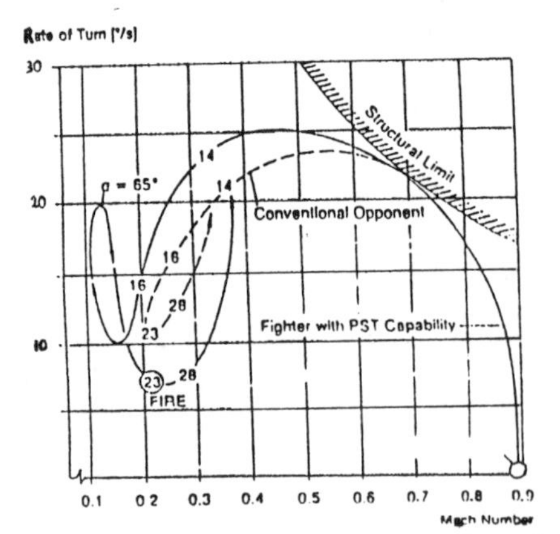

c. Maneuver characteristics of a typical air combat engagement. Elapsing time in sec. Result of computer simulations.

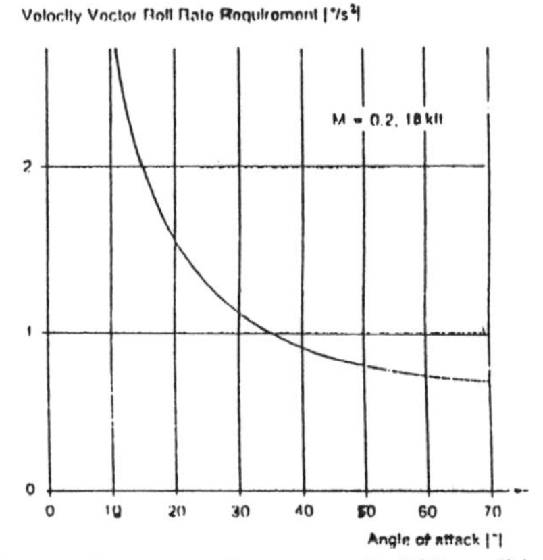

f. Control power requirements under PST conditions. Statistical results of manned combat simulations.

Fig. 11. **Herbst's simulation results (see text and Figs. 2, 3, 4).**

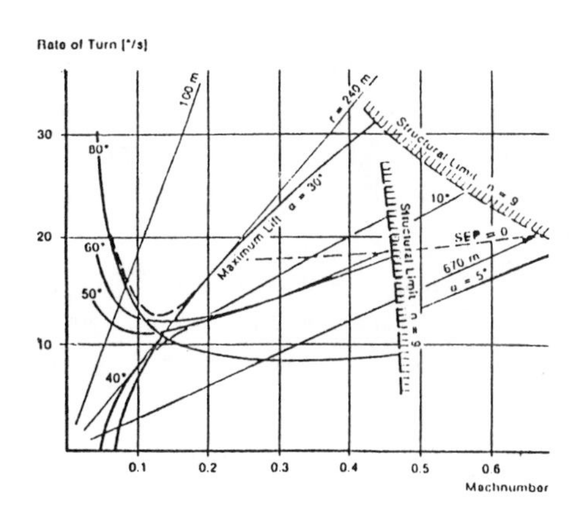

g. Maneuver states in a horizontal turn (aircraft dependent).

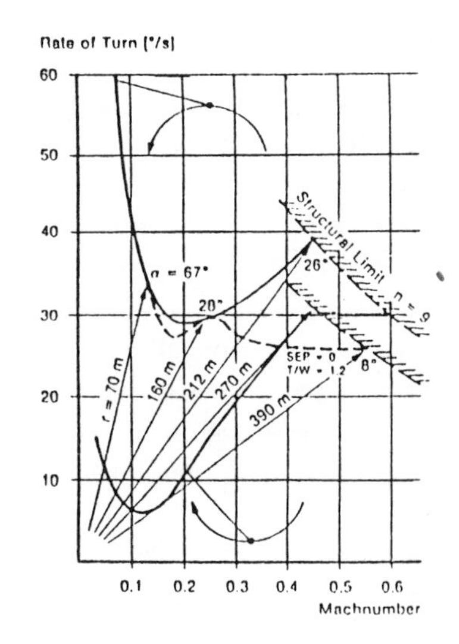

i. Maneuver states in a vertical turn (aircraft dependent). Upper curve for top, lower curve for bottom of a vertical maneuver.

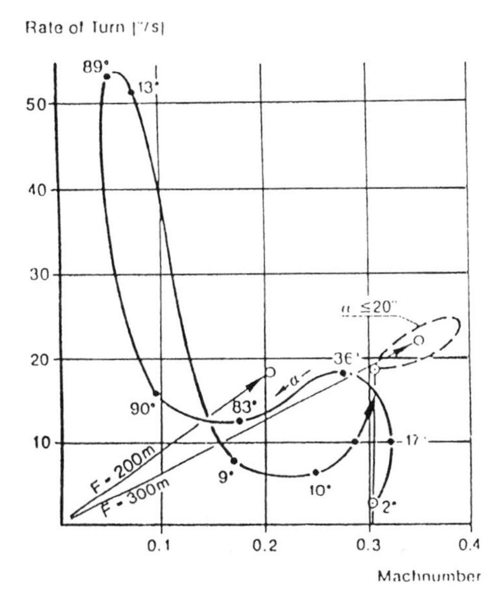

h. Comparison of maneuver cycles for minimum time maneuvers. Results of trajectory optimizations.

Fig. 11. (continu.)

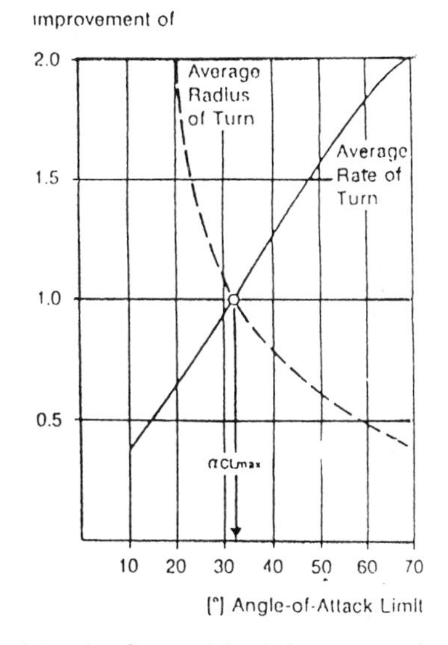

j. Results of manned simulations representing real aircraft and control dynamics. Standardized PST maneuver with 180° change of heading. Control augmentation by means of a ± 10° vectored nozzle in pitch and yaw.

For a conventional fighter such a high demand for roll cannot be met by aerodynamic control surfaces. Hence, the requirement for thrust yaw-pitch vectoring becomes a necessity to improve future, close-air-combat effectiveness to a degree unachievable by conventional performance.

II-7.3 A 180° Change of Heading Maneuvers

A 180° change of heading, with the additional constraint of returning to the point of departure at the same speed and altitude, is shown in Figs. 2, 3, 4, 11.

Fig. 11b shows trajectories of a typical initial phase engagement depicting two opponents with equal conventional performance. However, the aircraft moving to the *left* is *unvectored* and limited to a given value of maximum lift, while *the second fighter has yaw and pitch vectoring, PST-capability of up to 70° AoA.* Both aircraft are equipped with the same all aspect weapons.

The associated time history of turn rate vs. speed and AoA is plotted in Figs. 11h and 11i. Starting at high speed and the same altitude, both aircraft pull to *maximum load factors*, and then, *slow down to achieve the best instantaneous turn rates by means of gaining altitude.*

At the same time, *a smaller radius of turn helps to get the opponent into own weapon-off-boresight-cone and keeps the opponent from achieving a similar objective.*

Thus, a properly performed penetration into the PST-regime enhances both minimum radius and speed reduction, thereby *giving the vectored aircraft the first firing opportunity.*

II-7.4 High-Alpha Supermaneuverability & 2 Flight Envelopes

Let us re-examine now the high angle-of-attack (AoA) flight in combat by highly agile fighters. According to current thinking a high AoA, beyond stalling, will be the inevitable key element in enabling fighters to survive and win in battles both beyond and within visual range of the enemy. Thus, *future fighters will have 2 flight envelopes; one for beyond visual range and one for close-in combat.* The aircraft would be required to perform rapid transitions between them, for some beyond-visual-range engagements will inevitably turn into close-in combat.

The 2 envelopes would be governed by four categories:

Category 1:

Use of vectoring and IFPC to gain increased turn rates for beyond-visual-range engagements.

Category 2:

Use of vectoring and IFPC to gain improved turn performance at high and very low load factors, and at post-stall and transient overshoot under maximum lift conditions (fighter survival vs. missiles, defensive break-away, etc.).

Category 3:

Use of vectoring and IFPC to reach a high pitch rate to large AoA, followed by a velocity-vector roll, or a PSM, to achieve a rapid pointing-angle change.

Category 4:

Use of vectoring and IFPC to momentarily lower speed at very high AoA, and very high heading-angle rates.

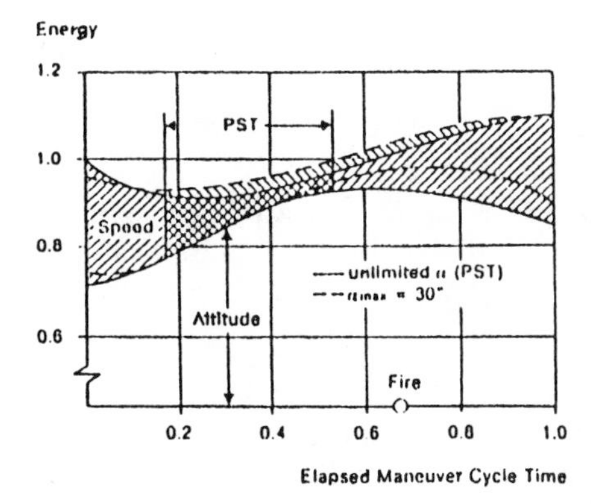

Fig. 12. **Energy management in a typical air combat engagement.** Result of computer simulations. Cf. Figs. I-2, 3 and 4. Herbst [154].

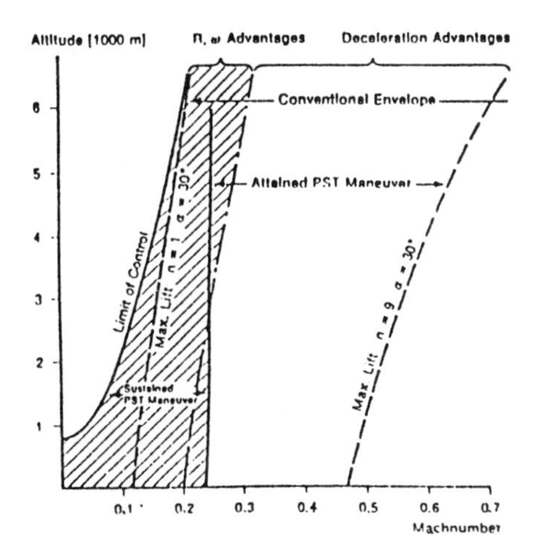

Fig. 13. **The PST flight regime (aircraft dependent).** Herbst [154].

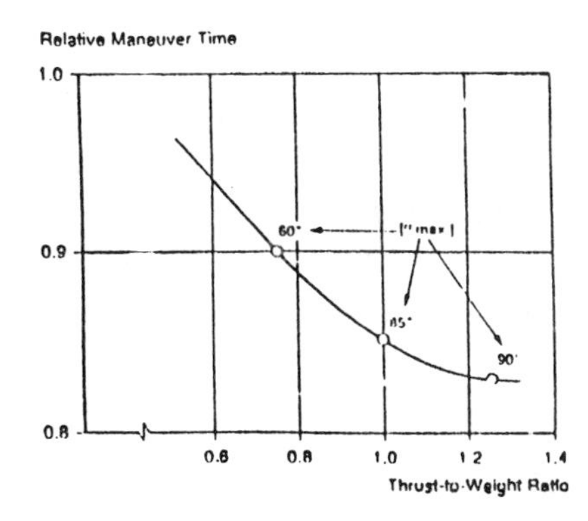

Fig. 14. **Thrust dependence of the time advantage in a minimum time maneuver with limited PST capability.** Results of trajectory optimizations. Herbst [154]. Consequently, vectored propulsion adds maneuverability to fighters having $T/W > 0.6$. Many old aircraft may, therefore, be upgraded.

These categories may be used quantitatively to obtain dominance over the threat, in terms of maneuverability.

However, these categories convey a glaring contradiction to some prior tendencies to design the fighter as a **'stand-off missile launcher', with no extra need to maneuver, and, therefore, with no need of vectoring, high AoA, and agility**. Nevertheless, the unfortunate experience with the F-4 during the Vietnam War has put serious doubts into the latter mode of thinking. *There is also a semi-intuitive resistance, especially among pilots, to slow down, even momentarily, in close-in combat.* However, the only alternative to properly-calculated, and performed combinations of the aforementioned categories may be to stay out of the "future air combat". Consequently, there is a well-defined need *'to expand envelopes'* by a proper *interaction* between *external flow, engine intake, vectoring nozzles,* and multiple control means such as ENJV, or canards *and the 2D-nozzle flaps themselves.* These should be interrelated to other advantages, such as stealth and all-aspect missile maneuverability. **Combined, they supply the vectored aircraft pilot with unique maneuvers to rapidly point, and, during the turn, to allow the computers to calculate and, then, to lock-on weapon (either gun or missile), put it within its fire envelope, and fire. It is the respective time delays of each 'unit operation' which must be properly minimized and well-integrated (cf. 3.5 in the Introduction).** *Thus, RaNPAS time for PSM (or PST) must be shorter than the combined computing, locking, releasing and (high-speed, high-g) turn times of the next-generation, all-aspect missiles in offensive thrust-vectoring engagements. And it is here that PSM by "no-tail" PVA would emerge as the pace-maker, determining technology.*

Moreover, with the possibility of rapid sideslip, or "jump", with or without thrust-reversal, while commanding very high nose rates, one can put the aircraft on such a "zig-zag" trajectory which will make it less predictable to adversary missiles, and confuse the opponent in defensive engagements.. Similar conclusions apply to future agile (vectored) RPVs, SAMs, and ACMs (cf. Lecture IV).

II-7.5 A Major Propulsion Problem: The Nonavailability of a PST Inlet

A major obstacle to gain the aforementioned advantages is *not wing stalling but engine stalling* (at very high AoA and turn rates, low speeds and high altitude). Engine controllabiltiy must first be improved in comparison with current engines – both by improving *compressor stall margin* and by use of *HIDEC* or IFPC. However, special *inlet doors, ducts and ramps are now needed.* Since other requirements may dictate a special inlet design, the industry requires the development of a variable inlet which may, uniformly, pass and compress air both from the wing's upper and lower surfaces and, in some designs, from inlet sidewalls. This can be done, alternately, or simultaneously, depending on the type of mission, take-off, maneuverability and landing needs. Such inlets are presently being designed and tested in various laboratories, including this one.

LECTURE III

THE MATRIX OF UNKNOWN VARIABLES OF VECTORED AIRCRAFT

III-1 Reservations and Precautions

Fundamental concepts, components and unit operations associated with vectored propulsion and supermaneuverability have been introduced and illsutrated in the Introduction and in lectures I and II.

However, a matrix of *unknown variables* emerges as one tries to proceed and *interconnect* them into cost-effective, fully, or partially-integrated systems.

This difficulty may be best examplified by any attempt to define and analyze *IFPC variables* for pure or for partially-vectored aircraft.

Thus, in seeking realistic design goals for vectored aircraft, one may first identify the main parameters to be considered, evaluate their relative importance within each 'engineering discipline', as well as their proper 'groupings' and 'subgroupings' for analysis and/or synthesis of their coupling effects to each other.

Traditional subgroupings into 'disciplines' involves at least 22 categories; e.g.:

1) PSM, PST/Stealth inlet variables and technology limits.
2) Thrust-vectoring nozzle variables and technology limits.
3) Engine technology limits (Appendix B).
4) Electro-Aero-Thermodynamics and Low observability technology limits.
5) Aero-thermodynamics, *IR* and heat-transfer variables.
6) Wing/body/propulsion system integrated variables.
7) Environmental variables.
8) Expected threat variables.
9) STOL, V/STOL, VTOL, STOVL variables.
10) The basic and revisited variables of flight mechanics and flight control.
11) Redefined and revisited mission variables and analysis.

●12) Payload/weapons/Propulsion/airframe coupling effects.
●13) Structural and materials technology limits.
14) Stability, maneuverability, and controllability technology limits.
●15) Reliability, durability, safety, and maintenance.
●16) Hardware manufacturing constraints.
●17) Cost-Effective objectives.
●18) R&D funding/management through realistic milestones.

●19) Time-table for a realistic program with respect to the threat.
●20) Governmental approval/budget/control effects and delays.
●21) Constraints in restructuring engineering and pilot education and training (including the delayed-time optimized vectored trainers will be available to the services).
●22) Proprietary and classified constraints.

* * *

By examining this list one realizes that there is yet no verifiable methodology regarding the following questions: How shoud one structure these categories? Or how should one evaluate their relative importance, as well as their coupling effects to each other?

Moreover, thanks to the *subdivision* of engineering science into *fragmented* 'disciplines', we may fail, *a priori*, to perceive the *'interconnectedness'* between these *'self-centered'* categories, to judge their *collective priorities*, and to estimate their *inherent structure and ordering* in conceptual and preliminary design.

In trying to overcome these problems, one may proceed with the problem of ordering the rest of these lectures and appendices.

Indeed, at this stage of the course, one is faced with empirical, logical, and subjectivistic discourses; empirical, because a great number of pioneers have already produced an enormous amount of empirical data on components, unit operations and partially-integrated thrust-vectored systems; *logical*, because all rational engineering singles out regularity and logical order; *subjectivistic*, because all thought is, to some extent, ordered by personal experience.

All combined, the feedbacks of conceptual design and experimentation, *as outlined in the illustration shown in the Preface, must be considered first and foremost.* What order we choose next, it remains only a minor link within the feedback networks of this illustration.

●III-2 PSM/PST/Stealth Inlet Variables and Technology Limits

A brief introduction to this broad subject is given in Appendix F: *"Limiting Engine-Inlet Envelopes"*

'Classical' and 'Non-Classical' intake-design approaches are discussed therein, including the near-term possibility to employ *inlet-distortion-insensitive cold-jet powerplants for VRA* (cf. Figs. II-1i and 1f).

PSM/PST/Stealth engine-inlet-compatibility will be enumerated in Volume II together with the variables required for IFPC systems of future PVA.

Volume II will also stress the inherent difficulty associated with the *'undefinable matrix of intake distortions'* and the different design options available for PSM/PST-tailored inlets.

●III-3 Engine Variables and Technology Limits

A brief introduction to this central subject is given in Appendix B. This appendix stresses *RCC-technology potentials and present constraints.*

In fact, much of what is discussed in this volume, will become feasible when RCC materials would dominate the performance of powerplants, *pushing their T/W ratio from present 8–12 values to 20* and beyond.

The matrix of variables associated with this category is well-defined in standard textbooks on gas-turbine theory, and is, therefore, not to be repeated here, unless associated with the following:

- IFPC trim and engine margins (Appendix F and Volume II)
- Thrust-vectoring nozzle (Lectures I, II, III and appendix C)
- RCC electro-aero-thermodynamics (volume II)
- Non-RCC cooling and heat transfer rates (see below).

●III-4 Engine Variables and Technology Limits

Some of these variables are defined and identified below in connection with inlet/wing/nozzle integration and PST-flight conditions.

●III-5 Electro-Aero-Thermodynamics and Low Observability

The material available in the open literature is to be collected and published within the framework of Volume II.

●III-6 Jet-Aero-Thermodynamics and Heat-Transfer Rates

The variables associated with this subgroup are well-defined theoretically and experimentally, as will be discussed below.

●III-6.1 Subsonic and Supersonic-Flow Heat Transfer

A major limit to the development of advanced powerplants is the current non-availability of Reinforced Carbon–Carbon (RCC) composite materials for hot-sections of powerplants.

A distinct and unique advantage of RCC is *the elimination of present-technology performance penalties to cool turbine blades, disks, A/B ducts, nozzles, etc.*, as well as weight and part-number reductions. RCC is also an excellent absorber of electromagnetic waves.

Faced with such potentials, this laboratory has been co-working with another laboratory in this institute, on the development of improved industrial processes to manufacture well-impregnated RCC materials. At the same time, an effort has been

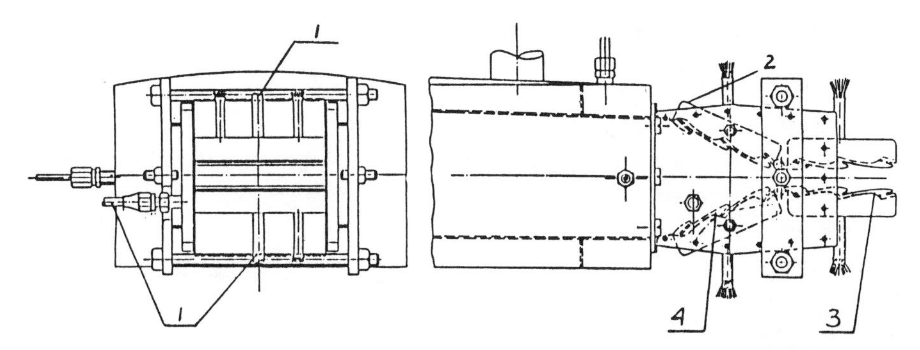

Fig. 1. **An early design of a film-cooled, *2D–CD*, vectoring nozzle [tested in this laboratory during 1980–83 for the evaluation of temperature distributions during subsonic and supersonic, hot-gas vectoring, up to 900°C, and up to $\delta_v = \mp 23°$].**

a. (Right) The converging and diverging flaps of this vectoring nozzles were produced from the (film-cooled) combustor liner of a *GE* F-404 engine (3). δ_v-variations: ±23 degrees. Hot-gas temperature: 900°C. (Left) Rear view of nozzle at $\delta_v = 0$. Throat $AR = 6.5$. Option: Thrust reversal.

1. Cooling air inlet; 2. Cooling-air flap cover; 3. Diverging-vectoring flap; 4. Converging flap.

Notes: The truncated sidewalls are also being film-cooled. The duct upstream of the nozzle is a water-cooled transition duct between an Allison-GM T-56 combustor and the hot, air-cooled, vectoring nozzle. [Vectoring nozzle angles were varied during the tests.] Cf. Fig. 1, Appendix E.

made to develop methods *to cool vectored nozzles made from non-RCC materials.* This effort is described below and in Figs. 1, 2, and 3.

The vectoring nozzles shown are of the early type employed in this laboratory in 1980/3. They are limited in size to 0.61 kg/sec airmass-flow rate, and in aspect ratio to less than 7. However, this was one of the first published studies conducted about subsonic and supersonic cooling effects on *2D–CD* wall-temperature distributions (δ_v was variable up to ± 23 degrees).

Another study, to be described next, was published in 1982 by Straight and Cullom of the NASA Lewis Research Center [48].

●III-6.2 The Straight-Cullom Heat-Transfer Correlation for Vectored Nozzles.

These authors stress the fact that while many tests have been reported on vectoring-nozzle payoffs, *cooling performance data, in the full-scale engine environment, are scarce. Accordingly, they present baseline thrust and cooling data for a 2D–CD* nozzle mounted on a J-85 turbojet engine in an altitude chamber.

Straight-and-Cullom tests cover a broad range of pressure ratios, nozzle throat areas, and internal expansion area ratios (see Figs. 4 to 7). Their nozzle configuration is shown in Fig. 4, while the cooling methods are shown in Fig. 5.

Vectoring was conducted by two methods:

- Nozzle-flap vectoring (using interchangeable nozzle-flap cooling panels).
- Pivoting the nozzle throat.

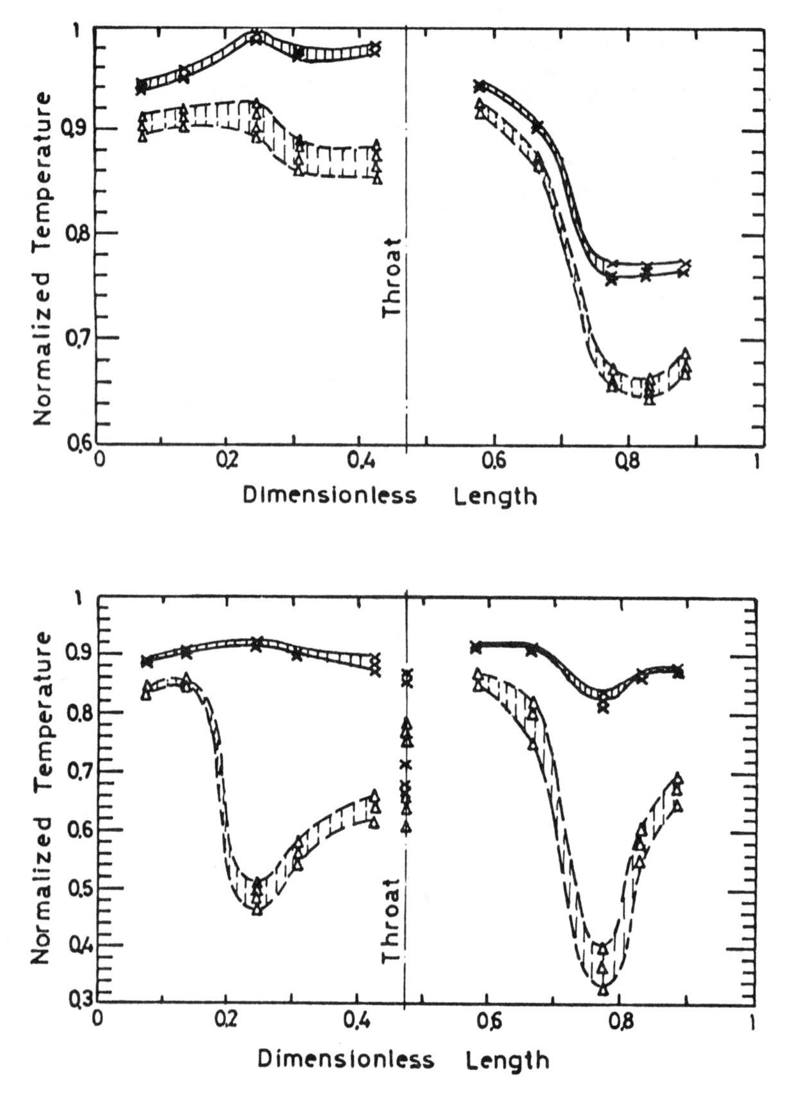

●Fig. 2a. **The effects of 2.5% cooling flow on the temperature distributions in the *2D–CD* nozzles shown in Fig. 1 (film-cooled) and Fig. 3 (uncooled).**

The 'sideline' distribution is a 'corner' temperature distribution (upper figures). The lower figures are for nozzle's center-line.

Cooling-air temperature was 26.2°C; Tg = 683°C; A_t = 11.2 cm^2 A_e/A_t = 1, δ_v = 0, $\dot{M}a$ = 0.669 kg/Sec (in the uncooled model); $\dot{M}a$ = 0.612 kg/sec (in the cooled model). The test facility is described in Fig. IV-7.

To keep costs down, their design philosophy was to use simple boilerplate structures, but keep realistic flow paths.

Simple, leaf-type, low-cost seals were used throughout to minimize leakage.

●III-6.3 Cooling and Leakage

The maximum performance potential of the nozzle was determined by correcting the measured thrust coefficients for seal leakage and cooling airflows that bypassed the nozzle throat. *Their data corrections indicate that substantial thrust penalties could result from coolant bypass flow and leakage, and that all significant leakage occurs upstream from the nozzle throat.* Moreover, they report that effective leakage area is only a function of differential pressure (independent of whether the leak is subsonic or sonic flow), and that the leakage area is equivalent to a simple-flow orifice (and not to a series of flow restrictions).

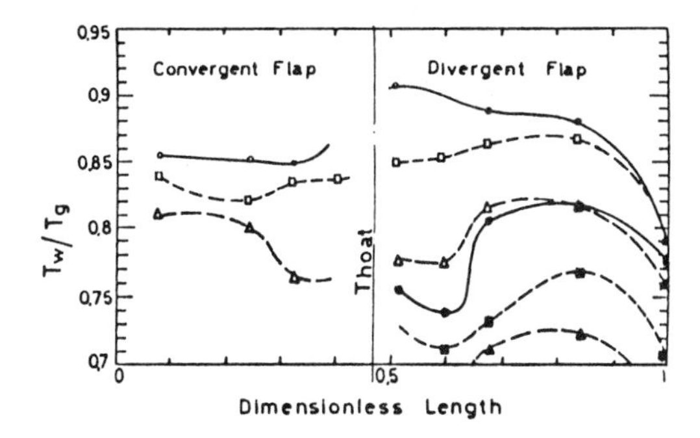

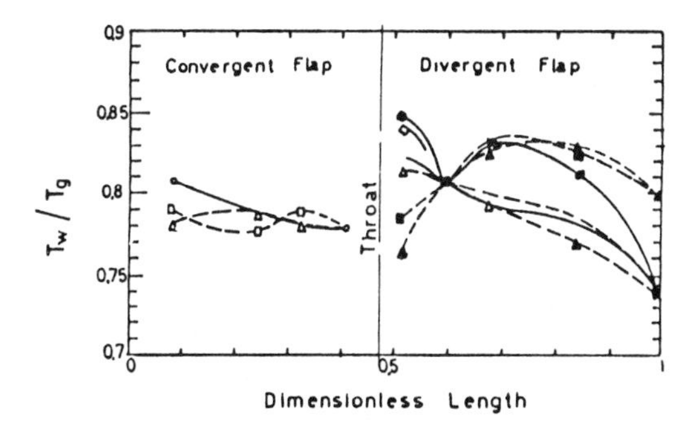

●Fig. 2b. **Temperature distributions in the *2D–CD* vectoring nozzle shown in Fig. 1 during supersonic vectoring at $\delta_v = 23°$ and Tg = 900°C, (upper figures) and during subsonic vectoring at $\delta_v = 21°$ and $T_g = 900°C$ (lower figures).**

Notes:

(1) The throat is located close to the normalized length-value 0.5.

(2) Black points represent test results for the lower flaps, while the open marks those measured on the upper flaps.

The experimental results demonstrate that there is a considerable difference between center-line and nozzle-corner temperature distributions, and between heat-transfer rates in the subsonic and the supersonic domains. This difference increases with NPR and δ_v [103].

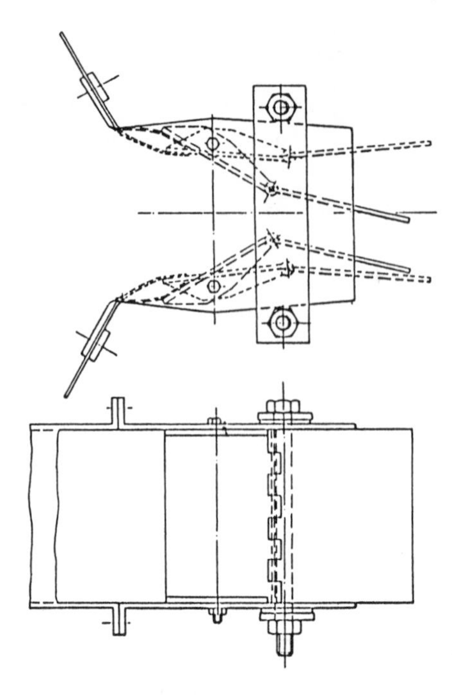

Fig. 3. **The first *2D–CD* vectoring nozzle tested in this laboratory in 1980 was uncooled.** Its dimensions are similar to the film-cooled nozzle shown in Fig. 1. Hot-gas temperatures for this model were limited to a maximum of 1300°C. Cf. Fig. 2a

●III-6.4 Flow Discharge Coefficient – C_{D8} [Eq. 7 below].

This coefficient was determined by ratioing the effective area of the throat to the actual measured area (cf. Eq. III-7), after correcting the data for leakage and coolant flows that by-passed the nozzle throat (Fig. 5).

The effective area was computed from the corrected values of throat flow rates and mixed-hot-gas temperature and the measured values of total pressures. Similar effects of cooling and leakage on C_{fg} have already been illustrated in Fig. I-2.

●III-6.5 Supersonic Hot-Gas Heat Transfer

The hot-gas-side heat-transfer coefficients were computed from the average bulk-to-wall temperature differences for each cooling panel, and the heat flux from the temperature rise of the coolant. The data are plotted in Figure 6 as a function of hydraulic diameter of the flow path between the two divergent flaps, and also compared with predicted values on the right-side of the figure. The data covered a range of Reynolds numbers from 7.0×10^5 to 2.0×10^6, computed on a hydraulic-diameter basis. Film-property values were determined from the average of hot-gas bulk and wall temperature. The equations used for the predicted values are:

$$Nu = \frac{h D_H}{k_f} = 0.023\, Re_f^{0.8}\, Pr^{0.4} \tag{1}$$

where

$$Re_f = \frac{D_H V \rho_f}{\mu_f} \tag{2}$$

The experimental values are as much as 40 percent higher than predicted at the higher heat-transfer rates. It is apparent from these data that a more rigorous analytical approach is needed to improve the predictions (which should include the effects of shocks).

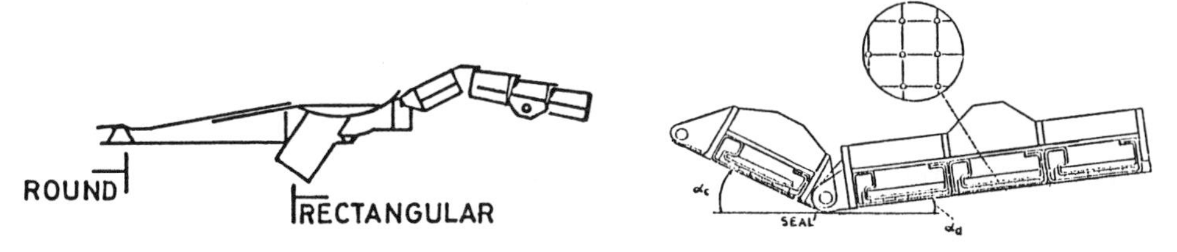

Fig. 4. **The cooling methods employed by Straight and Cullom [48] using a research turbojet engine (the *GE* J-85) equipped with a 2*D–CD* pitch vectoring nozzle.**

III-6.6 Accelerating Subsonic Heat Transfer

When the standard equation (1) was used for predicting heat transfer on the converging flap, the correlation was poor. Significant variations of heat transfer coefficient occurred with change in the angle of the convergent flap to the hot-gas stream. *The data were then re-plotted as a function of the flap angle in Figure 6, where it grouped about a single line.* The empirically derived equation for the line shown is:

$$Nu = 0.023\, Re_f^{0.8}\, Pr^{0.4} \left(1 + \frac{\bar{\alpha}_c^{0.8}}{15} \right) \tag{3}$$

This equation is the standard equation (1) with the addition of a term containing the *angle of the flap relative to the hot-gas direction* entering the *convergent* portion of the nozzle. The form of the added term was chosen such that when the flap angle is zero (parallel to the hot-gas flow), the equation reduces to the form of equation (1).* *No attempt has been made to date to explain the results analytically.*

* Stoll and Straub [J.F. Turbomachinery, 110, 57 (1988)] have recently reviewed some previous studies in film cooling and heat transfer in nozzles while adding their own studies of a water-cooled, low *AR*, 2*D–CD* nozzle constructed from copper.

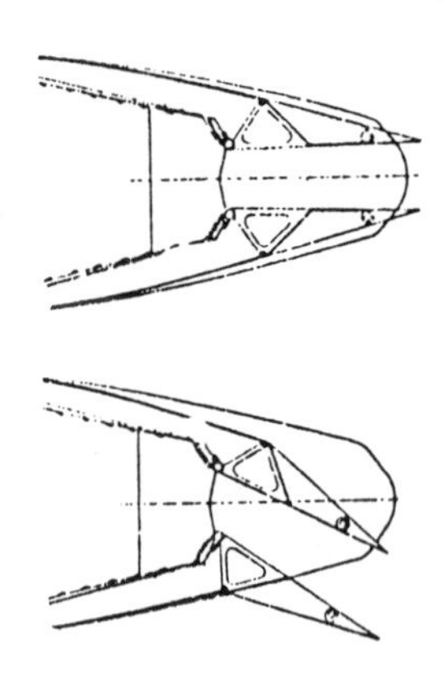

Fig. 5. **Straight and Cullom vectoring, *2D–CD* nozzle during cruise or vectoring.**

III-7 Wing/Body/Propulsion System Subintegration

III-7.1 The Primary Effects of Vectoring Nozzle Aspect Ratio

While NAR emerges as the most important parameter in pure vectored aircraft, other parameters must be considered as well. Eqs. (3) to (31) below sum-up the main parameters which must be defined for pure vectored flight-propulsion.

Let us first examine the effects of NAR on some preliminary design considerations.

As NAR is increased, the relative 'heights' of the nozzle exit components decrease, in proportion to the increase in width (cf. Figs. II.1a, 1b). *However, total exhaust system's length may increase*, because the circular-to-rectangular transition duct may have to become *longer* to maintain uniform flow distribution and minimum distortion/separation effects. *Moreover, it may also include a rectangular afterburner (an unorthodox, but feasible design!)*

A whole-system-design-approach to the entire matrix of problems involved with high NAR systems includes:

1) *Use of streamlined-flow-distributers/struts inside high NAR nozzles.*
2) *Use of these distributors as structural elements in the wing-structure design for wing weight reduction.*
3) *Use of the wing's skin and internal structure to reduce the weight penalty of high aspect-ratio nozzles.*
4) *Use of upper and lower uncurved panels to reduce production costs and improve integration with aft-wing sections.*
5) *Calculate transition-duct cooling loads due to increased heat-transfer area.* This duct may also form a new-type of rectangular afterburner.
6) *Improve cooling loads and IR signatures by use of internal wing mixers and inlet passage for cooling airflows.*

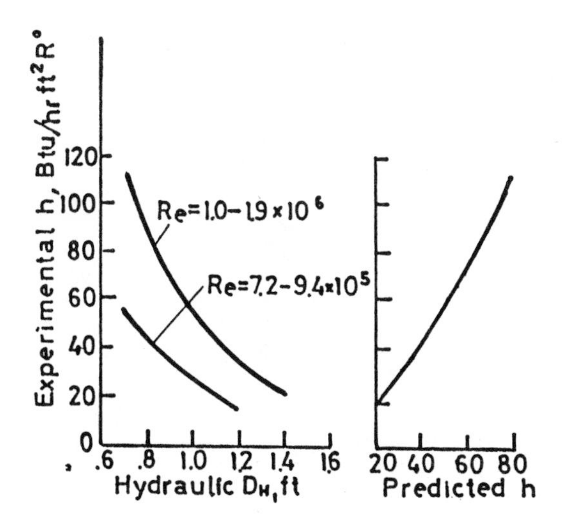

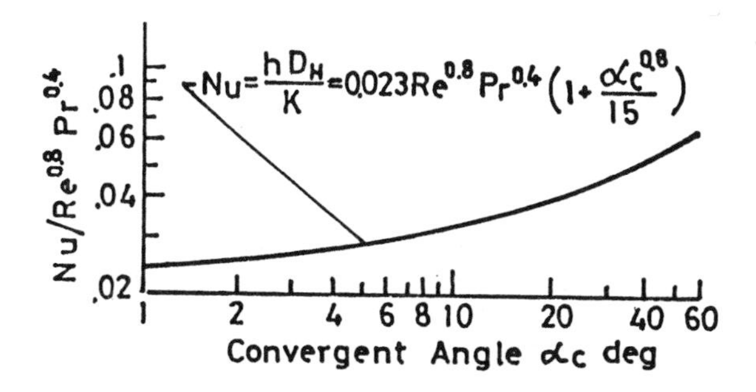

●Fig. 6. **Straight and Cullom impingement cooliing heat-transfer coefficients (upper drawing), supersonic, hot-gas, heat-transfer coefficients for cooled divergent flaps (drawing at center), and a heat-transfer correlation for vectoring, *2D–CD* nozzles, including the effects of converging flap angle in accelerating hot-gas flow (cf. eq. 2) [48].**

7) *Use RCC materials (see below).*
8) *Reduce boattail drag by high NAR.*
9) *Use sidewall truncation* (which does *not* reduce C_{fg}) – see Figs. I-15 and I-16.

III-7.2 Aerodynamic Effects; Regrouping the Main "60-Variables"

Equations 2, I-11, 12, 13 and 14 specify and define the following variables: h, D_H, k_f, V, ρ_f, μ_f, C_p, $\bar{\alpha}_c$, δ_v, δ_y, C_{fg}, M_{act}, T_T, Pa, P_T and γ. Aside from NAR, and the various nozzle-wing variables defined in Fig. 10, one must also evaluate the dependence of engine inlet, and C_{fg}, C_{D8}, C_A and C_v (see below), on engine, and on *'external flight conditions'. The result is an unknown matrix of coupled variables.* Thus, in trying to reduce the confusion associated with this unknown matrix, one may first group the corre-

●Table 1. **2D–CD configurations tested by Straight and Cullom [48].**
For further details see Appendix C.

Configuration designation	Nozzle throat area, A_t		Nozzle area expansion ratio A_e/A_t	Nozzle throat aspect ratio	Convergent wall angle °C deg	Divergent wall angle °C deg
	cm^2	$in.^2$				
A1	707.1	109.6	1.20	4.10	28.75	2.19
A2			1.50			5.66
A3			1.79			8.87
A4			2.29			14.60
B1	902.0	139.8	1.20	3.19	21.36	2.86
C1	1130.4	175.2	1.20	2.55	13.15	3.58
C2			1.50			8.91
C3			1.80			14.31
C4			2.30			23.72

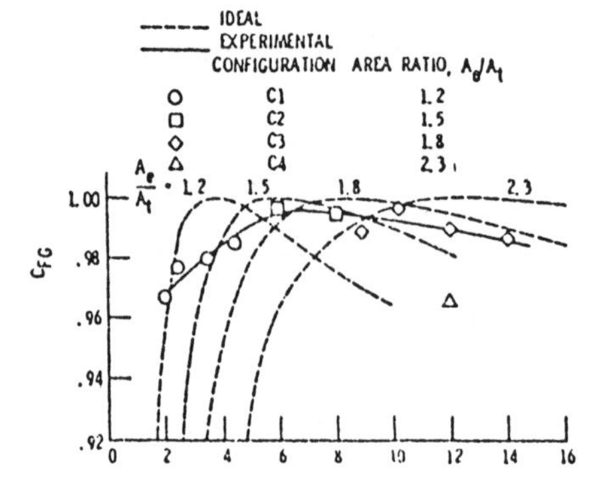

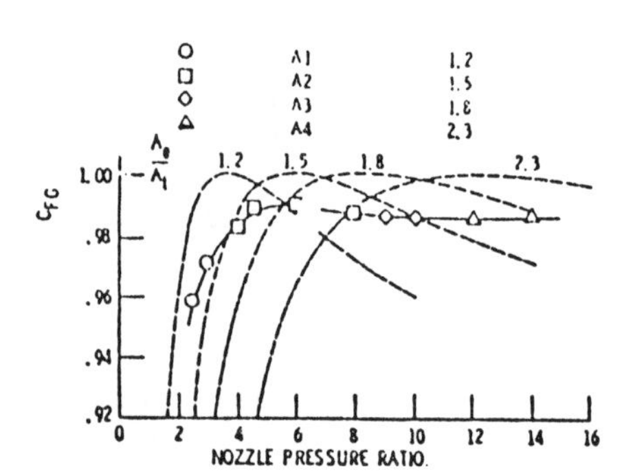

●Fig. 7. **The NASA-Straight/Cullom test results for the 2D–CD vectoring nozzle show how C_{fg} depends on NPR and A_9/A_8 (A_e/A_t) values [48].**

Note to the designer: Unlike the theoretical predictions, the actual test values presented have demonstrated a negligible dependence of C_{fg} on the expansion ratio A_e/A_t.

Provided these results apply to other vectoring nozzles (the data presented here are for $AR < 4.1$ and the *GE* J-85 engine), these results may have a significant effects on installed (integrated nozzle-wing structures) systems, especially in terms of reduced drag values associated with (lower-than-theoretical) A_e/A_t values at high NPR values. For further detail see Appendix C.

sponding variables, and, then, try to identify the more important ones, and/or select those subject to experimental verification in wind tunnels, in the jet propulsion laboratory, or in flying vectored RPVs (specially, and systematically designed to investigate certain unknown variables). There are, nevertheless, a number of variables which do not lend themselves to simple experimental (or theoretical) evaluation. One of them is defined by Fig. 10 and the following equation.

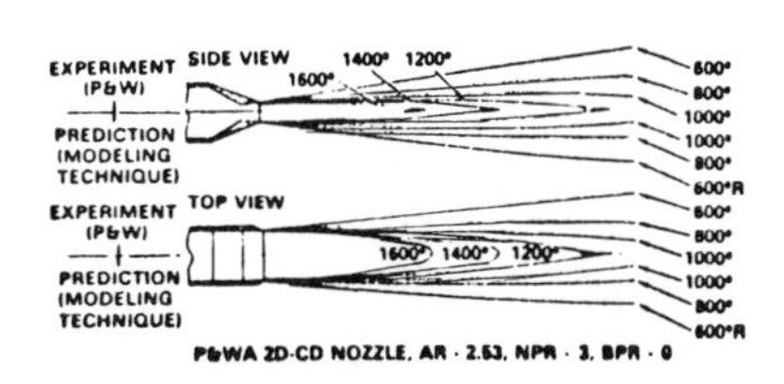

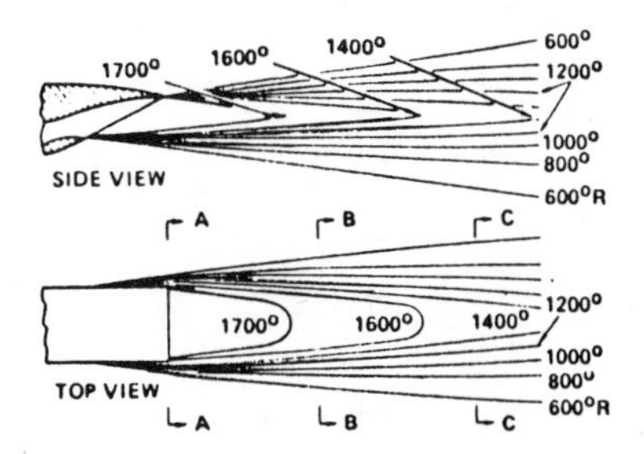

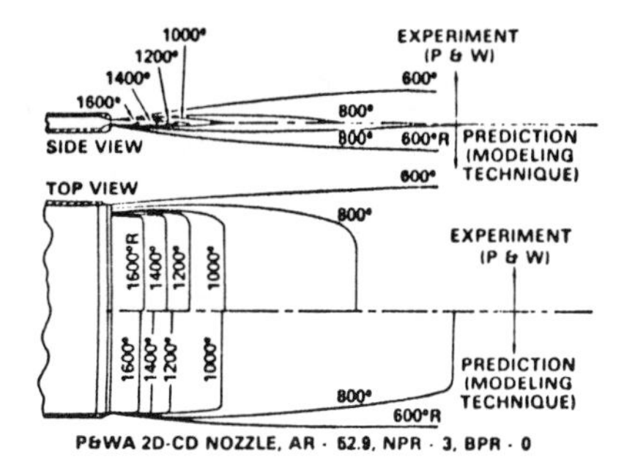

Comparison of Total-Temperature Contours for 2D-CD Nozzles.

Total-Temperature Contours for ADEN, NPR = 2.76, BPR = 0

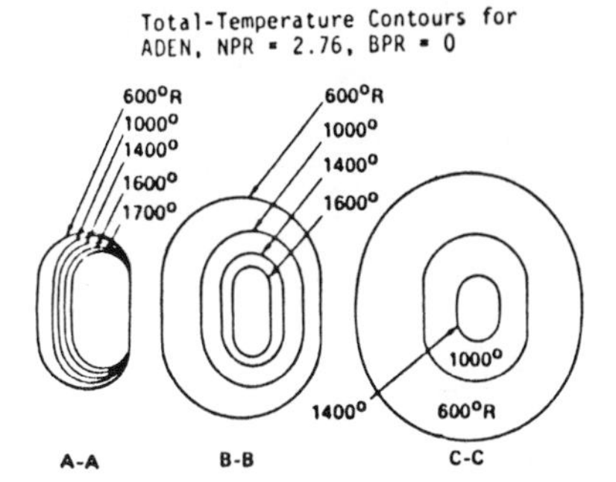

Cross-Sectional Total-Temperature Contours for ADEN, NPR = 2.76, BPR = 0.

●Fig. 8. **Chu and Der [91] calculated the temperature distributions inside high-aspect-ratio vectoring exhaust jets and compared their results with experimental data (lower left).** Their corresponding distributions for low-*AR* nozzles are demonstrated on the upper left corner of this figure.
As expected, the high-AR nozzle lowers IR signatures. A similar study for low-*AR ADEN nozzle is demonstrated on the right-hand side of the figure.*

$$C' = C + C_f \cos\delta_v + C_v^* \cos\delta_v \cdot \cos\delta_y \tag{3}$$

However, instead of the unknown variable c_v^* (Eq.5), one may use the experimental results of Figs. 18 and 19 below to estimate the main contributions to supercirculation and direct lift. *Yet, this would be a mistake, for one must first examine the whole integrated system together, e.g.:*

$$(\Delta C_L)_{\max} = (\Delta C_L)_{sc} + (\Delta C_L)_v =$$

$$= f(\text{NAR}, C_v^*, \alpha, \gamma_{\text{dist.}}, \delta_y, \delta_v, C_\mu, \beta, \dot{\beta}, M, \text{G.E.}, H, C_f, \text{BLS}, \text{SSW}, \text{T}_{\text{T8}}, \text{Ta}, \text{Re}, \text{IFPC}, \text{FCE}, \ldots) \tag{4}$$

* In the x-direction. Unless otherwise specified we assume that the engine (i.e., its center line) is installed with $\delta_v = 0$ with respect to the aircraft zero lift angle of attack. Cruise drag considerations, for instance, may add a few degrees in engine installation and the calculations given here should be adjusted accordingly.

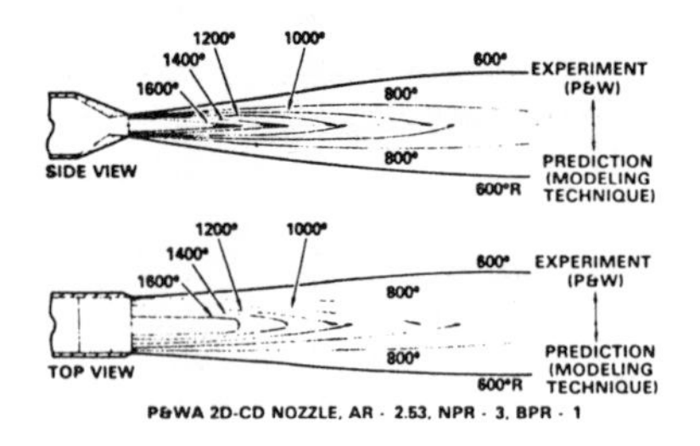

Comparison of Total-Temperature Contours for a 2D-CD Nozzle with Bypass Air.

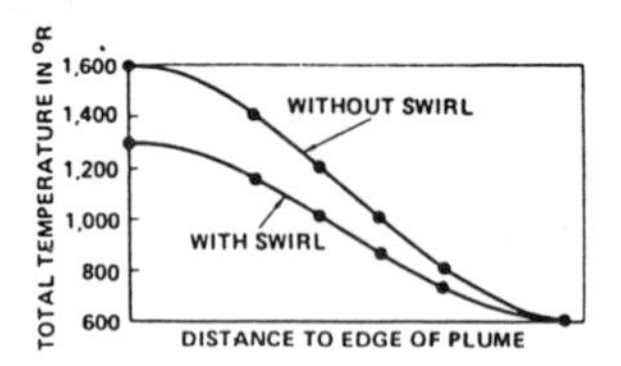

Illustration of a 30% Reduction of Plume Total Temperature Difference Due to Engine Swirl.

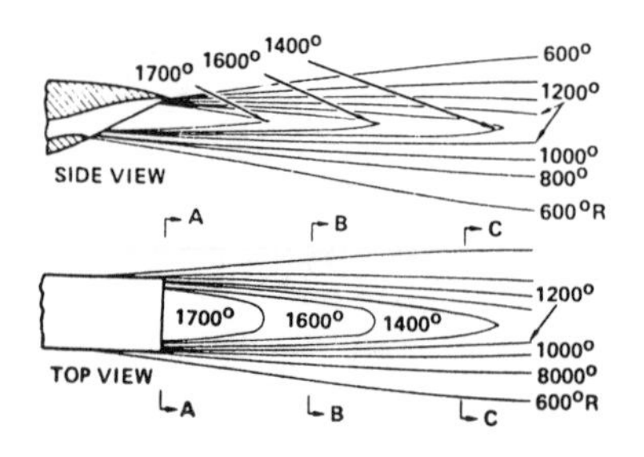

Total-Temperature Contours for ADEN, NPR = 2.76, BPR = 0.25.

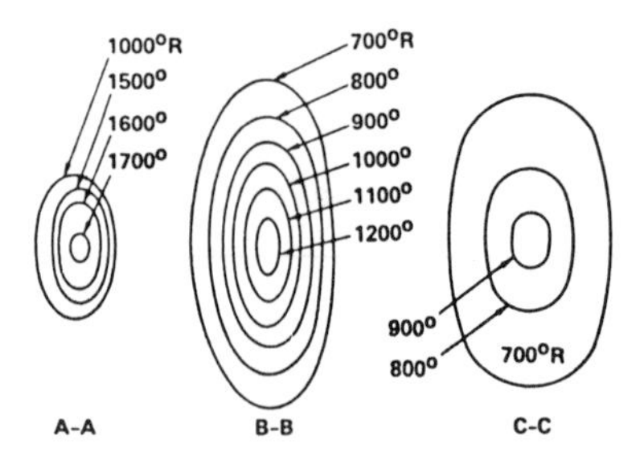

Predicted Total-Temperature Contours for ADEN (NPR=2.76, BPR=0.25) Under the Influence of Engine Swirl

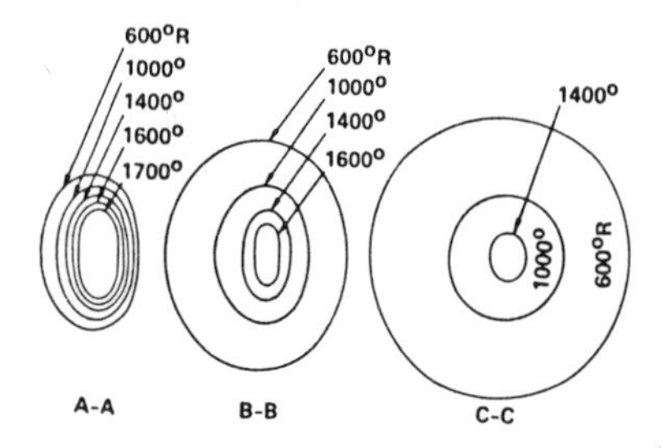

Cross-Sectional Total-Temperature Contours for ADEN, NPR = 2.76, BPR = 0.25.

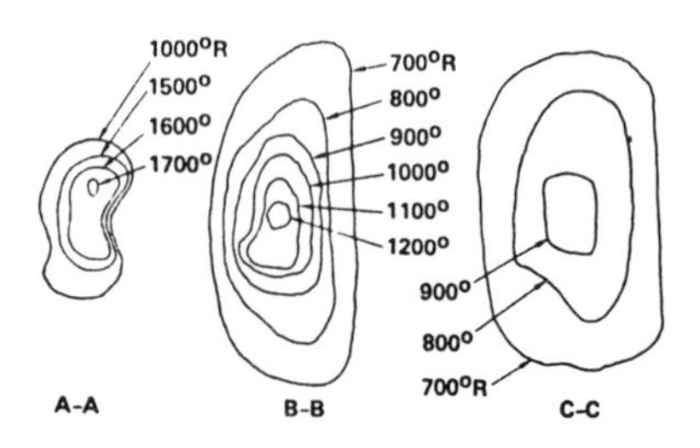

Measured Total-Temperature Contours for ADEN (NPR=2.70, BPR=0.25) in the Presence of Engine Swirl

●Fig. 9. **Various effects of engine swirl, bypass air and *AR* for *2D–CD* and ADEN nozzles were also reported in Ref. 91, by Chu and Der.** Note that the *IR* signatures depend on the fourth power of the absolute temperature.

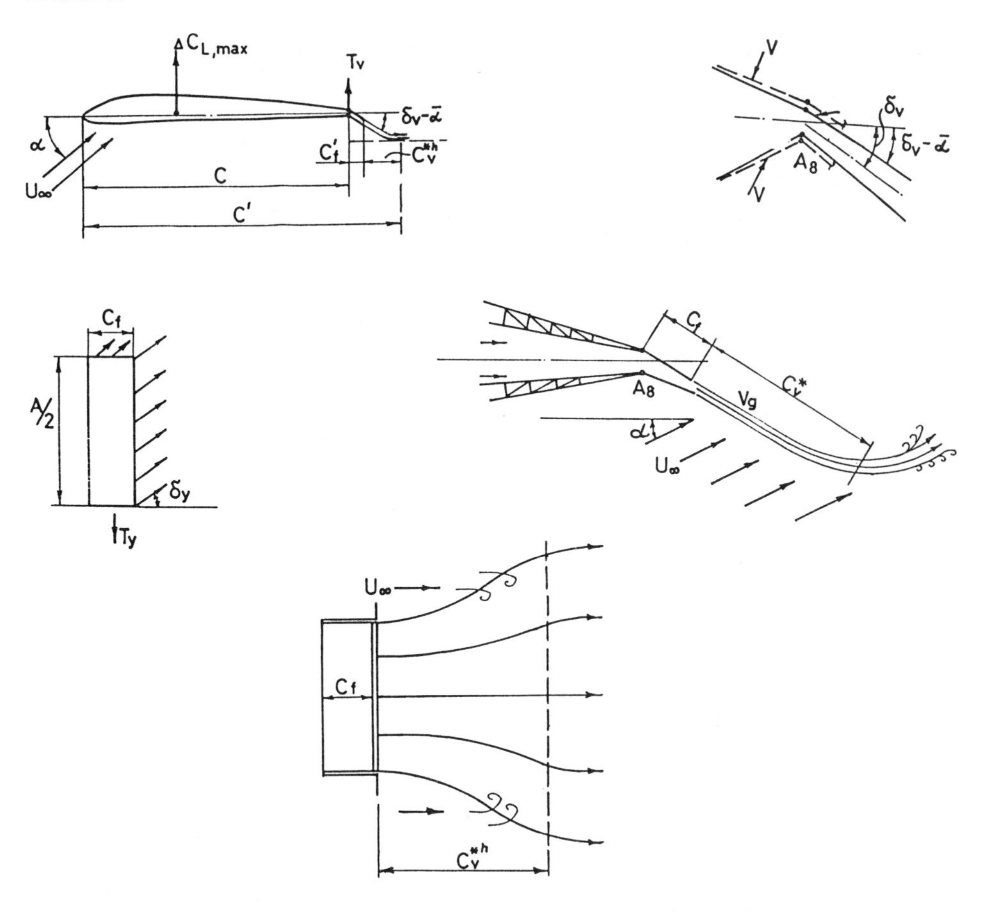

Fig. 10. **Preliminary definitions of an integrated wing/exhaust vectoring nozzle.**

Notes: V-adjustments require $(\cos \delta_v)^{-1}$ and $(\cos \delta_y)^{-1}$ increases in throat area (cf. Eqs. I-5, and Fig. I-5) and A_9 (Eq. I-6). *Analytic research in propulsion and aerodynamics must be integrative.*

Thus, vectored aircraft design calls for full integration of engine exhaust nozzle, the wing, aerodynamic control surfaces and the internal/external streamlined flow dividers (which become structural nozzle-wing struts).

$$c_v^* = \psi(\alpha, \text{NPR}, C_{fg}, C_{D8}, C_A, C_V, t^o, \text{NAR},, \text{G.E.}, \delta_v, \delta_y, M, H,$$
$$T, \beta, \dot{\beta}, \dot{\alpha}, \text{BLS}, \text{NG}, \text{Turb.}, \ldots) \quad (5)$$

$$C_{fg} = \Psi(\text{NAR, NPR}, \delta_v, \delta_y, C_f, C\text{–}R \text{ ducting and splitting, flow dividers/}$$
$$\text{Structural struts, IFPC}, \ldots) = T/Ti =$$

$$= \text{Actual thrust/Ideal thrust} = \quad (6)$$

$$= \frac{C_v C_A M_7 \,\text{actual} \cdot V_{9i}/g + (P_{s9i} - P_a) \cdot A_9 - \Delta C_{fg}(\text{Leakage+Cooling})}{V_S \cdot M_{7\text{actual}}/g}$$

(cf. Fig. 12)

$$C_{D8} = \frac{M_7 \,\text{actual}}{M_8 \,\text{Ideal}} = \frac{Ae_8 + Ae[\text{Leakage}] + Ae[\text{Cooling}]}{A_8} \quad (7)$$

(cf., e.g., Figs. 11, 16 and 12)

C_A = Thrust Loss due to the nonaxial flow from the Nozzle (see eqs.

17, 22, 23 and 24 below and Fig. 13). (8)

$$C_v = C_{fg}\,\text{peak}/C_A \text{ (see also eq. 6)} \quad (9)$$

$$(\Delta C_L)_v - \frac{T_v}{\tfrac{1}{2}\rho U^2 S_w} = \frac{C_{fg} P_{T8} A_8 \cdot \sin\delta_v \cdot \cos\delta_y \cdot \Phi(\gamma_8, P_{T8}/P_a)\, V_s}{\tfrac{1}{2} U S_w T_{t8}}$$

(10)

$$\text{NPR} = f'(\text{RPM}, P_{T2}, \alpha^*, \text{EGT}, t^o, \delta_v, \delta_y, TR, AB, M, \text{IFPC}, T_a, P_a, NG, \text{Degr., Trim}, ACC\text{–}D, FQ, \text{Cooling, Leakage, fan-core mixing}, \ldots) \quad (11)$$

Note: The thermodynamic stations are different for subsonic (converging), or supersonic (converging–diverging [C–D]) nozzles. Thus, for a supersonic nozzle,

$$T = \varphi(M, H, t^\circ, Inlet, \alpha, \beta, C_{fg}, C_{D8}, C_A, C_v, \text{NAR, NG}, \alpha^*, V_s, A_8, \text{IFPC}, AB, TR, T_a, Degr., Trim, ACC\text{–}D, FQ, \gamma_8, P_{T8}, P_a, \ldots)$$

$$= C_{fg} \cdot P_{T8} A_8 \varphi'(\gamma_8, P_{t8}/P_a)\, V_s / \sqrt{T_{T8}} \quad (12)$$

Substituting, one obtains

$$(\Delta C_L)^{\max} = f'(\text{NAR}, C_f, \delta_v, \delta_y, \alpha^*, M, H, t^\circ, \text{Inlet}, C_{fg}, P_{T2}, P_a, C_\mu, P_{T8}, \gamma_8, \bar{C}, C_{D8}, C_A, C_v, \text{RPM, NG, TR, EGT, IFPC}, V_s, \gamma_{\text{dist.}}, \text{G.E.}, \alpha, \dot{\alpha}, \beta, \dot{\beta}, \text{BLS, SSW, Re, FCE, Fan-core Mixing, Turb., Degr., Trim}, ACC\text{–}D, FQ, M_8, T_{T8}, A_8, \ldots)$$

(13)

One must also stress that the location of CP, for instance, varies with the degrees of vectoring. Moreover, the evaluation of FCE is a broad subject in itself (cf. Lecture V).

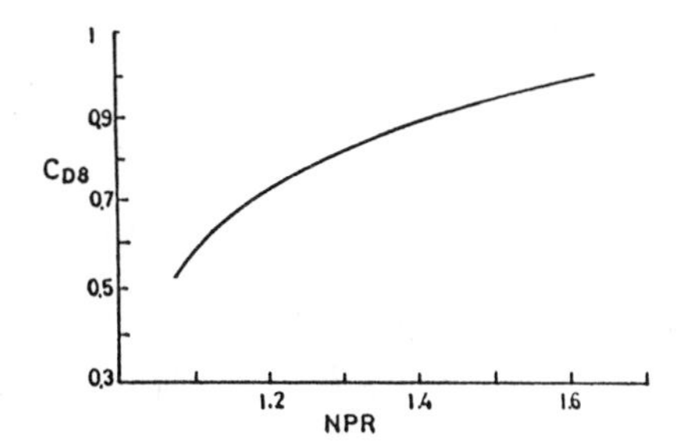

Fig. 11a. **C_{D8} as a function of NPR for a subsonic, circular nozzle.** (Data from the Turbo and Jet Engine Laboratory-Technion-IIT, 1987).

Notes:

1) In developing and testing subsonic or supersonic vectoring nozzles, one must first compare their performance to a *circular* (axisymmetric) nozzle, in order to asses their efficiencies with and without vectoring.

   Hence, such a Laboratory Data Base (LDB) should contain the dependence cf C_{fg}, C_{D8}, and C_v on NPR for an axisymmetric nozzle *isntalled on the same type of mission engine, and having the same A_8 and A_9/A_8 values.*

   Facilities required for such LDB tests are shown in Figs. IV-6 and IV-7.

2) The next LDB should include a comparison between vectored-system performance with a standard bellmouth inlet and that with high-alpha-beta-PST inlet designs that characterize subsonic or supersonic, partial or pure vectored aircraft.

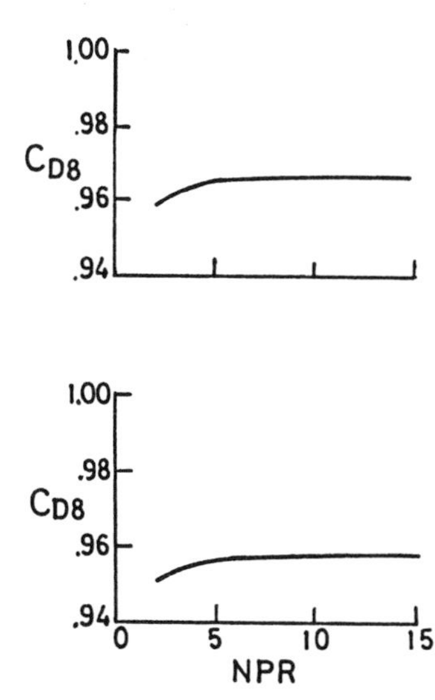

●Fig. 11b. **Variations of C_{D8} with NPR for low-*AR* 2*D*–*CD* vectoring nozzles with various A_e/A_t ratios [48].**

There are at least 45 parameters in Eq. (13). *However, for preliminary design considerations, the most important ones are only ten, namely:*

$$(\Delta C_L)_{\max} \cong f(NAR,\ \delta_v,\ \delta_y,\ C_\mu,\ M,\ H,\ C_{fg},\ C_{D8},\ G.E.,\ TR) =$$

$$\cong (\Delta C_L)_{sc} + (\Delta C_L)_v \quad \text{[see also Figs. 18 and 19]} \tag{14}$$

Inlet and nozzle geometric designs, as well as BLS, β, $\dot{\beta}$, etc., are important in analyzing aerodynamic effects in PST maneuvers. *Thus, a new family of subsonic and supersonic inlets must be designed and flight tested for PST angles of attack.*

For NAR < 10 designs one may use (cf. Fig. 19):

$$\frac{(\Delta C_L)_{\max}}{(\Delta C_L)_v} = 4 \qquad (\delta_v \leq 20) \tag{15a}$$

$$\frac{(\Delta C_M)_{\max}}{(\Delta C_M)_v} = 2 \qquad (\delta_v \leq 20) \tag{15b}$$

$$\frac{(\Delta C_D)_{\max}}{(\Delta C_D)_v} = 2.5\text{–}3.5 \quad (\delta_v \leq 20) \tag{15c}$$

When A_e (Leakage) and A_e (Cooling) are negligible, eq. 7 reduces to

$$C_{D8} = \frac{A_{e8}}{A_8}\ , \tag{16}$$

Eq. (16) is the ratio of effective to geometric cross-sectional, throat area of the nozzle during unvectored or vectored operation (cf. eqs. I-5, 6, and Fig. 13, Introduction). Examples of C_{D8} variations with NPR are given in Figs. 11 to 12. The effects of cooling and leakage are shown in Fig. I-2 and also in Figs. III-2a and III-2b.

C_{fg} accounts for *all* nozzle losses, while C_A accounts for the thrust loss due to the nonaxial exit of the exhaust gases from the nozzle.

In axisymmetric (rounded) nozzles the exit angle α_i varies from zero, at the nozzle center line, to $\bar{\alpha}$ at the outer radius. Hence, for a *constant* M_i per unit cross-sectional area

$$C_A = \frac{1}{A_9} \int_{r=0}^{r=\varphi/2} \cos \alpha_i(r)\ 2\pi r\ \mathrm{d}r \quad (\delta_v = \delta_y = 0) \tag{17}$$

or, more generally, when swirl still exists,

$$C_A = \frac{1}{A_9} \int_{\theta=0}^{2\pi} \int_0^{r=\varphi/2} r \cos \alpha_i (r, \theta)\ \mathrm{d}r\ \mathrm{d}\theta \tag{17a}$$

However, whenever M_i varies with the unit cross-sectional area, as may be the case in some high AR nozzles, *its functional dependence should also be integrated.*

It is also beneficial, to include streamlined flow-dividers/structural struts in relatively high aspect ratio $2D$ nozzles. The streamlined struts may reduce overall weight

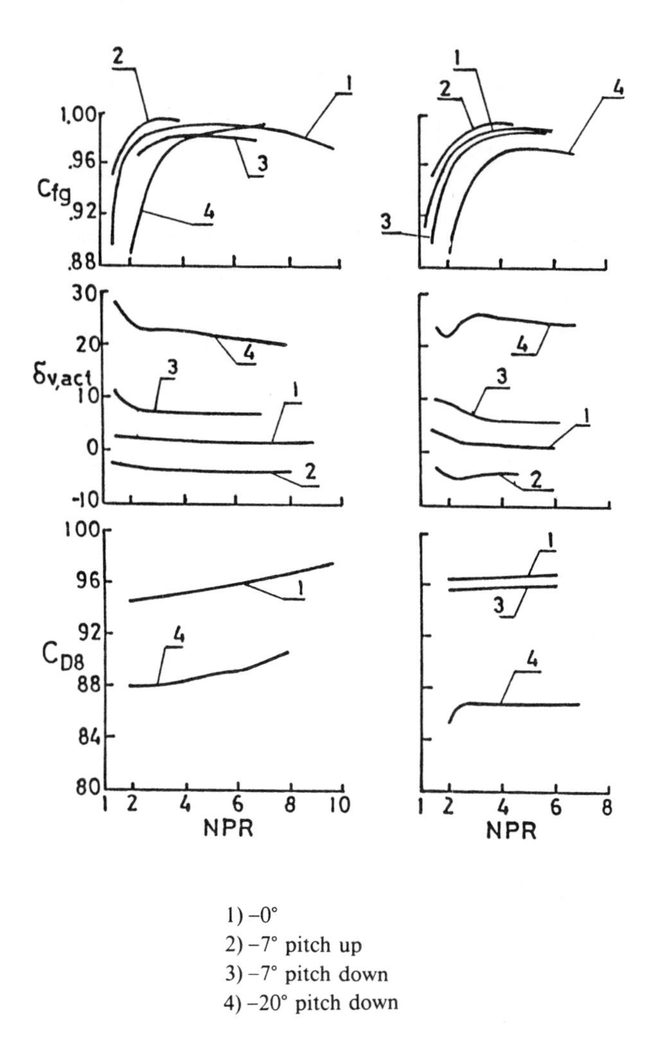

Fig. 12. **Capone and Berrier of NASA have demonstrated that there is, in principle, *a difference* between *geometric* flap vectoring,** δ_v, and *actual jet deflection* δ_v^{act}. 1–0°, 2–7° pitch-up, 3–7° pitch-down, 4–20° pitch-down. In SERN, and in other types of nozzles, this difference may be substantial. *However, for the 2D–CD type of nozzles the difference is quite small and, consequently, may be neglected in preliminary design estimations.* For instance, the mistake in δ_v^{act} varies *from* 4 *degrees, at internal subsonic conditions, to* 1 *degree in supersonic conditions.* The data presented here are for a 2*D–CD* vectoring nozzle. [NASA, Capone and Berrier, 1980, Ref. 56]. This figure also shows *a drastic reduction in* C_{D8} *at* $\delta_v = 20$ *degrees.*

Further data are given in Appendix C.

Notes for the Designer:

1) Constructing engine test rigs with multiple-degrees of freedom to simultaneously measure δ_v, δ_y, T_x, T_y and T_v is not only expensive, but time consuming.
2) Precalibration of δ_v^{act} with δ_v, and δ_y^{act} with δ_y, combined with detailed experimental data on C_{fg} during δ_v, δ_y vectoring angles (at each NPR value), allows the designer to employ engine test rigs which measure only T_x, and, then to use eqs.I-11, 12 and I-13 to calculate actual T_y and T_v values.

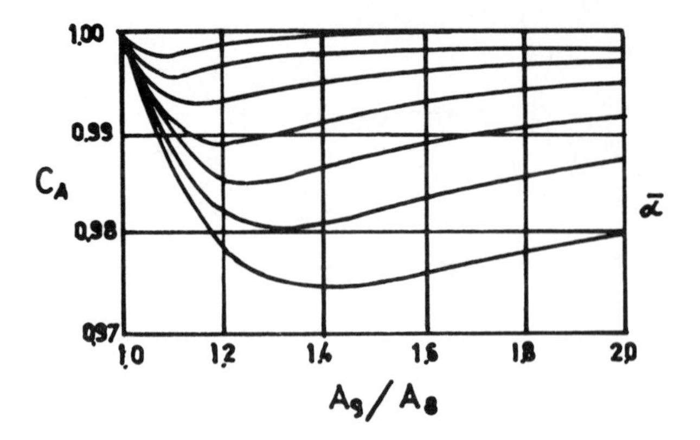

●Fig. 13. **An example of C_A (axisymmetric, C–D nozzle angularity coefficient) variations with the divergence angle $\bar{\alpha}$ (see eqs. 8, 17, 22, 23, 24 for details and explanations in regard to vectored, $2D$–CD nozzles).**

$\bar{\alpha}$ is defined in Figs. I-1 and I-5. Data taken from Kuchar's fundamental reference [138].

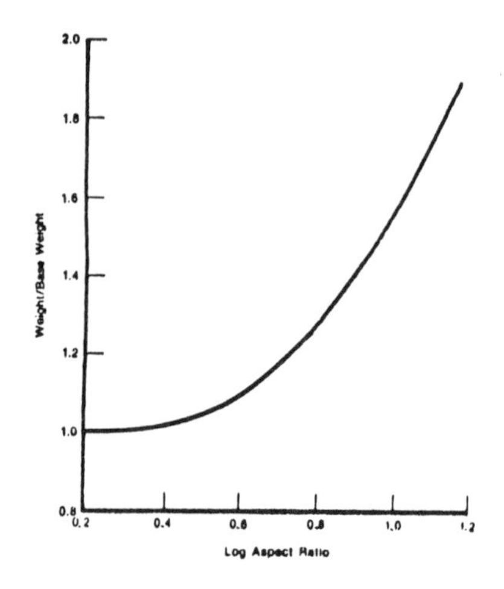

●Fig. 14. **Stevens-Thayer-Fullerton's estimated effect of aspect ratio on exhaust system weight [49].**

However, the actual weight may be *considerably reduced by full integration of high AR nozzles, such as shown in Figs. III-10 and II-1, with the wing structure.*

This was recently demonstrated in this laboratory using yaw-pitch vectoring nozzles having *AR* between 14 to 46.6.

of the nozzle in a well-integrated design with the external wing/body, while also reducing thermal signature and increasing C_{fg} and C_A values.

The streamlined struts may be so designed as to distribute M_i evenly and reduce swirl across the throat and exit areas, thereby allowing not only the use of the simplified equation (17) but, also, to increase C_A and C_v values (see, however, below).

Primitive transitions ducts may be designed by following eqs. 18–20 (cf. Fig. 17);

$$\frac{z'(x')^{\eta(x')}}{a(x')} + \frac{y'(x')^{\eta(x')}}{b(x')} = 1 \tag{18}$$

$$r'(x') = \left\{ \left[\frac{\sin\theta}{b(x')} \right]^{\pi(x'')} + \left[\frac{\cos\theta}{a(x')} \right]^{\pi(x')} \right\}^{-1/\pi(x')} \tag{19}$$

$$y'(x') = r'(x')\sin\theta$$

$$z'(x') = r'(x')\cos\theta$$

$$r'(x') = \left\{ \left[\frac{\cos\theta}{a(x')} \right]^{\pi(x'')} + \left[\frac{\sin\theta}{b(x')} \right]^{\pi(x')} \right\}^{-1/\pi(x')} \tag{20}$$

$$0° \leq \theta \leq 38$$

$$y'(x') = a(x')\tan\theta$$

$$z'(x') = a(x')$$

$$38 \leq \theta \leq 90°$$

$$y'(x') = b(x')$$

$$z'(x') = \frac{b(x')}{\tan\theta}$$

●III-7.3 Nozzle C_v, C_{D8}, C_{fg} Coefficients Revisited.

Being related to the viscosity of the gases, C_V is essentially a function of the "secondary nozzle surface area" at stations 7, 8 and 9, and of the velocity distributions near the solid walls, and, thus, of the Reynolds number.

However, experimental and analytical studies show that Reynolds number effects are less than $\pm 0.2\%$ for all exhaust nozzles and for all operating conditions, thereby allowing easier 2*D* nozzle scale-up and scale-down development efforts.

The main difficulty associated with the development of high-aspect-ratio 2*D* vectoring nozzles, is the fact that, while C_A and C_v may be estimated from analytical studies of 2*D* nozzle geometries, it is extremely difficult, and virtually impractical, to determine them by experimental methods (using, say, very precise flow angle measurements in subsonic and suersonic flow fields).

A practical solution to this problem may be the evaluation of either C_v or C_A from eq. (6) and the definition $V_{9i} = V_s$ when $P_{s9i} = P_a$, viz.;

$$C_{fg(\text{peak})} = (C_v C_A)_{\text{Peak}} \tag{21}$$

Since maximal C_{fg} values are readily determined experimentally for each vectored position (at any pitch, yaw and thrust-reversal angle), the problem may be reduced to calculating either C_A or C_v.

Comparing the difficulties and the associated reliabilities of various candidate theoretical approaches, one may realize that C_A estimation is a practical approach, especially when C_{D8} values are known experimentally.

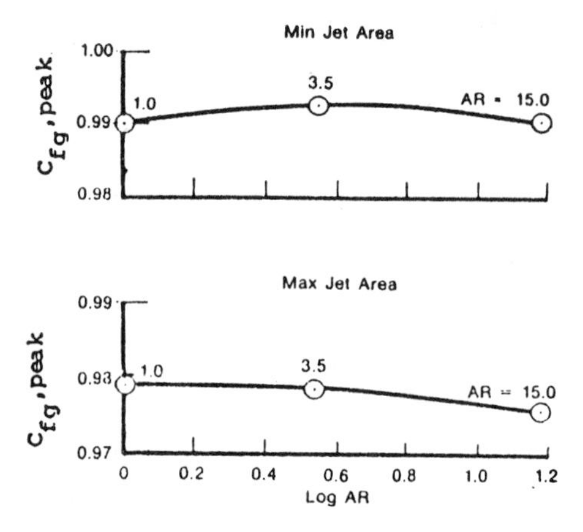

●Fig. 15. **A main design parameter is NAR (cf. eq. 14).** While it affects supercirculation (Fig. 18), it causes only minor changes in nozzle efficiency. Data taken from Ref. 49.

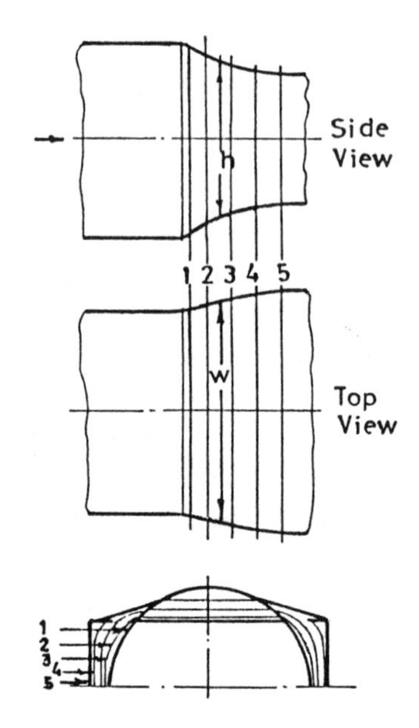

●Fig. 16. **An example of a *C–R* transition duct following the method demonstrated in Fig. 17 for low *AR* vectoring nozzles.** However, our recent laboratory test results with yaw-pitch vectoring nozzles having *AR* between 14 to 46.6 have demonstrated the need to replace this method by more advanced methods which do not add much weight, nor reduce C_{fg}, C_{D8} and other efficiencies of the HAR nozzles.

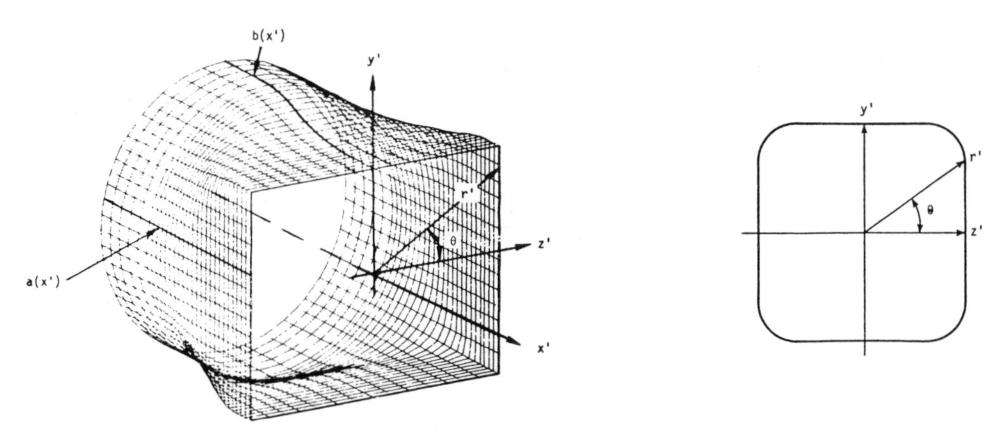

Fig. 17. **Internal geometry of transition *C–R* ducts for *2D–CD* vectoring nozzles characterized by *low* aspect ratios may first be estimated by the equations of a *superellipse*.** *However, for high AR vectoring nozzles this technique may not be suitable!* Cf. the Notes in Fig. 16.

The superellipse equations may be expressed as in eqs. 18–20.

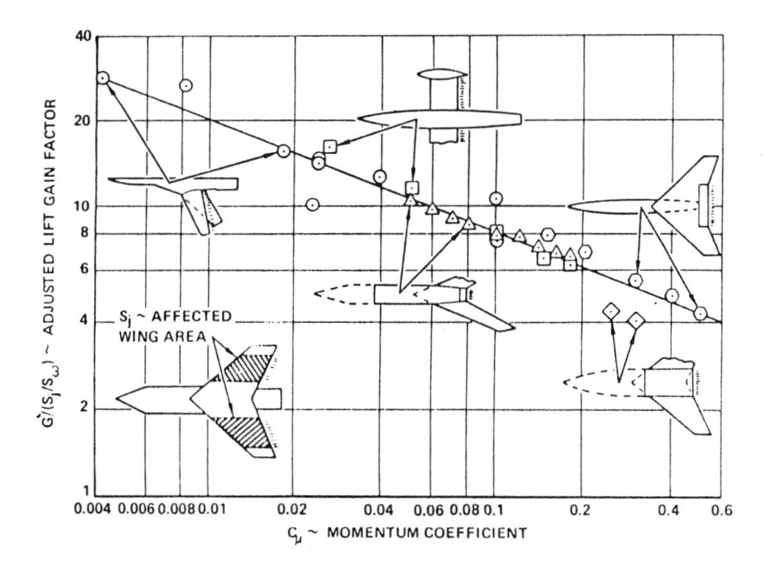

Fig. 18. **Supercirculation Lift Gain Correlation [164–167].** Cf. eq. 29 and Fig. 19.

Now, for RCC-type vectoring nozzles,

$$C_{D8} = \frac{M_7 \text{ Actual}}{M_8 \text{ Ideal}} \cong \frac{A_{e8}}{A_8}, \tag{22}$$

for no-cooling would be required and leakage losses may be substantially reduced (in comparison with the multiple flaps leakage of conventional nozzles).

An example of C_{D8} dependence on NPR for a small turbojet engine used in this lab

($T \leqq 350$ kg(f) at S.L. standard conditions) is shown in Fig. 11a. These results were evaluated from readily available experimental data on NPR, M_7, T_{T8}, and A_8. It should be stressed, however, that the evaluation of M_7 by a simple bell-mouth inlet excludes the possibility of testing [installed] vectored nozzles with various 2D [flush-mounted] PST-tailored inlets.

Now, introducing small δ_v- and δ_y-flight-controlled variations, eq. (17) may be employed to re-estimate the efficiency of vectored nozzles.

$$(C_A)_v = \sin\delta_v \cdot \cos\delta_y \cdot \frac{A}{A_9}\int_{z=0}^{z=B} \cos\bar{\alpha}_v(z)\,dz \tag{23}$$

$$(C_A)_y = \cos\delta_v \cdot \sin\delta_y \cdot \frac{B}{B_9}\int_{y=0}^{y=A} \cos\bar{\alpha}_y(y)\,dy \tag{23a}$$

In rectangular vectoring nozzles the ducts are normally converging (subsonic nozzles) or converging–diverging (supersonic nozzles) only with respect to the upper and lower surfaces.

Hence, when $\bar{\alpha}_y(y) = 0$, i.e., in ITV-nozzles with fixed, parallel sidewalls, eq. (23a) is reduced to

$$(C_A)_y = \cos\delta_v \cdot \sin\delta_y \cdot \frac{AB}{A_9} = \cos\delta_v \cdot \sin\delta_y = T_y/T \tag{24}$$

The split-up of C_A into z(pitch) and y(yaw) components may be required in overall moment calculations of vectored aircraft. (C_{fg}, C_v and C_{D8} may thus be split accordingly.) It may be noted, however, that T itself varies with δ_v, δ_y, α*, NPR, NG, NAR, etc.

Typical data for C_A are shown in Fig. 13.

III-7.4 The Range of Thrust Vectoring

Our laboratory tests show that C_{fg} does not decrease much with increasing NAR. This conclusion remains valid up to NAR $\simeq$ 50. However, special, internal, flow struts and turning curves must be designed for each NAR. *We also found that relatively small variations in δ_v and δ_y cause substantial effects on flight performance.* Simple δ_v-, δ_y-effects may be easily estimated from eqs. 1 and 2. However, the effects on C_{fg} must be evaluated on well-calibrated, thrust-vectoring test rigs. Such studies, as well as our flight experience with pure, vectored RPVs, show that, *with a variable canard:*

$$-20 \leq \delta_v \leq 20 \tag{25}$$

$$-35 \leq \delta_y \leq 35 \tag{26}$$

* The y and z-coordinates are the pitching and yawing axes, respectively.

However, most of the low altitude, high-turn-rates may be accomplished with

$$-10 \leq \delta_v \leq 10 \tag{27}$$

$$-10 \leq \delta_y \leq 10, \tag{28}$$

or less.

The range of Eq. (26) has been found useful *during runway taxiing without front-wheel-steering.*

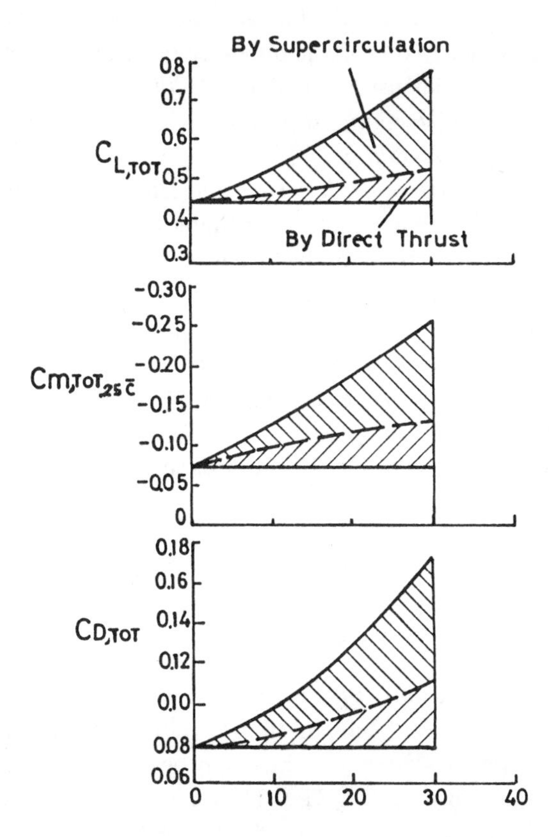

Fig. 19. **Supercirculation increases the maximal values of C_L, C_m and C_D.**
Upper shaded areas represent supercirculation, i.e., $(\Delta C_L)_{sc}$, while the lower shaded areas represent *direct lift generated by* $T_v = T \cdot \sin\delta_v \cdot \cos\delta_y$, i.e., $(\Delta C_L)_v$ (cf. eqs. 10 and 4). The "delta" represents values beyond maximal values obtainable with (unvectored), current-technology wings. For a lift gain correlation see Fig. 18 and eqs. 29 and 30.

The data represent computational and experimental data for relatively small NAR values.

Notes

1) Direct ΔC_M variations due to engine thrust are virtually the changes in pitch moment due to $\cos\delta_y \cdot \sin\delta_v$ engine variations (shown here only for $\delta_y = 0$).
2) Direct ΔC_D variations due to engine thrust are the changes in drag due to $\cos\delta_y \cdot \sin\delta_v$ engine variations. For additional effects see eq. 31, Fig. 18 and the appendices.

III-7.5 Supercirculation in Vectored Aircraft.

For subsonic flight speeds the additional, supercirulation lift may first be estimated by means of the 'adjusted gain factor G' of Fig. 18 defined by

$$G = G'\left(\frac{S_w}{S_j}\right) = \frac{(C_\mu)\,\mathrm{Sin}\,(\alpha + \delta_v) + (C_L)_{sc}}{(C_\mu)\,\mathrm{Sin}\,(\delta + \delta_v)} \tag{29}$$

where G is the usual lift gain factor, S_w the total (or reference) wing area, S_j *the superciruclation affected wing area* (which is defined as the portion of the wing area included between chordwise lines at the inboard and outboard ends of each exhaust nozzle, α the angle of attack (AoA), δ_v the pitch-vectoring angle (counted positive in downward rotations of the exhaust nozzle flaps), C_μ *the exhaust gross thrust blowing coefficent referred to the flight dynamic pressure and wing area*, and $(C_L)_{sc}$ *the increased lift coefficient due to supercirculation.*

Apparently all rounded (axisymmetric) exhaust nozzles are potentially incapable of supercirculation gains in lift. Hence eq. (29) is suitable only for rectangular-type (two-dimensional, or non-axisymmetric) nozzles which are fully integrated with the trailing edge of the wings as shown schematically in Figs. II.1 and III.10.

The available experimental data of references [72, 164–167] indicate that, for a given aircraft configuration, *G is relatively independent of Mach number over the subsonic range, but is strongly dependent upon the momentum (blowing) coefficient and the affected wing area* S_j.

The experimental correlation of Fig. 18 succeeds in coalescing the majority of the data points into a single line defined by

$$G' = G/(S_j/S_w) = 3.3/(C_\mu)^{0.392} \tag{30}$$

●III-7.6 Induced Drag Due To Supercirculation

Associated with the supercirculation lift is an induced drag force, D_Γ. The available experimental data on this effect (References 165–167) has been analyzed and correlated in terms of the efficiency factor e_Γ(which may be estimated as equal to 0.75 and is relatively independent of C_μ and S_j/S_w). Thus,

$$C_{D\Gamma} = \frac{[(C_{Lj0} + (C_L)_{sc})]^2 - C^2_{Lj0}}{0.75\,\pi AR} \tag{31}$$

where $C_{L_{j0}}$ is the 'jet-off' lift coefficient.

●III-8 Engine Sizing Problems

An interesting study has been reported in 1986 by Brooke, Dusa, Kuchar and Romine of the General Electric Co. While studying *2D–CD* ejector cooling systems for vectored propulsion they have raised the following design dilemma. Non-RCC, *2D*,

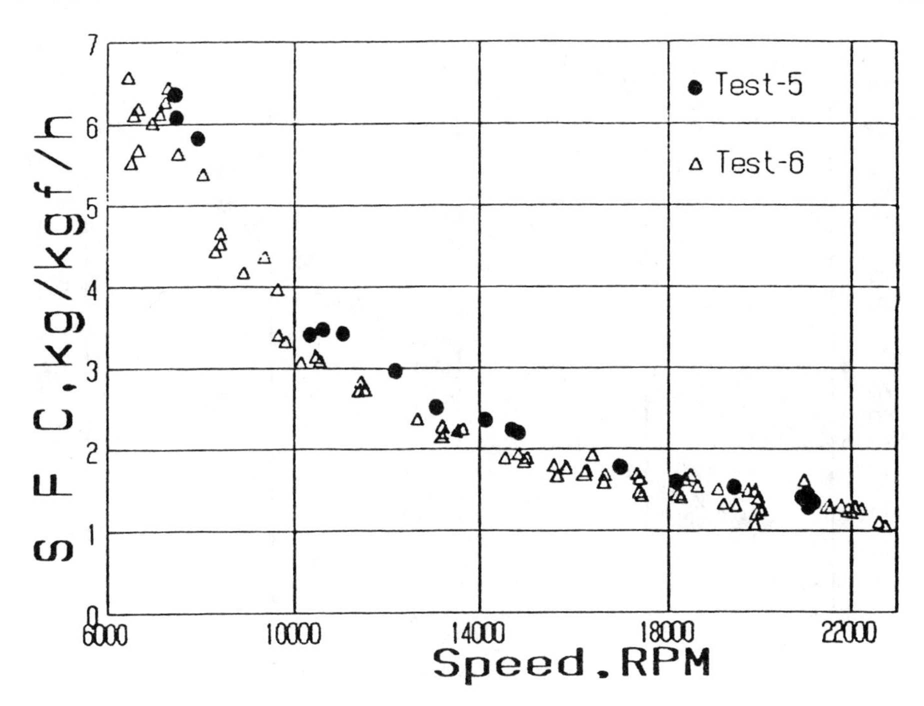

Fig. 20. **Engine SFC Variations (Data from this laboratory, 1986).** Turbojet Engine.
Test-5: with a fixed, two-dimensional, subsonic vectoring nozzle, with Aspect-Ratio = 26.6
Test-6: with a fixed, circular (axisymmetric), subsonic nozzle with the same exit are as in Test-5.

exhaust systems inherently have more internal surface area than conventional, axisymmetric nozzles, and, thus, *require more cooling air to maintain structural integrity during A/B operation.* The trend in relation to engines for advanced multi-mission systems is towards a low bypass ratio and high specific thrust cycles. This tends to increase exhaust gas temperatures and nozzle size, because of higher nozzle pressure ratios, and reduces the amount of exhaust nozzle cooling air available, thus making the nozzle cooling task more difficult. Moreover, increased emphasis on survivability is tending towards higher nozzle cooling flow needs. These increased cooling flow requirements have a direct impact on nozzle gross thrust during *A/B* operation, because the cooling air cannot be mixed with fuel and burned; thus, resulting in a lower *A/B* thrust. If the engine size is established at a max *A/B* design point (which is not likely for the new ATF for instance), then the exhaust system (which requires more cooling flow) can directly influence the engine sizing and weight. Also, from a system standpoint, it may become necessary to compromise engine cycle selection to provide adequate nozzle cooling air.

An ejector system may provide a solution to the cooling flow requirements of nonaxisymmetric exhaust nozzles and may result in an improved propulsion system. An ejector nozzle, however, significantly complicates the propulsion system installa-

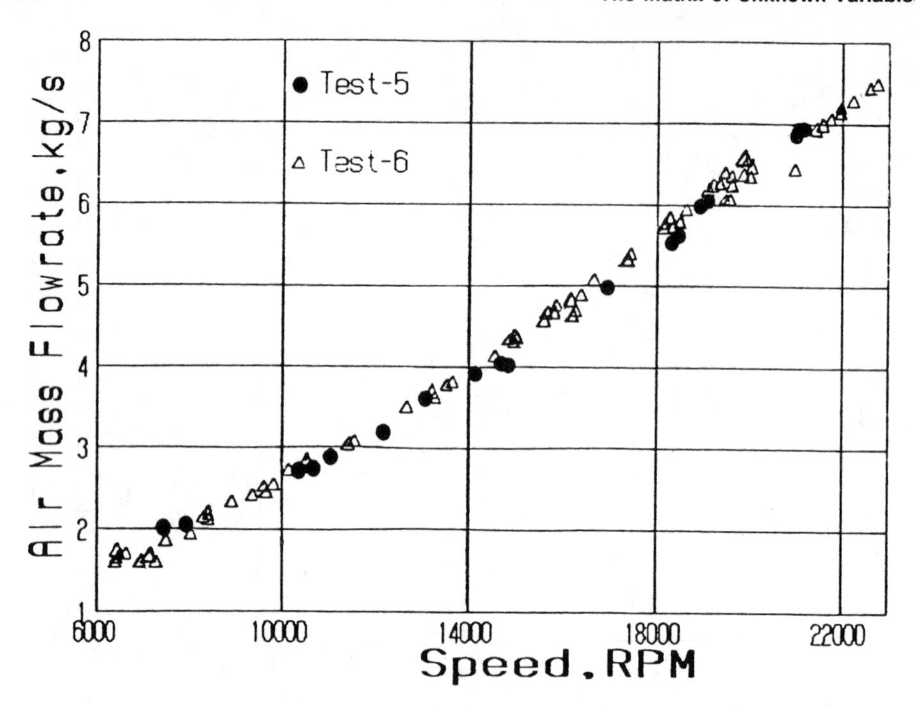

Fig. 21. **Engine Air-Mass-Flow Rate Variations.** (Data from this laboratory, 1986.) Tests 5 and 6 are defined in Fig. 20.

tion and requires careful evaluation of the entire inlet and nozzle installation. The impact on the installed propulsion system design and performance includes system net thrust, component weights, inlet, bypass ducts, nozzle-bay cooling, and life cycle cost. The system net thrust is impacted by changes in inlet ram and spillage drag, bleed and bypass system drag, nozzle afterbody drag, and nozzle internal performance.

III.9 Engine SFC and Air-Mass-Flow-Rate

Properly designed thrust-vectoring nozzles may cause only minor variations in engine performance (cf., e.g., Figs. 20 and 21). *Hence, transforming an unvectoring engine to a vectoring one, is, relatively, a low-cost, short-period program, involving conventional materials, cooling methods, and production means.*

LECTURE IV

VECTORED AIRCRAFT AS R&D TOOLS, OR AS SUPER–AGILE, ROBOTIC FLYING SYSTEMS

"The more we multiply means the less certain and general is the use we are able to make of them."

John Dewey

"Knowledge is one. Its division into subjects is a concession to human weakness."

Sir. H.J. Mackinder

IV-1 Which of the Available Jet-Powered RPVs, Cruise Missiles and Robot Aircraft May be Considered for Thrust-Vectoring Upgrading?

One of the most notable shifts in missile propulsion is currently from rocket, or rocket-ramjet power, to turbine power. Turbine power is now advancing in a number of technology fronts: RCC-Low-RCS (App. B); Counter-Rotating Propfans and propellers; advanced, yet extremely simple, low-cost, low-maintenance turbojet and turbofan engines (with and without TV); stealthy, high-range propulsion systems with very high terminal maneuverability (e.g., the anti-radar Tacit Rainbow missile).

Most of the RPVs available now are driven by piston engines. The minority are Jet-powered.

Jet-powered RPVs may be divided into "turbojet", "turbofan", or "rocket" types. Similarly, Cruise missiles (CM) and anti-radar missiles, may be divided into "turbojet" or "turbofan" types. Thrust vectoring may be considered an attractive option for the last two categories, especially when terminal supermaneuverability (Lecture II), low-cost, low-weight, flexible, stealthy, high-range, long-storage-time characteristics are required.

"Turbojet" Missiles

The "Turbojet CM" category includes the antiship CMs USN-McDD Harpon (surface-to-surface R/UGM/-84AC, or the air-to-surface AGM-84A/C), and the USAF-Boeing-SRAM-AGM-69A. These CMs are propelled by the Teledyne J402-CA-400 turbojet engine (a scaled-down, improved version of the Teledyne-J-69 family of engines, or of the Turbomeca Marbore older engines).

A (Microturbo-TRI-60) turbojet engine is the powerplant for the British ASM (Sea Eagle-P3T) antiship missile, a (Turbomeca-Marbore) turbojet engine for the Saab-08A-SSM-RB08A, 110-mi-range missile, and a (Turbomeca-Arbizon) turbojet engine for the French TH-CSF/OtoMelara-SSM-Otomat/Mk.2, 1540-mi-range, antiship missile.

Also included in this category are the Soviet AS-15 (Kent) air-launched, 1850-mi-range, CMs for launch by Blackjack, Backfire, Bear H, or by newer missile-carrying aircraft. It also includes the AS-1 (Kennel), 63-mi-range, antiship winged-CM, and the AS-2 (Kipper) CM with a range of 100 miles.

Thrust vectoring may add significant maneuverability, reliability and survivability merits to this category. It may also help to simplify the various launching methodologies, lower costs, signatures, and weight, increase range and improve flexibility and long-storage-time specifications (see below).

"Turbofan" Missiles

This category includes the (strategic) nuclear ALCM AGM-86B USAF-Boeing/Honeywell/ Litton, (1500+)-mi-range, air-launched CM (from the B-52, B-1, and B-2). These missiles are propelled by Williams International (WI) F107-W1 turbofan engines.

Thrust vectoring may add significant maneuverability, reliability and survivability merits to this strategic category. It may also help to simplify the launching methodology, reduce signatures, weight and cost (Fig. 1).. Here the TV nozzle may deflect only the core (hot) jet, or both the cold and hot jets.

"Turbojet" RPVs and Anti-Radar Missiles

This category includes the following subcategories:

1) *Surveillance/Target acquisition and "Multi-Purpose" (MP)*, such as the Boeing Brave 3000 (propelled by Noel Penny NPT 171 turbojet engine), the German/Canadian AN/USD-502(CL-289)-Canadair/Dornier (powered by KHD-117 turbojet engine).
2) *Target (T)*, such as the Fuji Heavy Industry BQM-34A (powered by Teledyne J-69-T-29 engine) and the J/AMQ-2 (powered by MHI TJIM-3 engine); the People's Republic of China Changkong IC (powered by WP-6 turbojet engine), the Australian/UK. Jindivik Mk. 4A (powered by R–R Viper Mk. 201 turbojet engine), the Aerospatiale CT-20 (powered by Turbomeca Marbore engine), and C-22 (powered by Microturbo TRI-60 engine); and the Italian/Meteor Aircraft & Electronics/Mirach 100, 300 and 600 (powered by Microturbo engines).

The US Target-RPVs include: the Beech Aircraft MQM-170A, MQM-107B/D, 999F, and BQM-126A (powered by Teledyne J402 and Microturbo engines), the Northrop AGM-136-A,* the Chukar-II-MQM-74C, and the air-launchable Chukar-III-BQM-74C (powered by WI-24-7), the Teledyne Ryan BQM-34A/S and –34E/T (powered by Teledyne CAE J-69 family of engines) and the larger RPVs QF-100 D/F and QF-5 of Tracor (powered by P&W J-57-P21 and GE-J85-GE-13 engines).

New turbojet engines are now being developed by various laboratories to fill the gap of low thrust levels required for small, long-range, reconnaissance/stealth/MP RPVs, or CMs. (E.g., the Teledyne CAE 305 engine [Fig. 1] with thrust levels in the range of 45 to 90 lb.t., a diameter of only 6.5 in. and a very low dry weight.)

"Turbofan" RPVs

New turbofan engines are now being developed by various laboratories to fill the gap of low thrust levels required for small, long-range, reconnaissance/stealth/MP RPVs, or CMs. The available candidates in the upper-level of this subcategory are WI F-44 (1800 lb.t.), the Garrett F-109-GA-100 (1330 lb.t.), and the WI-107 turbofan engine (600 lb.t.). Other engines have been proposed and/or developed in this relatively new category.**

---

* Jet power is mandatory for loitering and final maneuverability of such anti-radar missiles as the USAF/Northrop Tacit Rainbow AGM-136A, etc. It provides low cost, low weight, flexibility and high degrees of terminal manerability even without TV.

** E.g., propfans, turbine low-pressure stage and four-stroke rotary high-pressure section on co-axial shafts, cold-profulsion, etc.

IV-1.1 The Thrust Vectored Barak Missile

The rocket motor of the antiship TV Barak missile provides 3 thrust levels:

I – Low thrust for launch and turnover. (The missile is vertically launched and the aerodynamic fins are assisted during turnover by external thrust-vector control vanes in the rocket efflux. These are jettisoned after turnover, which occurs 0.6 seconds after launch.)

II – High acceleration boost.

III – Sustained thrust for short-range, high-maneuverability engagement.

The TV Barak missile has been developed jointly by Rafael-Israel's Armament Development Authority, and IAI (Israel Aircraft Industries). It is to become operational in the early 90's.

IV-1.2 Expanded Missions for Vectored RPVs and Robot Aircraft

Much has been written on the mission philosophies and on the design of various *propeller-powered RPVs and on (unvectored) jet-powered, RPVs and cruise missiles.* We do not intend to reproduce this material here, except briefly, when a comparison is required between those vectored and those unvectored.

To start with we note that (jet-powered) *vectored* RPVs may be subdivided into two types, namely:

IV-I.3 Advanced R&D Tools

The emergence of vectored RPVs as R&D tools replaces some of the more traditional roles of wind-tunnel testing techniques. This cost-effective methodology is capable of reliable simulation of manned vectored aircraft performance under realistic conditions involving, say, *PSM or post-stall maneuvers, STOL and V/STOL.* However, as in wind-tunnel testing, it encounters scale-up and scale-down problems. (One may first notice that materials and servos do not scale-up easily. RCS/IR signatures and aerodynamic/propulsion systems present special scale-up/scale-down problems. A few of these have already been discussed in II-3, and in Lecture III. A few additional ones will be reassessed below and in paragraphs IV-12 and V-1.1.)

IV-I.4 Unsurmountable VRA-Missions.

This class may be generally defined as *Vectored Robot Aircraft (VRA):* the term includes, in principle, cruise missiles, drones, decoys, TV-missiles, flying vectorable bombs, long-loitering target-designators, surveillance RPVs, etc.

However, the unique missions of VRA are not yet clear, nor well-defined. Some of these are to be discussed below, while others are postponed to Volume II.

Assertion I.

VRA-technology opens the way to new, expanded missions with unsurmountable performance in agility and launching or V/STOL capabilities (at low R&D, production, and storage-maintenance costs).

Assertion 2.

VRA-technology is one of the critical technologies required for high-performance, stealth, [cruise and/or terminal supermaneuverability (cf. Lecture II)] robot aircraft.

Assertion 3.

While the defence-role potential of VRA-technology has not yet been fully exploited, *it is already generating a revolution in the design philosophy of some unique, air, land and marine missions.*

Assertion 4.

The defence role potentials of vectored aircraft, in general, and of VRA-technology in particular, *depend on a new branch of research and development involving new vectored components, new powerplants, new PST inlets, novel unit operations, and, as yet, unknown, integrated flight/propulsion control theory.*

* * *

In trying to define the practical meaning of the aforementioned assertions, one may first examine the present state-of-the-art of *unvectored*, jet-powered, decoys, drones, RPVs, and advanced cruise missiles (ACM), and compare it with that of VRA. For this purpose one may first concentrate on conventional, *jet-powered-drones*, and, then, extend the comparison to other, less-known, applications.

Some of the more sophisticated missions of *target RPVs*, decoys, or *drones*, is in *simulating subsonic or supersonic "stealthy" or "unstealthy" aircraft, operating under a variety of different EW conditions*, (using such devices as Lunaberg lenses, corner reflectors, *IR* and shock-wave detectors, etc.). Indeed, *in the general air-defence training role, the decoy-drone is sub-scale to the threat that it represents, and, consequently, must impart performance and detection characteristics of the threat to be simulated.* Thus, if the threat is an advanced, vectored, cruise missile (AVCM), or a fighter, the simulator must also be vectored.

Yet, the most important uses of VRA are to be found in the *new roles that reflect their own inherent advantages.* These include:

1) *Helicopter Killers* (high closing speeds with final PSM-PST/RaNPAS vectoring agility) and *Anti-Radar Missiles* (ARM).
2) *AVCM and RPV Killers* (high closing speeds with final, PSM-PST/RaNPAS vectored agility).
3) *Ship Protectors* (V/STOL or VTOL with vectoring agility and very low speeds).
4) *Unmanned Rescue* (V/STOL or VTOL, stealth).
5) *Multirole (land)* (V/STOL or VTOL with vectoring agility).
6) *Multirole (Air)* (V/STOL or VTOL with vectoring agility and stealth).
7) *Penetrators* (low signatures).
8) *Advanced Vectored Drones (AVD) and decoys.*

9) *Advanced Vectored Cruise Missiles (AVCM).*
10) *Vectored RPVs armed with hypervelocity all-aspect missiles.*

Further classifications apply according to size and cost:

IV-2 Mini RPVs, mini AVCM, or mini-VRA.

Such vehicles are increasingly built now with stealth or enhanced-signatures configurations and characteristics, IFPC (including autostabilization systems), sophisticated avionics and weapon/surveillance subsystems. *However, because of their small size, and relatively low cost to develop and fly, these vehicles are often erroneously associated with model aircraft.*

A subdivision of this group inclues:

IV-2.1 Hot (Jet-Powered) VRA.

The penalties associated with this technology are:

1) *Jet-engine flame-out due to inlet distortion at PSM or during post-stall-maneuvers.* This limitation results in highly-limited agility, performance and mission characteristics, unless special, PST-tailored inlets are employed.
2) Use of high-temprature jets and materials increases cost, weight, *IR* signatures, cooling requirements, and complexity. It also reduces availability and reliability.

IV-2.2 Cold (Jet-Powered) VRA.

The payoffs associated with this technology are:

1) *Supermaneuverability capability* (cf. Lecture II) *due to thrust vectoring uninhibited by engine-intake distortion at high vehicle incidence* (high-α-β, PSM and PST-maneuvers).
2) High *T/W* ratios afford the development of a *bona fide* VTOL, or V/STOL capability.
3) Low *IR* signatures.
4) Low-weight materials increase the effective-payload capacity.
5) Low Specific Fuel Consumption.
6) Damage to launch-area/space equipment is minimized.
7) TBO, TBF, endurance, controllability and safety are maximized, while life cycle costs and storage expenses are substantially decreased.

These systems may be based on newly investigated, combined-cycles powerplants involving *multiple-stage axial or centrifugal compressors* (powered by, say, Wankel, or by smaller types of beyond -20,000-RPM, 5 to 150 Hp piston engines). The compressor supplies subsonic, or supersonic cold propulsion for unvectored, or vectored vehicles.

[For small turbojet engines one may use here the new Teledyne, 6-inch-diameter,

305 engine, or the SCAT, or the WR-24–7, 8, KHD 317/117, NPT, or Microturbo series.]

IV-3 Midi-RPVs, Midi-ACM and Midi-VRA.

This size-cost category is limited, still, to hot-jet propulsion systems. Otherwise, it is similar to the mini-class in terms of performance/missions. Jet engines for this category are now available from Teledyne or *WI*.

IV-4 Maxi-RPVs and VRA.

This category includes *fighter-size drones and flying bombs.* The large drones are often based on old fighters converted into target RPVs which employ various autostabilization, *EW* gear and other specific systems.

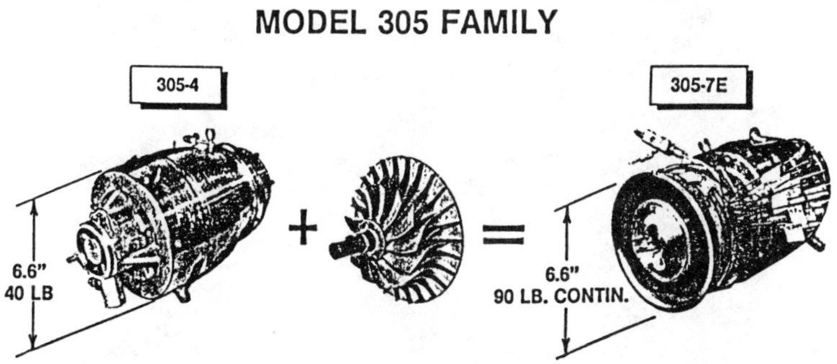

MODEL 305 FAMILY

305-4

305-7E

6.6"
40 LB

6.6"
90 LB. CONTIN.

Fig. IV.1. **One of Teledyne's proposals to power small robot aircraft.**

IV-5 Cost-Effectiveness of flying Vectored RPVs vs. Limited Wind-Tunnel Testing

Our integrated laboratory tests and flight-testing efforts with pure vectored RPVs are aimed at *reducing the risk decision regarding propulsion/flight designs.*

A near-term goal associated with these integrated studies is to demonstrate the *lower costs, time-condensed milestones, and high reliability-efficiency of this methodology, namely, of designing, constructing, lab-testing and flight-testing pure vectored RPVs, rather than relying only on the more expensive, scale-up methods associated with vectored-models-suction-ejection in wind-tunnel model testing.*

Thus, we proceed with the following integrated test stages:

STAGE 1:

Although we start the design of PVA with a wind-tunnel model testing, we refrain

Fig. IV-2. **Control Rooms No. 3 and 5 at the Jet Propulsion Laboratory (TIIT) help to substantiate the proposed laboratory/flight-testing methodology (cf. Figs. 6 and 7).** The lower one controls Components Research Test Room, while the upper one regulates the Engine Altitude–Speed Test Rig and R&D Facilities shown in Figs. 6 and 7.

from testing these small models with internal suction-injection systems (so as to try to simulate actual thrust-vectored conditions in flight).

STAGE 2:

New ideas are first tested on a component subscale test rig (Fig. 7). (Up to about 1 kg/sec engine air-mass flow-rate-testing, and up to 1,000°C.) This small-size, component category simulates supersonic and subsonic thrust-vectoring nozzles. The results of these tests are employed directly, or are scaled-up for the full-scale testing of STAGE 3.

STAGE 3:

For up to 7 kg/sec air-mass-flow-rate-testing, we use the laboratory's altitude-speed, *full-scale*, engine research system (cf. Fig. 6 and 7). This size category simulates various fighter engines, *provided certain guidelines and precautions are employed in the scaling-up process*, e.g., successfully passing first the component-test category ("STAGE 2"). The results of these tests are also useful for the next stage.

STAGE 4:

Laboratory and flight-testing of radio-controlled, pure vectored RPVs in the mini-size category of 7×4 ft to 9×4 ft. The propulsion systems of these TV–RPVs are scaled-down versions of optimized systems developed first through STAGE 3. These flight tests are employed to get very high AoA and extremely sharp turns such as those shown schematically in Fig. II-2. The tests include short takeoffs and landings and engine-out landing tests. Employing these results one may return to STAGE 1 and proceed again to STAGE 3 (sometimes without passing again through STAGE 2) [cf. the diagram in the Preface].

The experience gained by this laboratory through STAGES 1 to 4, has given verifiable proofs to a new, cost-effective methodology in the design, testing, development and flying of vectored propulsion/flight systems. *So far the cost and time, in comparison with the more traditional wind-tunnel methods, appear to be an order-of-magnitude lower (depending on the design, size and relevance of the laboratory and flight-testing data required).*

IV-6 Preliminary Phases of Vectored RPV Design.

In current design methodology the integration process is often limited to steps involving simple combinations of well-proven components with new ones; e.g.;

- An existing engine with a new nozzle.
- An existing engine with a new inlet.
- A new nozzle with an existing wing structure.
- A new inlet with existing, aerodynamic surfaces, and flight-control systems.

The results of such processes may only lead to partially-vectored aircraft, which are characterized by less significant gains in performance and survivability.

Alternatively, one may first study the following:

- *Expected technology limits of radars, IR and optical sensors, etc.*
- *Expected technology limits of powerplants, including PST inlets and thrust vectoring nozzles (cf. Appendices B and F)..*
- *Expected technology limits of supermaneuverability and PST-controllability.*
- *Expected threats.*
- *Expected missions.*
- *Financial limitations.*

Fig. IV-3. **Early windtunnel models (see also Fig. 11, Introduction).** The one on the left was flight tested (see Fig. IV-5). See also Fig. IV-4.

Then, as a first iteration, one may examine the methodology presented schematically in the Preface, and decide on a program which involves the relevant variables. Next, one may gradually reduce the number of variables to the minimum required to end up with a *first prototype of a proof-of-concept, vectored/stealth RPV.* Thus, shape, *RCS, IR*, materials, NAR, supercirculation, limits of yaw-pitch-reversing vectoring, thrust-to-weight ratio, maximum engine intake distortion, cold or hot propulsion powerplant, supermaneuverability, takeoff and landing factors may first be integrated to adapt to a tentative mission. Next, a few *tentative shapes* may be constructed *around* the best propulsion system available (including its optimal, laboratory tested, inlet-core-engine-vectoring nozzle configuration). *Trimming for low signatures, and proper wing loads, payload, performance, weight, range, etc., one may fix the wing aspect ratio, and the wing total surface area, sweep and range of incidence variations, etc.*

A range of (unvectored) wind tunnel models may then be built and tested to evaluate some preliminary variables, with or without canards, vertical stabilizers, tail, etc. The data may then be fed to a highly simplified computer simulation program.

By evaluating the location of the *aerodynamic center* (and the center-of-pressure variations), one may use simulation programs to decide on *static stability limits, etc., for the first flying vectored RPV to be tested.* A 5% positive static stability may be a good starting guess in case the simulation is not sufficiently reliable.

A mock-up model may then be built in the laboratory, and all actual servos, batteries, fuel vessels, inlet, engine, nozzle, telemetry–metry/flight computer installed in it, till the required static stability limit is met. The mock-up, or the flyable prototype should be constructed according to one-to-one scale drawings and in the following order:

1) Install (STAGE-3-tested) engine intake(s), and the selected (STAGE-3-tested)

exhaust vectoring ducts and nozzles, with proper flaps for engine-out emergency controls and internal, streamlined flow dividers/structural struts (cf. Fig. 1 in the Introducton) in such a design that the inlet/engine/nozzle becomes the main carry-through structure for control loads!

2) Following the low-signature, external skin lines, build the aircraft structure first around the (laboratory-tested) 2-*D* nozzles shape, following each nozzle top and side cross sections and integrating the wing structure with the exhaust system internal structural struts.
3) Repeat stage 2 for the engine intake. This should be a PST intake for hot propulsion, or a simple intake for cold propulsion.
4) Introduce retractable landing/takeoff gear, engine auxiliary equipment/instrumentation/control servos, fuel tanks, fuel and servo lines, on-board computer, etc.
5) Construct and install the canard (if required) and its control servos.
6) Construct and install vertical stabilizer(s) (if required), and integrate them with the nozzle-wing structure.
7) Install batteries, telemetry–metry data-acquisition/FBW computer, servos, amplifiers, sensors, analog-to-digital converters, and other accessories, as required for each particular prototype.
8) Hang the vehicle, without completing its external skin, and find its center of gravity.
9) Trim the center of gravity location to meet predetermined static stability margin, moments of inertia, etc., usually by moving the batteries and the computers (fuel vessels must stay centered around the vehicle center of gravity).
10) Balance the vehicle around its lines of symmetry.
11) Complete vehicle's skin covering job.
12) Operate the engine(s) and evaluate static takeoff thrust, with and without vectoring, using the remote or FBW control commands.
13) Trim propulsion system and servos for remote, radio-controlled or FBW operation.
14) Measure the empty weight of the prototype and calculate thrust-to-weight ratio with fuel vessels full and empty, depending on the mission/research project aims.
15) Evaluate T/W ratios for various pay-loads.
16) Use the computer simulation program to estimate runway distance till rotation, at various α and δ_v settings and timing.
17) Final trimming of wheels height/propulsion-system control/δ_v and δ_y, etc.
18) Calibrate the on-board, data-extraction computer and freeze its software, mode of operation and number and type of variables to be recorded during the flight tests.

IV-6.1 Flight Tests

One may now proceed to *runway testing and trimming, and eventually to a well-designed flight program, with proper monitoring of performance, including data acquisition.* At a given stage of this program one may also evaluate *IR* and *RCS* signatures,

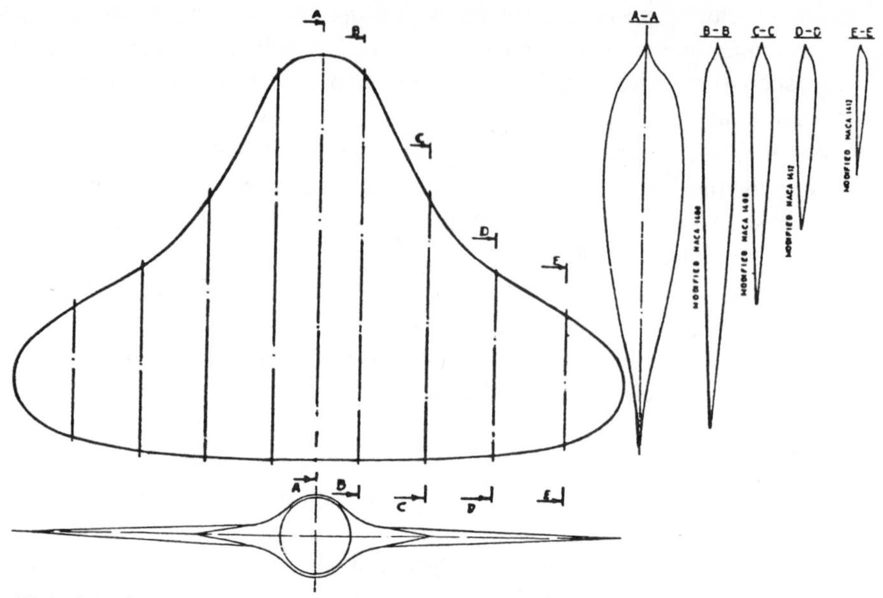

Fig. IV-4. **One of the early prototypes tested in our subsonic wind tunnel during STAGE 1 of our integrated vectored propulsion/flight-testing program** (cf. also Figs. II-1a to II-1h, IV-3, IV-5).

and, accordingly, make variations in shape and materials of the vehicle, and in its propulsion system.

To obtain PST maximum-agility comparison-maneuvers the propulsion system should first employ *inlet-distortion-free, cold-jet vectoring*, and, therefore, low-temperature materials. Then, gradually, as the program proceeds into *higher flight speeds*, a proper switch to *hot-jet vectoring* may be required. In turn, this requires new materials and PST inlets to be integrated with the propulsion system, and repeating the whole process from the very beginning at a much higher cost, lower program pace, and much higher safety risks.

A cost-effective program should also stress the reliability of *inflight data acquisition, including reliable video-computer recording.* In any case, low-speed maneuverability and STOL/VTOL characteristics should be evaluated first.

Here the simplest and lowest-cost program uses only advanced PC computer "cards" reconstructed in a new mode, *which readily stores in its advanced PC RAM all flight data during a few critical minutes of, say, flying in well-defined horizontal and vertical sharp turns/loops involving PST conditions.*

Then, following landing, the flight-borne computer is connected for data feeding to another, ground-based computer, near the runway. Using suitable preprograms, the ground computer may plot the results, compare them with previous flight data, and with flight objectives, and, finally, may demonstrate the best conditions for the next flight to be followed. The flight crew may then discuss the results to determine the next flight procedures.

* * *

The flight test data may once again be examined in the laboratory for incorporation of some possible modifications in the basic design. For instance, *following the analysis of rolling flight data of a given prototype during high angles-of-attack maneuvers, a decision may be made to improve roll rate by increasing, say, the rolling moment arm Y* (cf. Fig. 1a in Lecture II), and, accordingly, *to change the geometry of the exhaust nozzle, and, consequently, also that of the entire shape of the vehicle.* This may lead to the construction of modified wind-tunnel models. The entire process may then be repeated, ending up with a super-agile, vectored-stealth RPV.

* * *

Low-speed agility of vectored-stealth RPVs must then be compared with that of a similar, unvectored RPV, both having the same thrust-to-weight ratio, moments of inertia, shape, size, etc. The incredible sharp turns obtained with the former are not usually the objective, but only the baseline for proper comparisons with unvectored baseline performance and agility metrics.

* * *

Scale-up problems are the next important problems to be treated. For instance, during flight studies of engine-intake distortion (of a model RPV of the new STOL F-15), an extensive use is made of various *scale-up analogies involving similar Reynolds numbers in both model and actual systems.* However, many design parameters do not scale-up, e.g., materials and servos.

●IV-7 Synergetic Studies As Mid-Course Adjustments

Figure I in the Preface describes the feedback mechanism employed in this laboratory during the stages described previously.

Some of the early conceptual design options are related to the *iteration process* involving the emergence of *"figures of merit"* and configurations, and their testing in *wind tunnels*, the jet propulsion laboratory, and in *actual flight.*

Some of the early configuratons tested in our subsonic wind-tunnel during STAGE I of our vectored propulsion/aircraft program are shown, without canards, in Figs. 4 and 5. *These are subsonic, single-engine prototypes.*

However, whenever the entire vectored propulsion system is to be *integrated* with such configurations (including, say, vectored nozzles and inlets as shown in Fig. II-1f), *a new family of "figures of merit" and configurations must be developed and tested.*

Probably the best method to start such a testing is as follows:

PHASE-1

a) Determine the *general aim and scope* of the entire vectored aircraft program (i.e., R&D, or a given RPV–RA mission, or RPV simulation of available aircraft to be

Fig. IV-5. **GAL-3; one of the early pure vectored RPV designed by this laboratory (1987).** Cf. Fig. II-1. Its first flights were conducted by (from left-to-right,) Erez Friedman, Mike Turgemann and Tzahi Cohen, in Megiddo Airfield during 1987, using radio controls. It is now equipped with on-board computers. Its wind-tunnel model (Fig. IV-3) is shown on top. The (cold) jets emerge from the high AR (=46) yaw-pitch-roll thrust vectoring nozzles in the back trailing edge (marked with strips). No rudders, ailerons, flaps, etc. have been used. The vertical stabilizers will be removed during future flight tests with IFPC. Using such RPVs, PSM/PST/RaNPAS maneuvers are being tested since 1989.

transformed into partially-vectored aircraft, [e.g., the F-15 transformed into an F-15 STOL fighter]).

b) Determine the overall level of *observability and reflectivity* required (using different nozzles, inlets, shapes, materials, etc.).

c) *Determine the overall level of high-α-β, STOL, VTOL, or V/STOL requirements.*

d) *Determine priorities, e.g., synergetic studies of high α, β-inlets, vectored nozzles, supermaneuverability, low observables, STOL, V/STOL, control links, or PST–IFPC.*

e) Use the material of this design course to evaluate:

e-1: Simplified *computer simulation programs* for takeoff and landing for various values of wing loads, T/W, L/W, $-D/W$, S, C_L, C_m, C_D, δ_v, δ_y, δ_c, stability data, wing shape, W, α_m, Y, D, TR, C_{fg}, C_{D8}, C_A, C_v, ground effects, inlet distortion limits, emergency sizing of C_f (Fig. III-1), etc. (Cf. Lecture III for the matrix of unknown variables involved).

e-2: *Computer simulation programs for observables.* (The physico-thermodynamic fundamentals associated with this subject are to be treated in Volume II.)

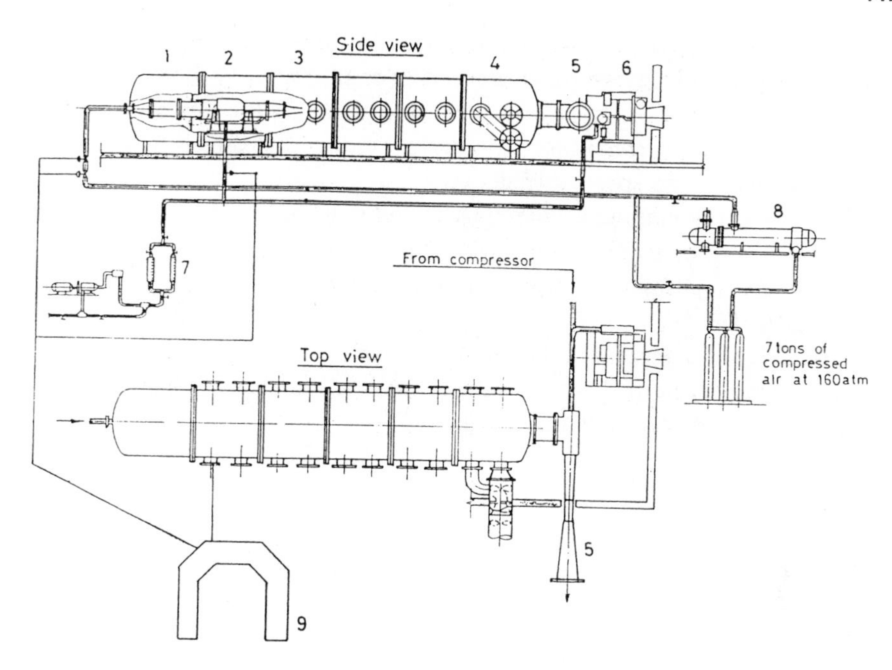

●Fig. 6. **The proposed integrated methodology of laboratory/flight tests is based on "STAGE 3", i.e., on the altitude-speed, full-scale, engine *R&D* system of this laboratory (cf. Figs. 2 and 7).** 1) Engine inlet section, 2)Engine section, 3) Vectoring nozzle section, 4) Exhaust section (includes fans for S.L. tests), 5) Ejector system, 6) Gas turbine for low-altitude simulation, 7) Fuel system, 8) Large heat exchanger for heating or cooling inlet air during altitude simulations, 9) Control Room No. 5. **Note:** Engine inlet section can be modified to fit R&D needs of various PST/Stealth engine inlets operating at various α, β, $\dot{\alpha}$, $\dot{\beta}$ conditions. Thrust vectoring nozzles tested are those scaled-up from component tests (cf. Fig. 7).

e-3: Early candidate *'figures of merit'* to be tested in the *wind tunnels*, but without vectoring means, canards, gears, etc.
An example of such a preliminary family of wind-tunnel models is shown in Figs. 4 and 11 in the Introduction.

f) Use the preliminary wind tunnel test results to *verify and re-evaluate* the computer programs and to define *preliminary PST–IFPC limits* for various *missions and control methods.*

PHASE-2

a) *Determine Statement of Work (SOW) and Program milestones*, using the initial data of Phase-1.
b) Determine *limitations and proper interactions between computer simulation, wind-tunnel tests, jet-propulsion tests, RCS–IR* tests, telemetry–metry methods of flight-control tests, flight tests, analysis, and conclusions.
c) Determine the flight-testing program methodology, including the landing proce-

dures, telemetry–metry feeding of flight data into a portable computer near the runway, video-recording/computer analysis of results, etc.

d) *Re-examine the selected type of* (radio, cable, fiber, or *IR*-laser beam) *flight control system, safeguards, emergency procedures,* number of channels, number of flyers, altitude, maximum speed, airfield, etc.

e) *Re-examine the materials, landing gear and propulsion method employed.*

f) *Re-examine scale-up and scale-down design limitations.*

g) *Choose the type of core engine required* (cold-jet, or hot-jet propulsion) (cf. Figs. I-1i, 1f).

h) *Choose engine sizes and thrust levels required for* laboratory testing of the entire inlet-engine-nozzle propulsion system, or of its individual components, using proper scale-up/scale-down techniques.

PHASE-3

1) Fix the overall *size and shape of the vectored-stealth RPV*, including its T/W, L/W, C.G. – A.C. criteria, etc.
2) Determine the *type, AR and shape of the PST-inlet* to be tested first in the lab under realistic operating conditions (i.e., *flush, top or bottom mounted, mode of variation, mode of control, etc.).*
3) Repeat sub-stage 2 so as to fit the *inlet AR*, type of propulsion and shape of RPV *and its size and space,* and method of PST–IFPC.
4) *Determine the degree of supercirculation and RCS/IR signatures, and, accordingly fix the AR and shape of the vectoring nozzle.* (CF. Fig. III-10).
5) Repeat sub-stage 4 till exhaust nozzle fits with the configuration, and, simultaneously, determine the type of vectoring, i.e.:

   - *pitch vectoring only,*
   - *pitch and yaw vectoring.*
   - *pitch, yaw and TR vectoring,*
   - Simultaneous yaw-pitch-roll vectoring.

6) Design and construct a number of *full-scale, laboratory models* of *inlets* and *exhaust nozzles.*
7) *Perform full-scale laboratory tests of inlets and nozzles* (cf., e.g., Figs. 6, 7. Note, however, that this sub-stage may take a number of years).
   E.g., install a *bell-mouth inlet* on a conventional type engine equipped with rounded, *unvectored* nozzle and *calibrate* the propulsion system using *full-scale, laboratory test rigs*, including the evaluation of C_{fg}, C_{D8}, C_A, C_v, T, SFC, pressures, temperatures, air-mass-flow rates, etc. (cf. Figs. I-18 and I-19 and Appendix A).
8) *Replace the bell-mouth inlet* by the required *new inlet*, and repeat the tests to verify *inlet-engine compatibility*, low-level of distortion, and the expected limits of PST maneuverability (cf. the Introduction and Appendix F).
9) *Replace the rounded exhaust nozzle* with the required *vectored nozzle* and repeat

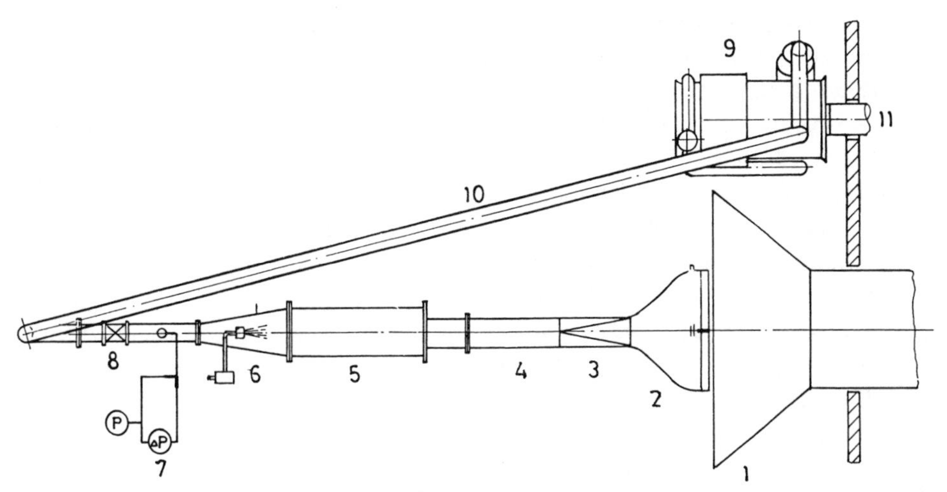

●Fig. 7. **The proposed integrated methodology of laboratory/flight tests is based on "STAGE-2"** (§ IV-5), i.e., on the component, subscale system for testing new ideas. 1) Exhaust system. 2) *2D–CD*, simultanous yaw-pitch-roll thrust vectoring nozzle with high aspect ratio undergoing component tests by a highly-instrumented, test rig before scaling-up to the "full-scale" tests (Fig. 6). 3) Circular-to-rectangular duct. 4)No-swirl/uniform flow duct. 5) Combustor. 6) Fuel system controlled by the instrumentation of Control Room No. 3. 7) Air-mass-flow-rate monitoring. 8) Flow control valve. 9) Gas turbine suplying up to 1 kg/sec air at up to 2.8 atm(g). 10) A connecting, flexible duct. (Cf, § IV-6, IV-7).

the tests. (Note, the new nozzle should have the same A_8, A_9 as those of the circular conventional nozzle tested in 8).

10) Compare the test results, and, following additional iterations, *freeze the propulsion system design.* (*Note:* Before freezing the design one should take into account the main design criteria enumerated in the text and in the appendices.)
11) *Feed the final design data into the computer simulation program for sizing routines* (e.g., the overall length of the inlet + engine + *C–R* duct + exhaust nozzle fixes the *minimal overall length of the aircraft, or in scale-down calculations, of the vectored-stealth RPV to be flight tested*).
12) *Scale-down* the optimized propulsion system to fit the required size of the vectored RPV and, accordingly, *construct inlet(s) and exhaust nozzle(s).*
13) Test the scaled-down propulsion system in the laboratory, including trim and modifications. This includes two sub-stages, namely, testing the propulsion system and testing the propulsion system installed in the vectored RPV, as described below.

    Thus, *without changing engine-inlet-nozzle geometry, build the aft shape of the "RPV" around the exhaust vectoring nozzle, using maximum structural integration ribs and supercirculation benefits.* (Note: If impossible, modify the wind-tunnel model and repeat the entire process.)
14) *Install the vectored-nozzle servos, or actuators, and test them with the radio, cable,*

fiber or IR-lazer-beam control system as in actual takeoff, flight and landing conditions, with and without engine power.

15) *Install canards, or ENVJ, and variable inlet servos or actuators.* (*Note:* A subprogram for the evaluation of proper canard or ENVJ systems should also be conducted during the preliminary design phases).
16) *Install retractable (or non-retractable) landing gear in relation to C.G., and, takeoff and landing conditions, as chosen from the computer simulation program.*
17) *Complete the forebody and install the canards or ENVJ, while leaving the forebody upper skin open for further work.*
18) *Install batteries, flight and data acquisition computer, and the payload, for an initial C.G. evaluation.*
19) *Move the batteries, servos, computer and payloads, till the required stability margin, the required moments of inertia* (for similarity comparisons with the full-scale aircraft), *and other criteria, are met.*
20) *Install fuel vessels in symmetry around the C.G. of the vehicle.*
21) *Repeat C.G. tests with and without fuel.*
22) *Test axial symmetry balance of moments, and trim the vehicle accordingly.*
23) Complete and close the upper skin of the vehicle while installing all *inputs-outputs connections* (fuel, control, computer telemetry–metry, starters, etc.)
24) Start the engine and perform *laboratory tests* (including measurements of installed thrust at various throttle power conditions, thrust-vectoring controls, canard-flap controls, computer input-output commands and data extraction modes of operations, etc.).
25) Perform initial runway tests, including testing of the data-acquisition computer, feeding into another computer, as well as the controllability of vehicle during turning/taxiing, etc.
26) *Freeze the first flight conditions and limits, including emergency procedures*, and data extraction metrics.
27) *Place high-quality video camera(s), and a high-performance, portable computer near the runway for recording of turning radii, velocities, α, β,* high-α-β *maneuvers, takeoff distance, landing distance, canard/vectored flaps angles, etc., during or at the end of each flight* (i.e., telemetry or metry systems).

PHASE-4

1) Feed takeoff, flight and landing data into the stand-by computer and examine the relevant video-recorders links and coordination procedures.
2) Use the portable computer simulation programs and the flight data to freeze the next-flight procedures and conditions.
3) Repeat flights. Alternatively, one may perform mirror-like, sustained horizontal turns, of both vectored and unvectored RPVs, simultaneously.
4) Evaluate and compare the performance-agility metrics (Figs. 19 and 20, Introduction) and rethink and reassess the entire program SOW/milestones.

5) Design modified configurations and 'figures of merit' in light of the flight testing results and performance metrics conclusions.
6) Design a modified vectored propulsion system as before.
7) *Integrate the new design, as before, and reevaluate the control and data acquisition modes.*
8) Perform wind-tunnel testing with the new models.
9) Perform *Jet-propulsion testing* with the modified *components.*
10) Perform *Jet-propulsion testing* with the *entire propulsion system.*
11) Repeat Phases 3 and 4, starting from § 12 of Phase 3.

IV-8 Master Plan For Vectored RPVs and For Vectored Fighters

Funding for vectored aircraft may depend on individual initiations of new ideas and also on a master plan. Indeed, to consolidate the multiple military programs, one must first eliminate redundancy, obsolete designs and ineffective programs.

A number of existing programs would be curtailed and requests for proposals for a family of new air vehicles must be issued. In the specific domain of RPVs this includes three missions: close-air operations; short-range, shipboard operations; and deep operations. Other subdivisions of new missions have been enumerated previously in this chapter.

IV-8.1 New Programs

NATO nations have recently begun sea-trial demonstrations to assess the capability of RPVs operating from small naval vessels.

A few RPV builders have already agreed to demonstrate their systems on board NATO vessels ranging from 4,000-ton frigates to 1,500-ton corvettes and 500-ton fast patrol boats. Under the proposals, the companies have agreed to finance the demonstrations, with the navies of the participating countries providing the ships and range time. However, vectored RPVs have not yet been included in any of these programs, except the Barak missile program (IV-1.1).

The DoD is interested in an RPV for close-in operations for which the prime candidate is a VTOL system, or a system capable of operating from a small ship.

The US Air Force will test RPVs to evaluate the use of unmanned vehicles to assess runway damage.

Requests for proposals for a baseline family of RPVs would be issued in early Fiscal 1990 under the US plan, with testing and evaluation scheduled to begin in Fiscal 1992.

Under the plan, at least three airframes will be developed, one for each of the three mission areas. Some will have unique subsystems, such as a signals intelligence package for deep operations, and others that will be common, such as a television or a forward-looking infrared system. Another requirement could be development of two data links that would be common to all the RPVs, as well as a single mission planning and control system.

The US program would match off-the-shelf systems with the requirements to come up with the definition of the most cost-effective system.

IV-9 STOL RPVs

Takeoff and landing distances are drastically reduced in vectored RPVs, *depending on the thrust-to-weight ratio of the RA, and on the positions and angles of the front "downward-pointing vectored jet", as well as on its distance from the aircraft center of gravity.*

Automatic, or semiautomatic mechanisms, may then be designed and employed *for a gradual transition from V/STOL, or semi-hovering mode of operation, into regular flight at maximum speed, and vice versa. Such vehicles have been recently tested by this laboratory, using 3 engines (Prototype No. 4), and a single engine (Prototype No. 5).*

IV-10 Fan and Core Flow Vectoring

In using turbofan engines to propel, maneuver and control vectored aircraft, the hot and cold streams may be *separately diverted into hot and cold nozzles*, each with its material and operational characteristics (cf. Fig. A-2).

Alternatively, the two streams may be mixed, as in conventional engines. However, the elimination of cooling needs, and the possibility to introduce low-weight, low-temperature materials in the "separated flows" design, appears more attractive. Indeed, the evaluation of NAR, $(\Delta C_L)_{\max}$, T, δ_v, δ_y, C_{fg}, etc., *for each stream would then be different.* Theoretically, $(\Delta C_L)_{\max}$ may first be estimated by computer simulations programs. However, the errors involved in using this methodology should not be underestimated. Empirical, flight-testing studies appear, therefore, to be the most reliable methodology, at least for the near term.

IV-11 Nozzle Length and Shape

Decreasing the 2D nozzle length reduces weight and increases overall T/W ratio. However, problems involving BLS, swirl, and non-uniformity may become inhibitive. Cold fan air is preferably ducted to the outboard section of the wing, thereby involving different shapes & BLS/mixing problems than those associated with hot nozzles. NPR is also different for each vectored nozzle (cf. Fig. 2, Appendix A).

IV-12 Scaling-Up and Scaling-Down of Aircraft Dynamic Characteristics

PST/Stealth/Vectored aircraft encounter unfamiliar dynamic characteristics (cf., e.g., Figs. 2 and 3 in Appendix D). Traditionally, the general equations associated with maneuvering about an aircraft's axes are given by:

$$I_{yy} \cdot \dot{q} = \tfrac{1}{2} \cdot \rho \cdot v^2 S\bar{c} \cdot C_M \qquad \text{for the pitching moment}$$

$$I_{xx} \cdot \dot{p} + I_{xz} \cdot \dot{r} = \tfrac{1}{2} \cdot \rho \cdot v^2 \cdot S \cdot b \cdot C_{\alpha}, \qquad \text{for the rolling moment}$$

$$I_{xz} \cdot \dot{p} + I_{zz} \cdot \dot{r} = \tfrac{1}{2} \cdot \rho \cdot v^2 \cdot S \cdot b \cdot C_N, \qquad \text{for the yawing moment}$$

where I is the moment of inertia and

$\dot{p}$, $\dot{q}$, $\dot{r}$ represent the angular accelerations about the x, y and z axes, respectively;
C_M, C_α, C_N represent moment coefficients for pitch, roll and yaw, respectively; and
S, b, $\bar{c}$ are the wing area, span and average chord, respectively.

The latter two equations take into account the coupling between roll and yaw. However, none of the above-equations takes into account the gyroscopic effects presented by the rotation of the engines during aircraft maneuvering.

When scaling a model for similar dynamic characteristics one is first faced with two requirements:

I: To maintain a similar weight distribution.
II: To maintain the same relations between moment coefficients and moments of inertia, e.g., for the pitching moment one would have:

$$I_{yy} \cdot \dot{q} = \tfrac{1}{2} \cdot \rho v^2 \cdot S \cdot \bar{c} \cdot C_M,$$

or,

$$\frac{C_M \cdot S \cdot \bar{c}}{I_{yy}} = \frac{\dot{q}}{\tfrac{1}{2} \cdot \rho \cdot v^2}$$

To proceed, one may add frequency scaling-down functions, and also read Ref. 179, and the next paragraph.

●IV-13 On the Limitations of Wind-Tunnel Tests

Can reliable wind-tunnel tests of vectoring aircraft be conducted? Indeed, reliable simulation of, say, *PST-inlet suction and distortion*, combined with 2*D*-nozzle pressure ratio/integration/yaw-pitch-reversing/vectoring simulations of actual, dynamic, in-flight conditions, encounter *inhibitive technical and cost-effective problems*, some of which will be described below.

It is here that the designer of future vectored aircraft may prefer to switch to *actual flight-testing of vectored RPVs to obtain optimized, actual, cost-effective test data for scale-up design considerations.* Consequently, the flight testing of vectored RPVs, as described above, emerges as a highly cost-effective R&D tool.

Nevertheless, wind-tunnel testing remains an essential R&D tool during initial design iterations of both research RPVs and manned vectored aircraft. Thus, wind-tunnel tests, even without suitable suction/ejection flow simulations of PST-inlets and PST-vectored propulsion systems, may give the designer of vectored aircraft an initial estimation of C_L, C_D, C_M, AC, Strouhal No., Reynolds No., Nusselt No., etc.

Canard and other control surfaces can be evaluated in such tests to trim nozzle vectoring/reversing pitching moments. Indeed, various cold flow, performance tests of 2-*D* nozzle models have been conducted in the past decade or so in wind tunnels, and are reported in the open literature [1, 4, 6, 8, 13, 14, 17, 32, 34, 40, 45, 62, 67, 69].

Such tests results have indicated that:

- At transonic and supersonic speeds, the *2D–CD* nozzle exhibits higher thrust-minus-afterbody-drag performance than the conventional, axisymmetric nozzle, while at low subsonic speeds, the nozzle incurred up to 2 percent thrust-minus-afterbody-drag penalty (cf. Figs. I-17, 18).
- Thrust vectoring produced significant induced lift increases (cf. Fig. III-19).
- Thrust reversing is effective for static and in-flight operation involving high-α-β maneuvers. However, PST-deceleration is much more effective than *TR* (Internal Reports of this laboratory).
- At $\delta_v \leq 20°$ thrust vectoring, the thrust loss is as expected from $\cos(\alpha + \delta_v)$, depending on the nozzle design, i.e., on the difference between actual and geometric deflections during thrust vectoring.
- Other variables which can be evaluated by wind-tunnel tests include NAR, nozzle type, nozzle pressure ratio and area ratio, horizontal tail deflection, nozzle thrust vector angle and thrust reverser/tail interference, afterbody/nozzle force data, etc. [cf. Appendices C and D].

●IV-14 *IR* Signatures

Design criteria for vectored aircraft should permit estimation of internal propulsion performance, weight, mixing and cooling requirements, and the corresponding infrared signatures of various advanced concepts.

Such static ($M = 0$) subscale rig tests have been reported in the literature [6, 91, 103]. They provide the technical data base for the development of improved prediciton techniques for *IR* signature, mixing and cooling requirements [103] and also of internal performance of various nozzle types over a wide range of nozzle aspect ratio (NAR) and operating conditions.

LECTURE V

PARTIALLY VECTORED AIRCRAFT

"It is better to debate a question without settling it, than to settle a questin without debating it."

Joseph Joubert

"To find the meaning of an idea we must examine the consequences to which it leads in action; otherwise dispute about it may be without an end, and will surely be impractical."

C.S. Peirce (1878)

V-1 Canard-Configured Vectored Aircraft

Thrust vectoring may require a variable incidence canard or canard/flaps for *counter-balancing the large pitching moments which are incurred, especially during STOL and high turning rates.* During such PST-RaNPAS turns, the thrust vectoring nozzle, and the canard work as a *'force couple'*, rotating the aircraft *around its center of gravity and couple it properly with nozzle's dynamic vectoring deflections.* Obviously, the canard effectiveness declines sharply during PST maneuvers above, say, 70 degrees AoA.

Several wind-tunnel investigations have been conducted to examine this problem, including optimized locations, shapes, etc., for canard integration with super-agile aircraft (cf. Figs. 10 and 11 below and § VI-2.5.3).

V-1.1 Preliminary Design of Partially-Vectored, Canard-Configured Aircraft

The Sedgwick-Lockheed example (Sedgwick, 164) discussed below is highly instructive in arriving at some preliminary conclusions regarding canard-configured partially vectored aircraft. To start with, the analysis is based on a comparison with a datum fighter aircraft shown in Fig. 1 and detailed further in Tables 1 and 2.

Table 1. Dimensions of Wing and Control Surfaces (Sedgwick, 164)

Characteristics	Wing	Canard	Vertical Tail
Area – ft^2	400	87,5	54.0
Aspect Ratio	3.5	3.7	1.309
Span – ft	37.4	17.5	8.4
Root Chord – in.	216.9	88.9	118.5
Tip Chord – in.	39.7	31.1	35.5
Taper Ratio	0.183	0.35	0.30
Mean Aerodynamic Chord – in.	148.7	64.6	84.4
Leading Edge Sweep – degrees	35	29	50
Thickness Ratio – Root – %	5	4	4
Thickness Ratio – Tip – %	5	4	4

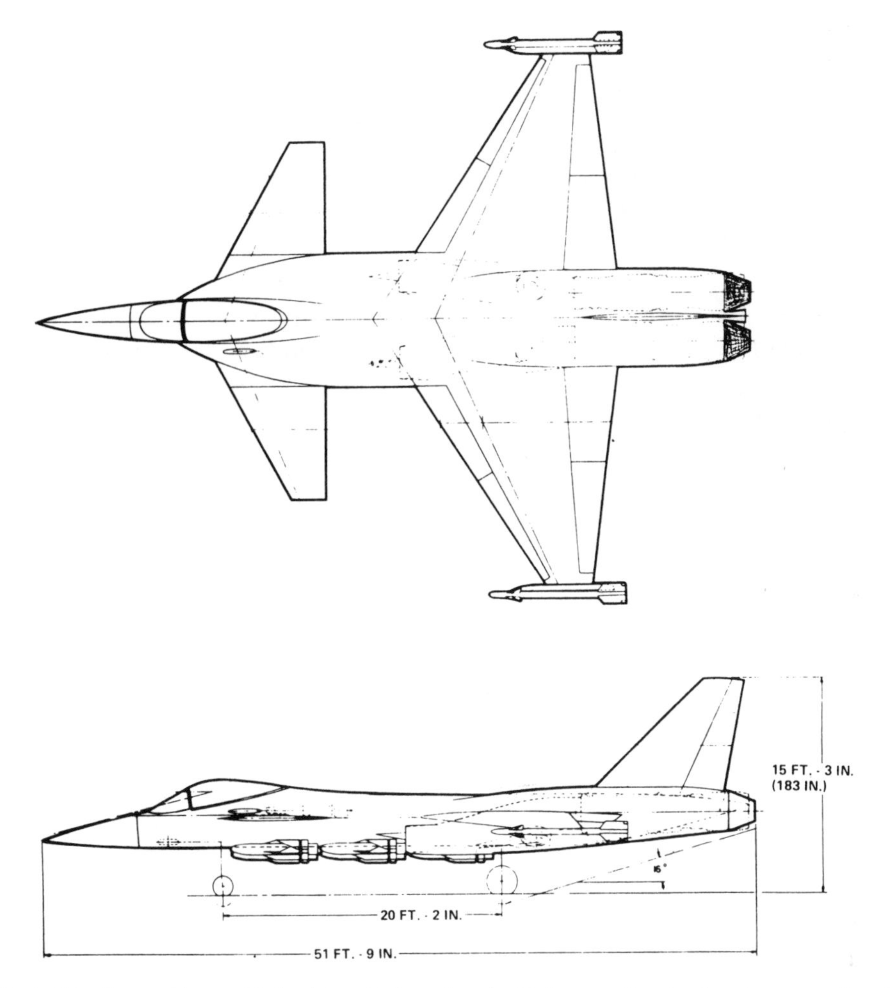

Fig. 1. **The Datum Aircraft (Sedgwick, 164) is equipped with 2 conventional exhaust nozzles of two GE J101 engines (cf. Table I for additional data).** No visible effort has been made to reduce outer-skin-shape/geometry *RCS* and *IR* signatures. Yet, this early design concept is highly instructive for this course. (Cf. Figs. 2 to 4.)

Secondly we note that *supercirculation effects have been incorporated in the preliminary design* (Fig. III-18).

Thirdly we note that the primary emphasis is placed on performance and/or weight, which are influenced by *nozzle-aspect ratio selection* and, accordingly, by *inlet-engine integration with exhaust nozzle and the wing structure* (Figs. 2 to 4).

Fourthly, we note that whenever significant differences occur, the most favorable design concept has been sought for each nozzle concept.

Finally we note that *all four prototypes use the same stores and forebody arrangements, have the same basic wing, canard, and vertical tail geometries as listed in Table 1.*

V-1.2 The Datum, Unvectored Aircraft

The datum aircraft is shown in Fig. 1. It has close coupled, all movable, horizontal canard stabilizer, and a single vertical tail. Its main characteristics are:

1) The canard is located forward, on the strake, so that the downwash from the canard does not significantly unload the wing [cf. Figs. 10 and 11 for other types of canard designs].
2) The inlets are partially shielded by the wing strakes from flow angularities produced during high AoA flight. However, the inlets may be highly visible to enemy radars.
3) The engines are installed inside the fuselage at a minimum nozzle-to-nozzle spacing consistent with small rolling inertia, structure, and engine maintenance considerations.

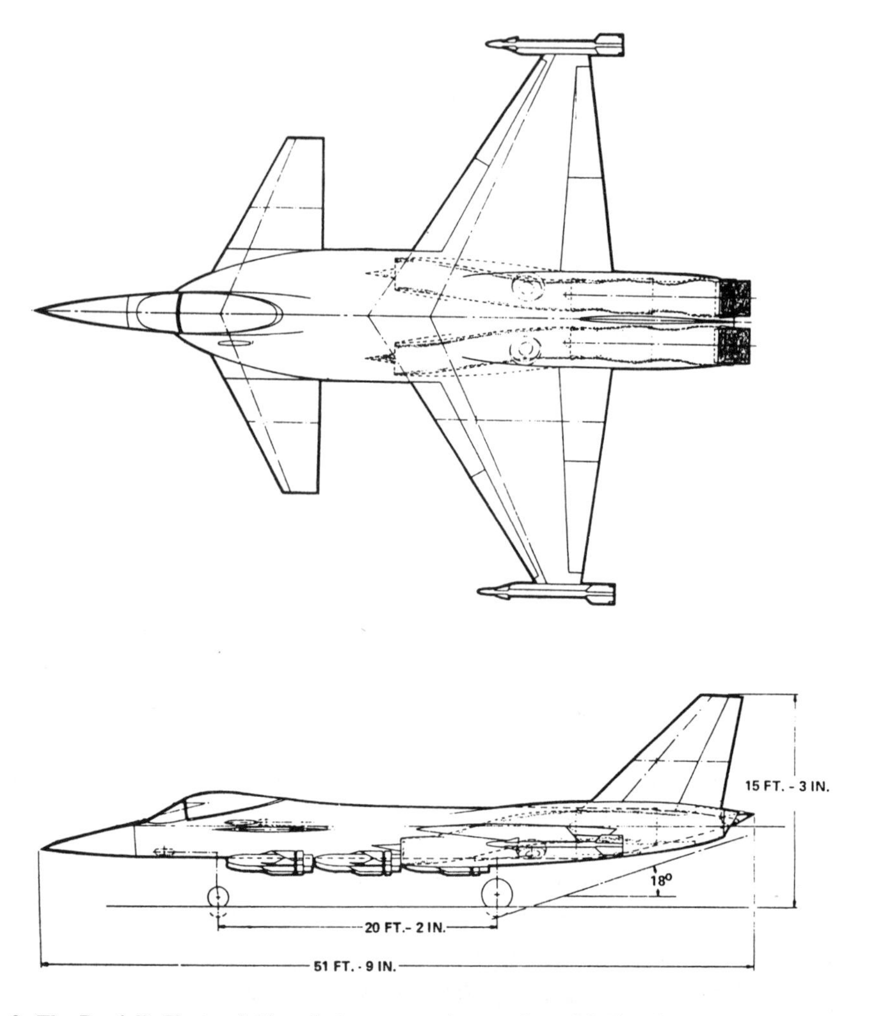

Fig. 2. **The Partially Vectored Aircraft thrust-vectoring nozzles.** (Cf. Figs. 3 to 4.) (Sedgwick, 164)

4) The narrow nozzle spacing, the nozzle interfairing configuration, and the single vertical tail, have been selected so as to yield the highest possible (installed) T–D force, following the experimental and analytical methods employed at Lockheed – California Co.
5) The main landing gears are located on the lower outboard side of the fuselage, and retract aft.
6) The wing has leading- and trailing-edge flaps.
7) No visible effort has been made to reduce outer-skin-shape/geometry *RCS* and *IR* signatures.

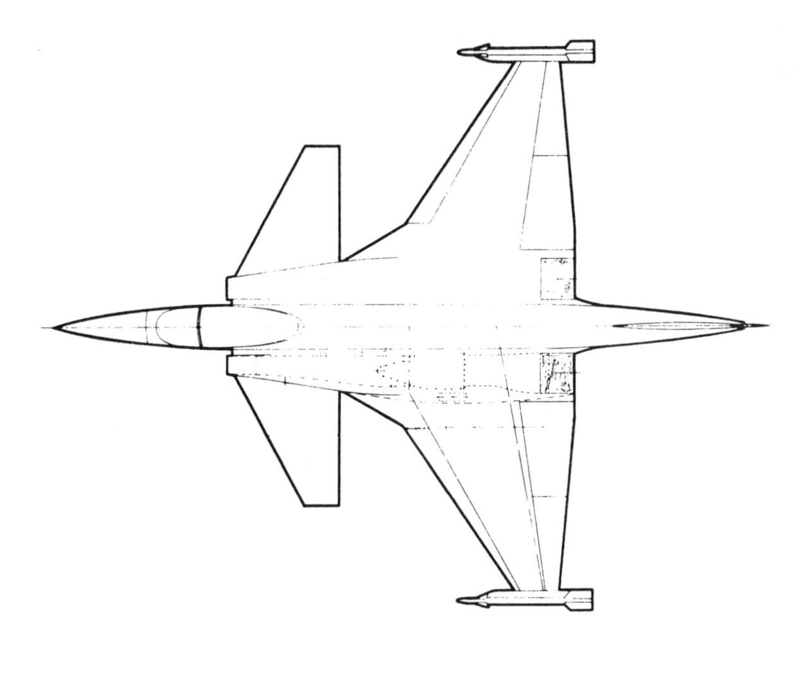

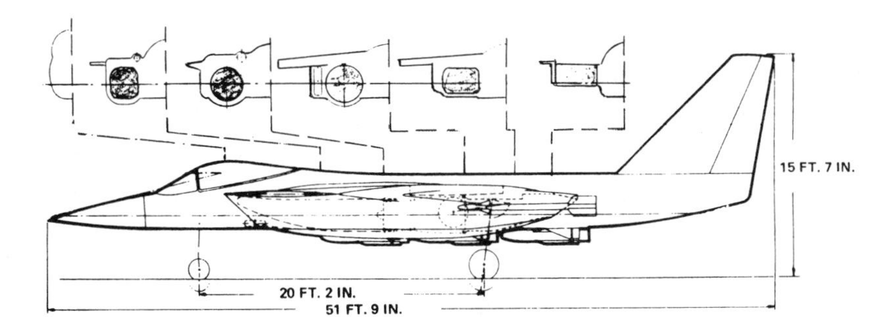

Fig. 3. **The Partially-Vectored Aircraft with *AR* = 6 supercirculation thrust vectoring nozzles** (see also Fig. 4) (Sedgwick, 164).

V.1.3 The Partially-Vectored Aircraft with NAR = 4

This fighter is shown in Fig. 2 and is identical with the datum fighter except for:

1) The replacement of the conventional, axisymmetric exhaust nozzles with the ADEN vectoring nozzles. (See, however, our reservations concerning the ADEN nozzle in Fig. 5 Appendix A, and in Chapter VI.) All other differences are a direct result of this replacement, and are concentrated in the aftbody.
2) *NAR* = 4 refers to the exhaust nozzle **throat** aspect ratio at the Maximum Dry Power Conditon (MDPC). It yields, with proper nozzle interfairing, a substantially lower afterbody drag.
3) While the required canard lift force for trim during maneuvering has increased with pitch thrust vectoring, the associated canard area requirement has not exceeded the takeoff rotation requirement (which is essentially the same for both fighters).

V.1.4 The Vectored Aircraft With *NAR* = 6 and Low Supercirculation

This fighter is shown in Fig. 3.

The major changes with respect to the prior two fighters are:

1) The engines are installed in semi-nacelles just outboard of the fuselage, and upstream of the inboard portion of the wing trailing edge.
2) A modified inlet is installed directly upstream of each engine. The modification icnludes two-dimensional, horizontal-wedge (overhead ramp), fixed-geometry, two-shock inlet (cf. Fig. 4, Appendix F).
3) To preclude adverse effects of the canard wake on inlet performance, the leading edge of the inlet external compression surface is placed just upstream of the canard leading edge.
4) The upper surface of the nozzle fairs smoothly into the wing upper surface, and the upper side acts *both as a thrust deflector and as a wing flap.*

V.1.5 The Vectored Aircraft with *AR* = 17 Medium Supercirculation Nozzles

This fighter is shown in Fig. 4.

The significant differences between this design and the previous *AR* = 6 fighter are:

1) *The cooling airflow limitations preclude installation of a conventional afterburner, and, consequently, a C–R transition section has been added.* Thus, *A/B* burning takes place in this *C–R* ducting. (It is not yet known whether or not burning in rectangular duct is feasible, especially with respect to cooling loads. Yet, new fighter engines may need no *A/B*.)

2) Five stiffening rings are employed on the outside of the *A/B, C–R* duct.
3) The outboard offset of the *A/B C–R* duct and the large change in cross-sectional shape necessitates *internal streamlined turning vanes.* These turning vanes and two

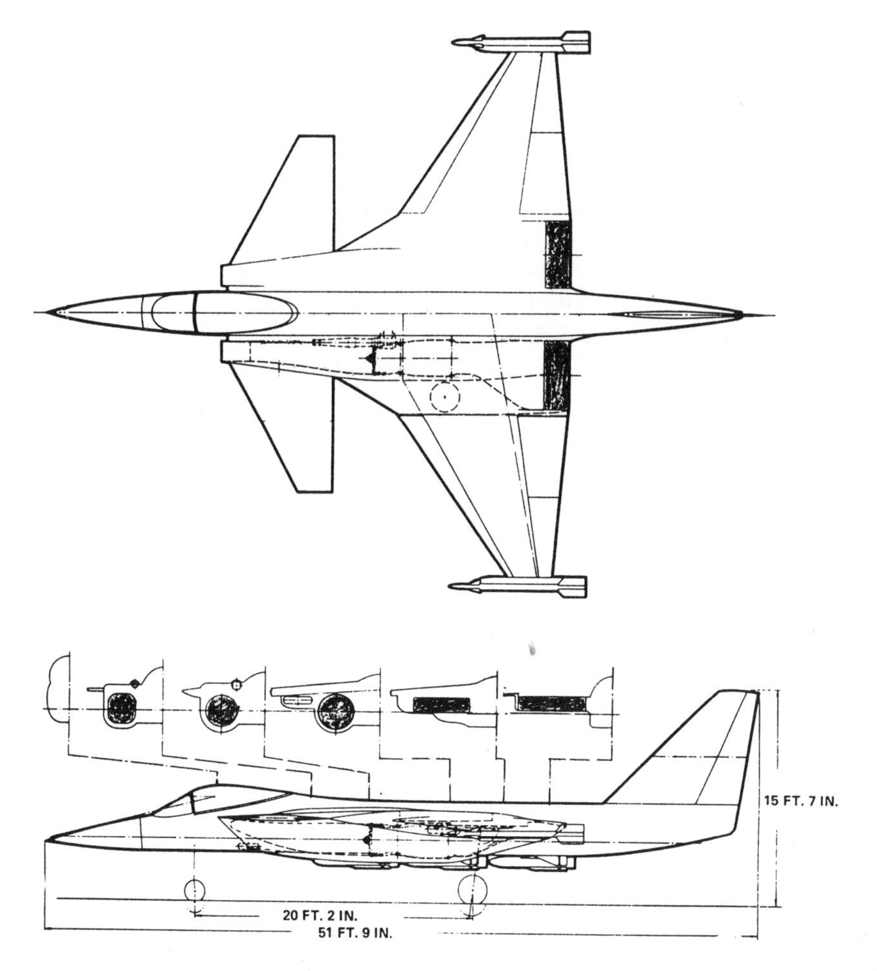

Fig. 4. **The Partially-Vectored Aircraft (Sedgwick, 164) with Aspect Ratio 17, thrust-vectoring supercirculation nozzles.**

Note the internal air-exhaust gas ducting for the vectoring propulsion system. Estimation of supercirculation effects may be done with the help of Fig. III-18. For the definitions of *pure vectored aircraft*, see Figs. II-1a to 1i, and the Introduction.

vertical struts upstream of the nozzle throat, are used to carry loads between the upper and lower wing surfaces.

4) *Increased wing area affected by supercirculation enhanced lift.*
5) Main landing gear design differences.

V.2 Mission Definition vs. Aircraft Design Constraints

The subsonic, close-air-support mission for the aircraft analysed here, as an instructive example, is illustrated below.

The estimated 5214 lb armament load includes (Sedgwick, 164):

- close-air-support missiles.
- air-to-air missiles.
- 20 mm gun.

The design requirements for 60% of takeoff fuel include:

- Level flight acceleration from 250 KEAS (Mach 0.46) to 494 KEAS (Mach 0.9) in 35 seconds at 10,000 ft altitude, with full armament load.
- 2 g's sustained maneuver at 150 KEAS at 5000 ft altitude, with full armament load.
- 5.5 g's sustained maneuver at Mach 0.9 at 30,000 ft altitude, with combat armament load (full armament load less 3276 lb of close air support missiles).
- Maximum speeds with full armament load and 60% fuel not less than Mach 0.9 at 10,000 ft altitude and Mach 1.6 at 20,000 ft altitude
- Structural design load factors are +8.00 and −3.0 at full armament load and 100% internal fuel.

V-2.1 Leakage Estimates

The vectoring ADEN nozzle requires increased coolant leakage with increased NAR. Since the lost flow is directly proportional to thrust, it becomes equivalent to a loss in C_{fg}. The overall result is therefore as follows:

Fighter	Leakage Flow/Power Loss, Percents	
	Dry Power	*A/B* Power
Datum	0.5	0.3
$AR = 4$	0.2	0.2
$AR = 6$	0.2	0.2
$AR = 17$	0.4	0.4

Similar estimations for cooling flow needs and pressure drops finally give the overall effect of AR on C_{fg} (cf. Fig. III-15).

V.2.2 Weight and Fuel Iterations

The computerized weight routine used by Lockheed (Sedgwick, 164) starts with an initial specified takeoff gross weight (TOGW), engine weight, and geometric data, such as wing area and control surface areas (Table 1).

The flight design gross weight is derived, and, along with other inputs, is used to calculate the group weights, which are summed to arrive at the aircraft operating weight.

The fuel weight is, therefore, the difference between TOGW and the sum of the operating weight and the payload.

If this fuel weight is insufficient to perform the specified mission, TOGW is increased, new groups weights are calculated, and a new fuel weight determined.

The computational loop is thus repeated until convergence takes place, as shown in Table 2.

Table 2. Weight Summary (Sedgwick, 164)

		INCREMENTAL WEIGHT		
	DATUM AIRCRAFT	ASPECT RATIO 4	ASPECT RATIO 6	ASPECT RATIO 17
WING	0	0	269	269
BODY	0	0	−411	−475
LANDING GEAR	0	0	208	208
NACELLE AND ENGINE SECTION	0	11	636	650
ENGINES	0	402	492	976
FUEL SYSTEM	0	−23	−67	−91
MISCELLANEOUS	0	−7	5	−3
TOTALS	0	383	1132	1534
OPERATING WEIGHT	17,284	17,667	18,416	18,818
PAYLOAD	3,716	3,716	3,716	3,716
ZERO FUEL WEIGHT	21,000	21,383	22,132	22,534
FUEL	7,000	6,617	5,868	5,466
TAKEOFF GROSS WEIGHT	28,000	28,000	28,000	28,000

V.2.3 Synthesis and Design Reservations

Aircraft synthesis results for all four prototypes are presented in Figure 5 as plots of TOGW vs. takeoff wing loading W/S (lb/ft^2) and T/W ratios.

While the values obtained by this instructive example are to be treated with caution, and reservation, the methodology employed here is highly instructive for the preliminary design cycles of partial, as well as pure vectored aircraft.

The reasons for caution emerge from the limitations mentioned earlier, and from the updated information about improved components, unit operations and testing methodologies as described in this course.

V-3 Upgrading Extant Aircraft to Partially-Vectored Fighters

Figs. 16 and 12 show feathers required to upgrade an extant fighter, like the F-15, into a partially-vectored one. So far, the F-15 modifications have not included yaw vectoring of the jets (13, 18, 62, 70, 71, 111, 128, 234). Yet, they provide some yaw vectoring through thrust spoiling of one engine; a highly non-inspiring design (Fig. 12).

So far, all proposed upgradings of existing aircraft have been limited to 2D nozzles with low aspect ratios, or to ETV. These limitations are due more to tradition and inertia than to *bona fide* constraints, such as limited nozzle space area, adverse interfer-

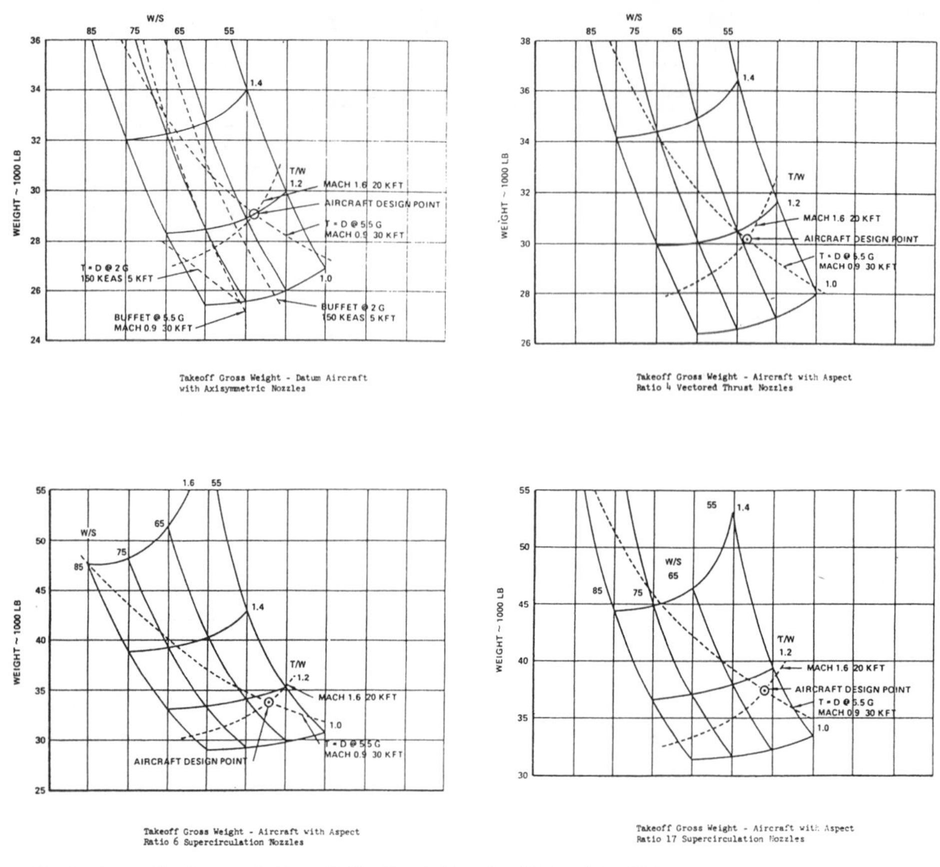

●Fig. 5. **Partially-Vectored Aircraft Synthesis Results (Sedgwick, 164).**

ence with existing wings, tails, rudders, actuators, fuel vessels and cooling requirements. Consequently, one may conclude that future upgradings of extant aircraft into partially-vectored aircraft, would be much better in comparison with the expected performance of the current F-15 S/MTD (13, 18, 62, 70, 71, 111, 128, 234). To realize this conclusion let us first examine the expected increase in performance of the F-15 S/MTD, in comparison with the standard F-15C (Table 3).

To start with, there would be a +78% increase in the maximum lift coefficient. This value may be considered low in comparison with the potential increases in C_L when properly integrated *2D* nozzle/wings are used (cf., e.g., Figs. III-18).

Secondly, one may examine Fig. 12 to realize why the yaw control power on the modified F-15 demonstrated would be quite poor. As shown it would be provided by spoiling axial thrust on one engine, while maintaining full axial thrust on the other.

The utilization of this kind of yaw control has a number of drawbacks:

First, engine thrust variations are relatively time consuming. Secondly, the yawing-moment arm in the proposed configuration is relatively small. Consequently, it is

TABLE 3
Improved Performance Expected for the Partially-Vectored (STOL) F-15

(The F-15 S/MTD fighter, in comparison with the standard F-15C)
(13, 18, 62, 70, 71, 111, 128, 234)

	% Change
Maximum Lift Coefficient	+78
Deceleration Rate in Flight	+72
Landing Run	−72
Roll Rate, Mach 1.4/40,0000 ft	+53
Pitch Rate, Mach 0.3/20,000 ft	+33
Take-off Roll Distance	−29
Approach Speed (at a constant α)	−16 Kt
Acceleration Rate, Mach 1.4/40,000ft	+30
Cruise range	+13

much less effective than in pure vectored aircraft configurations (cf., e.g., Fig. II.1), or in a newly-proposed, simultaneous yaw-pitch-roll thrust-vectored F-15 (Internal Report, USAF/WPAFB-Program: "Vectored F-15", This laboratory).

USAF specifications call for the STOL F-15 to operate at the F-15's normal combat weight and typical weapon loads from 1,500 ft × 50 ft runway at night in a 30 kt crosswind with a 200 ft eight-octa ceiling and 0.5 n.m. visibility with only passive ground-based landing aids. The *2D* nozzles will be capable of 20° pitch-up/down (either together or differentially) in full afterburner at sea level. They should also be

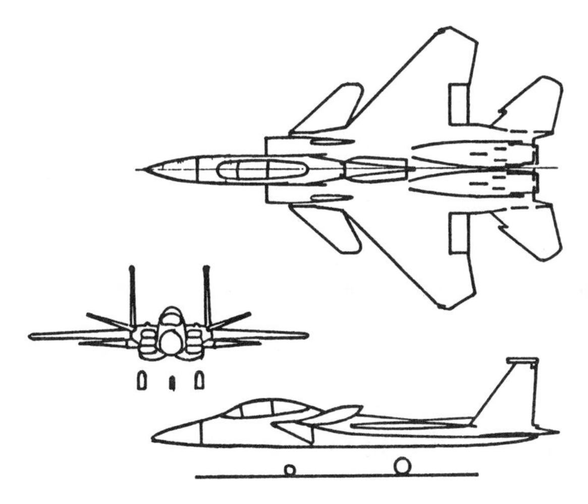

Fig. 6. **The Upgraded, Partially-Vectored (pitch-only) F-15 S/MTD (128).** (See Table 3 for details of the expected improved performance).

This may be the first manned vectored aircraft to fly in the "open history of aviation" (the Harrier's methodology is a forerunner to this technology. However, the Harrier does not employ supercirculation, nor rectangular, vectored nozzles.).

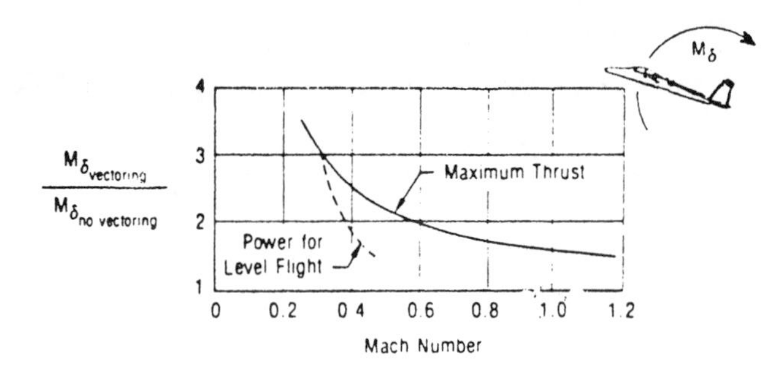

Fig. 7. **Vectoring moments variations during maximum thrust maneuvers at various altitudes and pitch-throttle settings (Mello and Kotansky of McDD, 128)** (Altitude = 30,0000 ft). (Thrust vectoring and enhanced agility are discussed in the Introduction.)

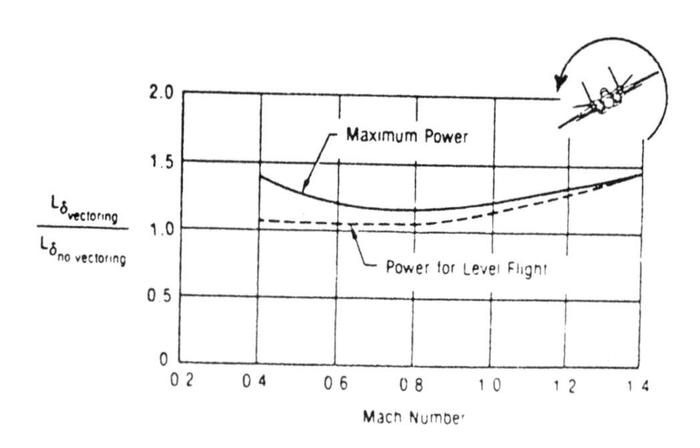

Fig. 8. **Thrust vectoring in the new F-15 STOL demonstrator is also expected to enhance roll performance (Mello and Kotansky of McDD, 128)** (Altitude in this example is maintaied at 10,000 ft. (See, however, the notes in Figs. II.1, 2,11–14.)

capable of full thrust-reversing at 50% military power. Its selected canards are actually F-18 tailplanes, and its pitch rate improvements may reach 100% (13, 18, 62, 70, 71, 111, 128, 234).

V.4 Upgrading Studies of F-15 by Vectored RPVs

Our jet-propulsion laboratory tests, as well as the vectored RPV program, including flying F-15C and F-15 S/MTD 1/7th models with similar propulsion/aerodynamic characteristics, high α, low speed performance and inlet distortion data are now being monitored, using our low-weight, on-board, computerized metry systems. Performance metrics are being evaluated for:

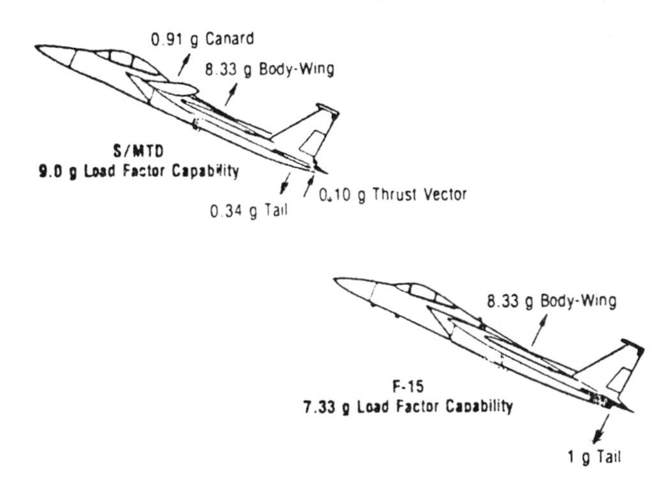

Fig. 9. **Structural load distributions on the new F-15 S/MTD and on the conventional F-15 baseline for an equivalent 8.33g body-wing load at a low supersonic flight condition (Mello and Kotansky of McDD, 128).** Target tracking of vectored F-15 should be much enhanced.

The conventional F-15 requires an equivalent counter productive 1g tail download, resulting in a net 7.33g load factor capability.

The new F-15 S/MTD, with the forward canard load to trim-out the wing load pitching moment, results in a much smaller tail loading and an increased 9.0g maneuver load factor capability.

The increase in load factor at supersonic speeds, as a function of vehicle gross weight, may be achieved with no increase in baseline aircraft structural weight.

1) F-15C (baseline-1).
2) F-15 S/MTD (Baseline-2).
3) F-15 S/MTD with additional yaw thrust vectoring.
4) A tailless F-15C equipped with our NAR = 46.6, simultaenous yaw-pitch-roll thrust vectoring system (cf. Lecture IV).

It should be noted that the Soviets are currently testing similar designs on the Su-27-1024 agile interceptor. Moreover, the unvectored Su-27 can maneuver (the "Pougachev's cobra maneuver") at up to 110–120° AoA. (Speed reduction is from 250 to 80 mph.)

●V-5 Upgrading F-18 and F-16 to Become Partially-Vectored Fighters

Capone and Berrier [56] have conducted a comprehensive study on partially-vectored F-18 performance with *2D–CD* vectoring/reversing nozzles using the Langley 16-foot Transonic Tunnel and a 0.1-scale model of the fighter aircraft. Higher *T–D* performance (untrimmed) has been demonstrated in comparison with current-technology F-18. Figs. 15–18 provide a few views on this type of model testing. ETV has recently been selected as means to investigate high-AoA performance, using ETV-F18. Similarly, ITV is employed by this laboratory for studying the PST performance of F-16, using ITV-RPVs.

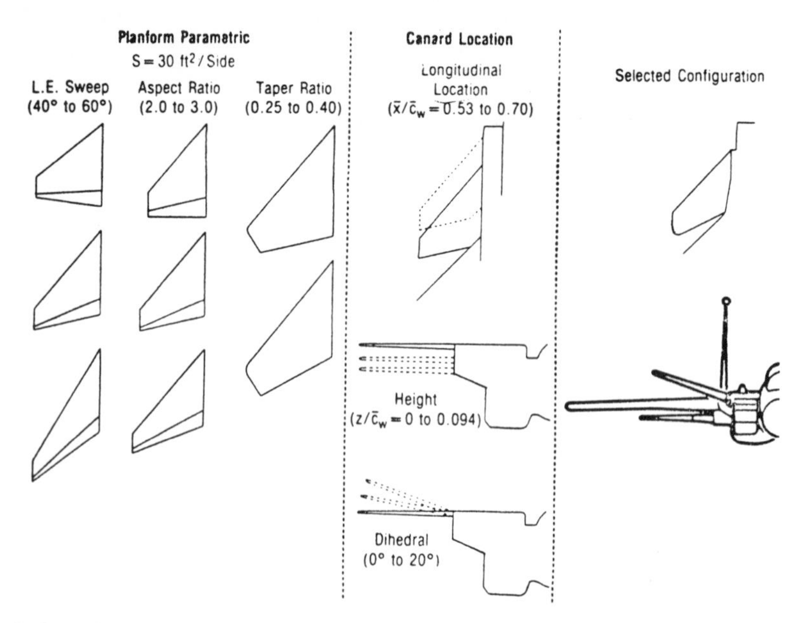

Fig. 10. **Independently variable, close-coupled, dihedraled canards, fully exploit the potential of vectored exhaust nozzles to enhance maneuverability of pure and partially-vectored aircraft (cf. Fig. II-1) (Mello and Kotansky of McDD, 128).**

The canards may be variable in pitch through an angular range of −35 degrees to +15 degrees.

Since no efficient thrust-yaw-vectoring is available for the new F-15 STOL fighter, the canards must be dihedraled to provide *direct side force (and additional yawing moment) through differential deflection in flight.*

In the subsonic and transonic speed range, the canard acts somewhat as a leading edge device, and is scheduled symmetrically with vehicle Mach number and angle of attack.

At supersonic speeds the canard provides additional lift, and because of its forward location, has a benefit in *the reduction of supersonic nose-down moments.* The forward location also results in a powerful positive trimming device to reduce negative tail and vectoring-down loads at supersonic conditions.

The canard geometry and positioning on the vectored aircraft may be evaluated for optimal lift, drag, stall characteristics, yawing moment and side-force generation.

The various planform configurations shown on the left side of the figure had been evaluated. The one selected for the F-15 STOL fighter is shown on the right. The one installed on the STOL F-15 is actually an F-18's tailplane. Different longitudinal locations, heights and dihedral angles, have also been considered (center) (128).

V.6 The Search for Design Philosophy in the Post-ATF Era

There are many wind-tunnel reports which examine, under subsonic and supersonic conditions, a matrix of inlet, nacelle, and exhaust nozzle configurations (cf. the Miller-review – 8, 9, 47). These include various external or internal propulsion components which are normally incorporated into a wind-tunnel model, insuring proper geometric simulation of aircraft-propulsion interference effects in a large range of Mach numbers. However, this method is limited from a number of viewpoints to be discussed below.

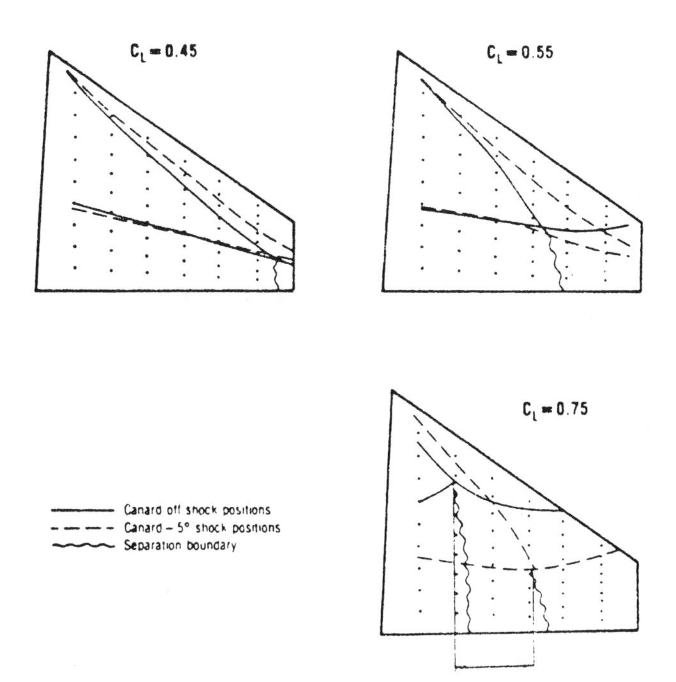

●Fig. 11. **Canard-Wing Interactions (Mello and Kotansky of McDD, 128).** The favorable interaction between the proposed canard and the F-15 wing upper surface is shown here for Mach number = 0.9. The canard-induced flowfield improves the wing-flow separation characteristics. The improvement is greatest at the higher C_L values, thereby further enhancing the maneuver performance of the aircraft (13, 18, 62, 70, 71, 111, 128, 234).

The canards offer a combination of relaxed static stability, lifting trim loads and favorable flowfield interactions on the wing drag-due-to-lift (cf. Fig. II-1).

Other interactions are demonstrated by Fig. 13.

Let us first consider the various high-performance propulsion systems which have been proposed for different V/STOL, subsonic and supersonic aircraft.

V/STOL proposals variously stress propulsive lift systems ranging from remote lift fans driven by rigid shafts, ducted high-energy gas, lift/cruise nacelles, swivel nozzles, remote augmentors, reaction control/compressor bleed systems (cf. Figs. B-7, 8, A-1-5), up to partial and pure vectoring propulsive systems (Lecture II).

The primary problem in the design and application of these systems is to retain their performance without robbing the powerplant of its ability to perform safely and economically. One of the highest interactions between airframe and these propulsion systems occurs in the transition from wing-borne flight to thrust-vectored maneuverability and controllabiltiy. Additional drawbacks are related to the slow responses of these TV systems, as well as to their complexity and low cost-effectiveness. **Accordingly, yaw-roll-pitch ITV remains the only cost-effective choice for vectored aircraft.**

Here the propulsion system must be designed as the primary flight control in terms

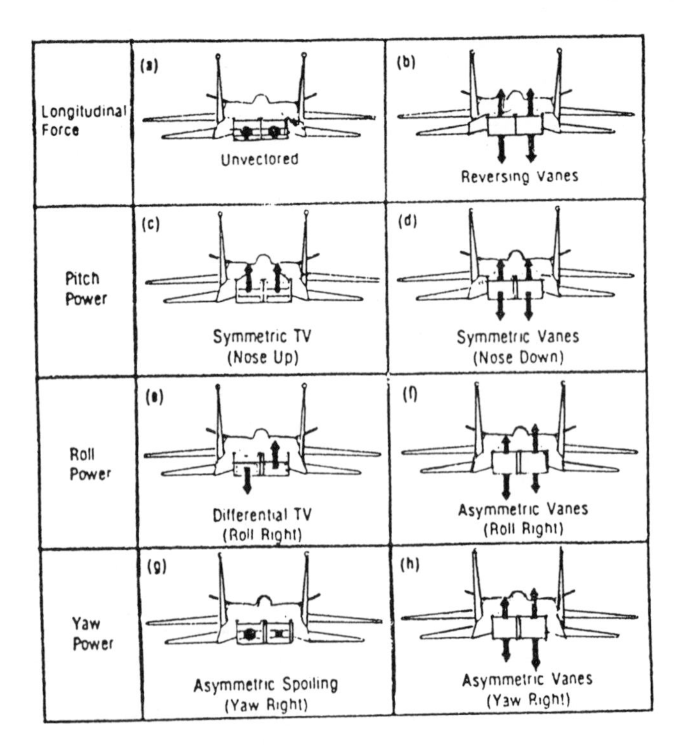

Fig. 12. **The thrust-vectored yaw, roll and pitch control modes of the new F-15 STOL Demonstrator (Mello and Kotansky, 128).** During yaw-thrust vectoring the F-15 S/MTD employs thrust spoiling of one engine, instead of direct yaw-vectoring of the jets themselves. The latter is expected, according to our recent experience with vectored RPVs, to be much more effective in terms of thrust efficiency and response times, as well as in terms of the low actuator forces required for yaw control of the jets (cf. Fig. II-1). Furthermore, it remains effective at the PST domain of supermaneuverability (cf. the Introduction and § V.5). It should also be stressed that, in general, the *2D* nozzle considerably improves **stealth** performance in comparison with circular nozzles.

of forces, moments, rates, responses, accuracy and reliability. IFPC technology must therefore integrate various design methodologies and employ them to enhance performance and revolutionize the design philosophy of fighter/attack aircraft in the post-ATF period. The benefits include the definitions of:

1) Force/moment envelopes in takeoff, transition and hover modes.
2) Thrust vectoring/modulation envelopes and cross-coupling coefficient ranges.
3) Reaction control demand effects on bleed requirement and, consequently, on engine performance degradation.
4) Definitions of IFPC variables, their ranges and their cross-coupling effects.
5) Verification of the aforementioned IFPC variables by flight testing manned and unmanned vectored aircraft.

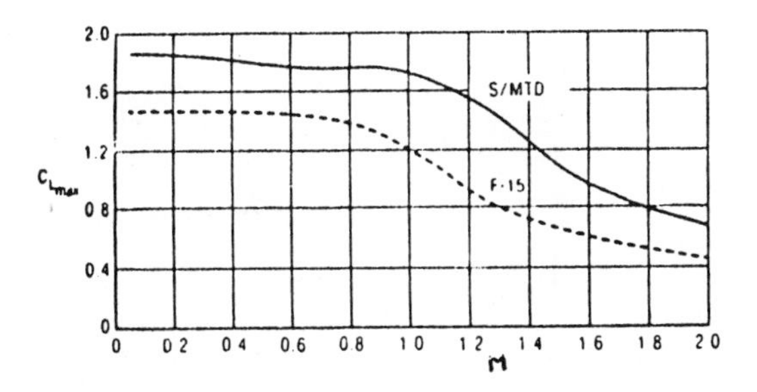

Fig. 13. **Blending of vectoring nozzles with close-coupled canards improves trimmed maximum lift coefficient of the STOL F-15 in comparison with the base-line F-15 (Mello and Kotansky, 128).** The gain is 27% at $M = 0.8$, and 58% at $M = 1.6$. Similar designs are being tested on the Soviet Su-27-1024.

The higher lift coefficient is due to synergistic combinations of increased lift and relaxed static stability afforded by the canard and an increase in control power from both the canard and the 2D–CD vectoring nozzles.

These increases permit improvements in instantaneous maneuver capability in many regions of the flight envelope.

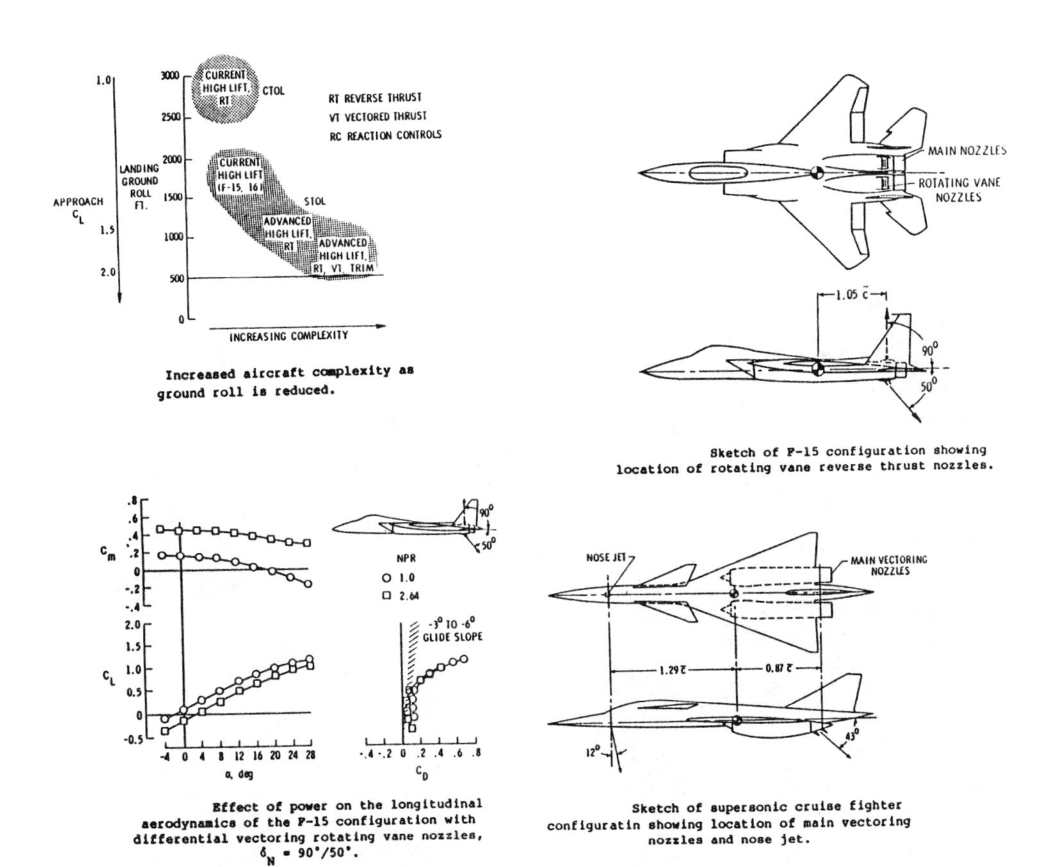

●Fig. 14. **NASA (Paulson and Gatlin, 155) has proposed various interesting methods to trim and vector aircraft, including *"canard-blowing" for STOL fighters.***

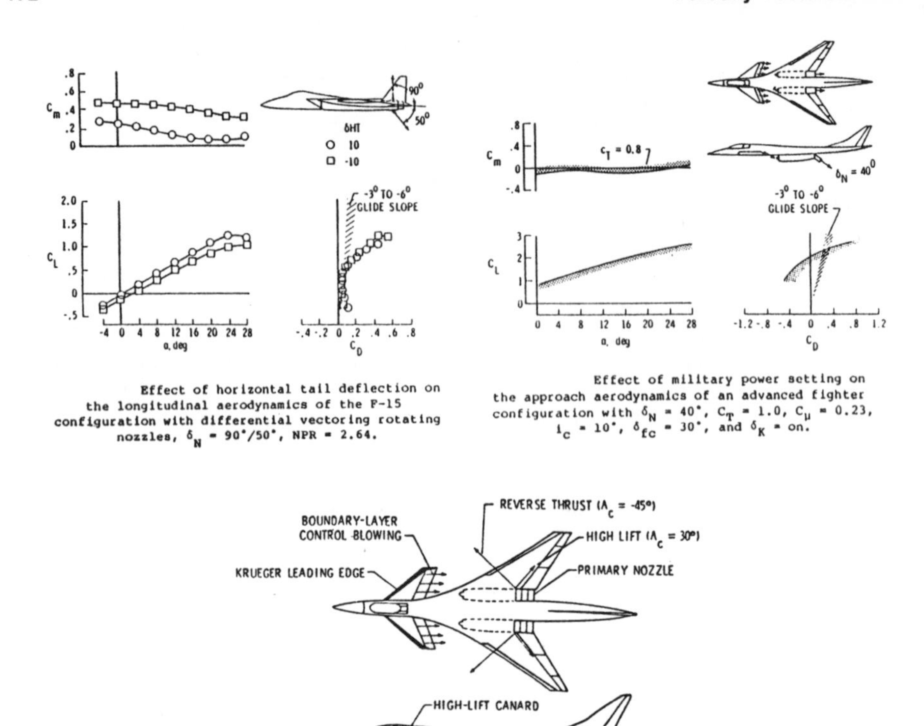

Fig. 14. Continued.

Fig. 15. **Thrust vectoring model proposed for the F-15, as tested by NASA's wind tunnels (Bare and Pendergraft, 234).**

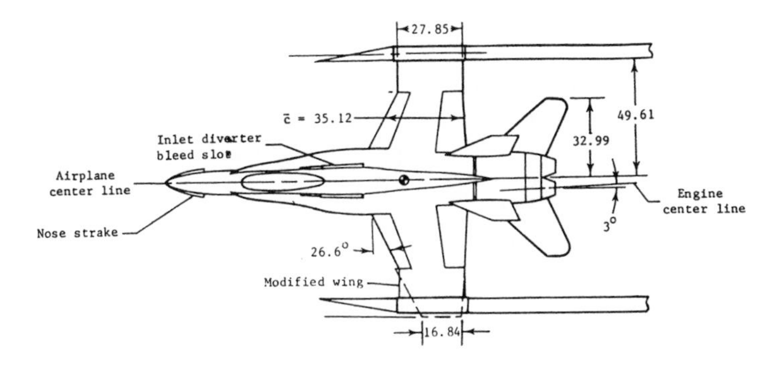

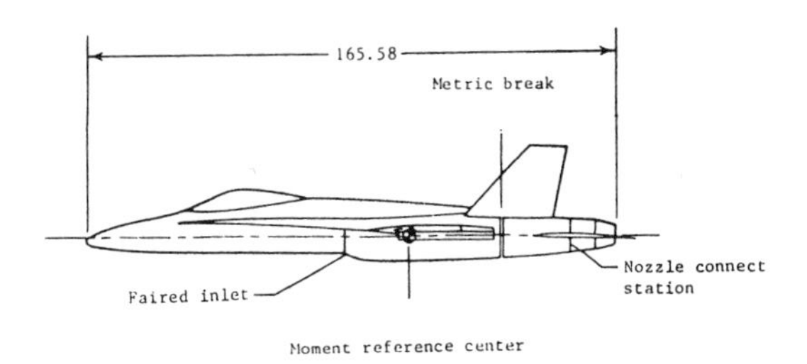

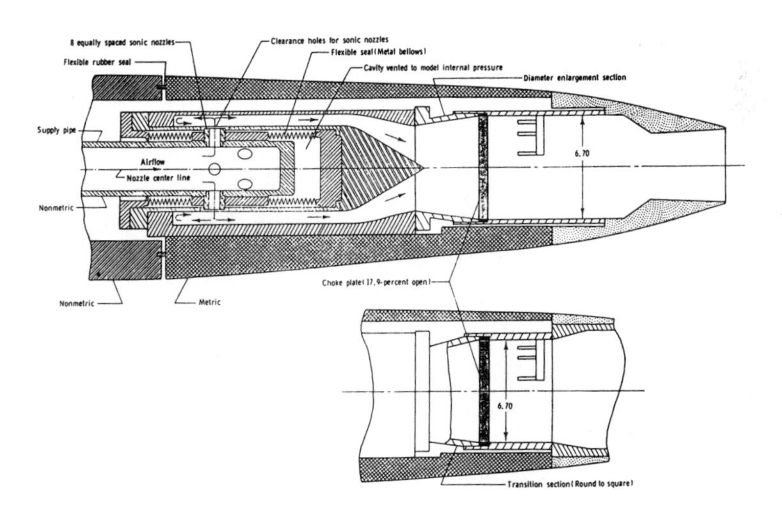

●Fig. 16. **The F-18 model during wind-tunnel tests [Capone and Berrier, 56].** The lower figure shows the details of the nozzles and instrumentation employed. The dimensions are in centimeters.

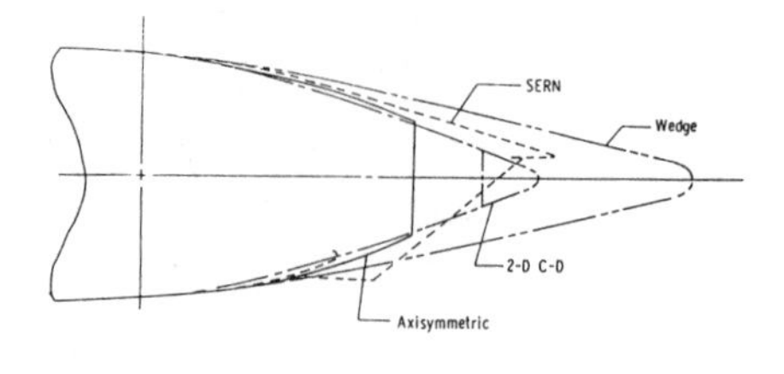

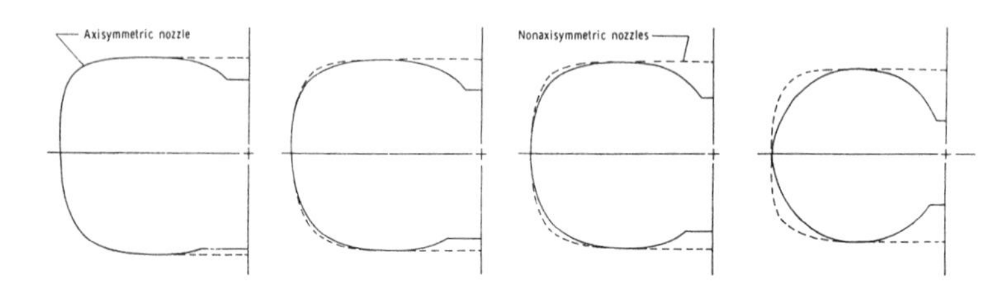

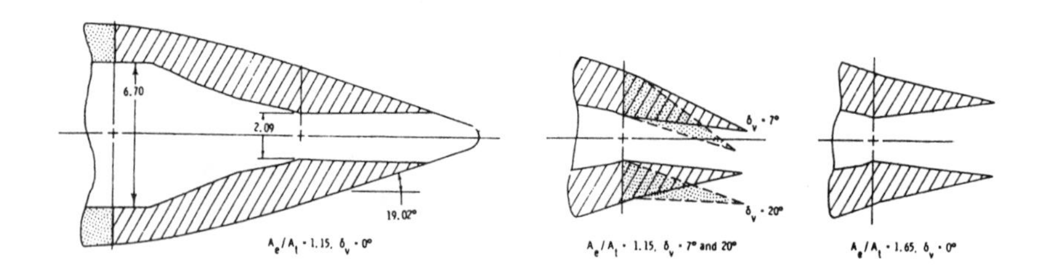

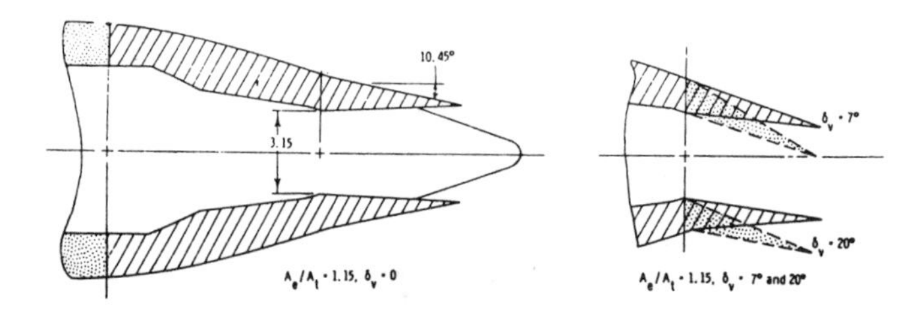

●Fig. 17. **Details of the nozzles tested by using the system and instrumentation shown in Fig. 16.** Sketch showing composite view of nozzles tested and some afterbody cross sections. Sketch of 2–*D*/*C*–*D* nozzle. Nozzle has diverging sidewalls

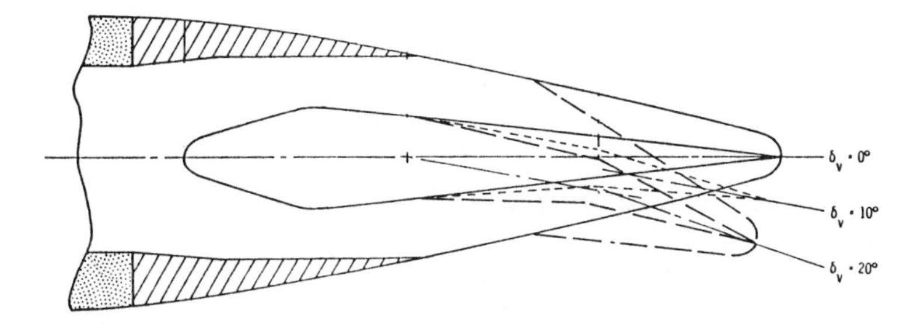

Fig. 18 **The wedge-type vectoring nozzle.** Picture shows the NASA wedge nozzle in a down pitching vectoring, $\delta_v = 20°$ (cf. Figs. 16 and 17).

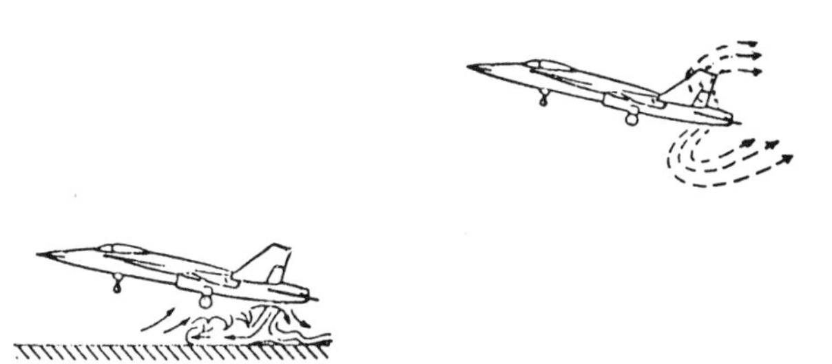

Fig. 19. **Hot-gas-ingestion into the inlet is encountered during thrust-reversal landing.** The ingestion begins below a minimal ground speed. Asymmetric flow-field interactions with the vertical stabilizers are also encountered during in-flight thrust-reversing (cf. Fig. 10, Appendix D).

Unfortunately no flight-tested IFPC variables exist today for pure vectored aircraft. It is imperative, therefore, to start flight testing programs that would evaluate these variables. We return to this subject in Appendix F.

V-7 The Need for Partially-Vectored Trainers (179)

Existing trainers, such as the F-5F and A-37/T-37, may be easily modified to become partially-vectored trainers *(without, however, expanding their flight envelopes).* These trainers are now powered by J-85 and J-69 engines for which a number of pitch-yaw-roll thrust vectoring nozzles have already been developed by this laboratory. Young, unexperienced pilots may thus employ these low-cost trainers to switch, in flight, from normal aerodynamic control to (weak) vectored control, and back. Thus, until a pure vectored trainer becomes available (probably not before the year 2000), a low-cost vectored trainer may serve to fill the extant gap in pilot education. Here the F-5F may also supply unlimited yaw thrust vectoring, while the A-37/T-37 may be somewhat limited by the fuselage during yaw thrust vectoring.

LECTURE VI

THE PROS AND CONS OF INTERNAL OR EXTERNAL THRUST VECTORING:

In this lecture we assess the pros and cons of two, post-Harrier, thrust-vectoring methodologies: Internal Thrust Vectoring (ITV) vs. External Thrust Vectoring (ETV). In trying to select the best technology, one must compare the technical methodologies, systems and conclusions of this lecture, with the general methodologies presented in the Introduction. However, this selection is not essential for the industry of future fighters, for ETV may be regarded only as a cost-effective tool for investigating RaNPAS–PST, high AoA performance.

1.1 IS ETV EFFECTIVE AT THE RaNPAS-PST DOMAIN?

Partial Jetborne Flight (PJF) has been defined in the Introduction as a flight in which elevons, ailerons, flaps, canards, or elevators, or rudders, are still being used in conjunction with a thrust-vectoring system. Most of the ETV-aircraft assessed below, as we shall see, must, therefore, be classified as partially-vectored aircraft, i.e., present ETV-aircraft are generally restricted to PJF.

Practically this means that the maneuverability and controllability of ETV-aircraft are still affected, to a degree, by the external-flow regime. Consequently, high AoA and/or high sideslip flights in the PST domain may still suffer from some of the classical aerodynamic limitations. To comprehend this subtle point one may have to refer to a few specific examples. This will be done next.

1.2 ETV-F-14 AS THE SIMPLEST EXAMPLE OF THE TECHNOLOGY LIMITS OF ETV

An F-14 flight with a single, post-exit, yaw vane was first conducted in 1987. As the yaw vane is deflected into the free jet, it provides external thrust-vectoring force/moment for yaw control at AoA up to 40 degrees, and at very low subsonic speeds.

During this flight, it was observed, as might be expected, that beyond a limiting AoA, the improved ETV-roll rate encounters a limit. This limit is, in part, due to the aerodynamic interaction of the external air flow, at high AoA, with the two engine exhaust-jets, and with the yaw-pedal deflector.

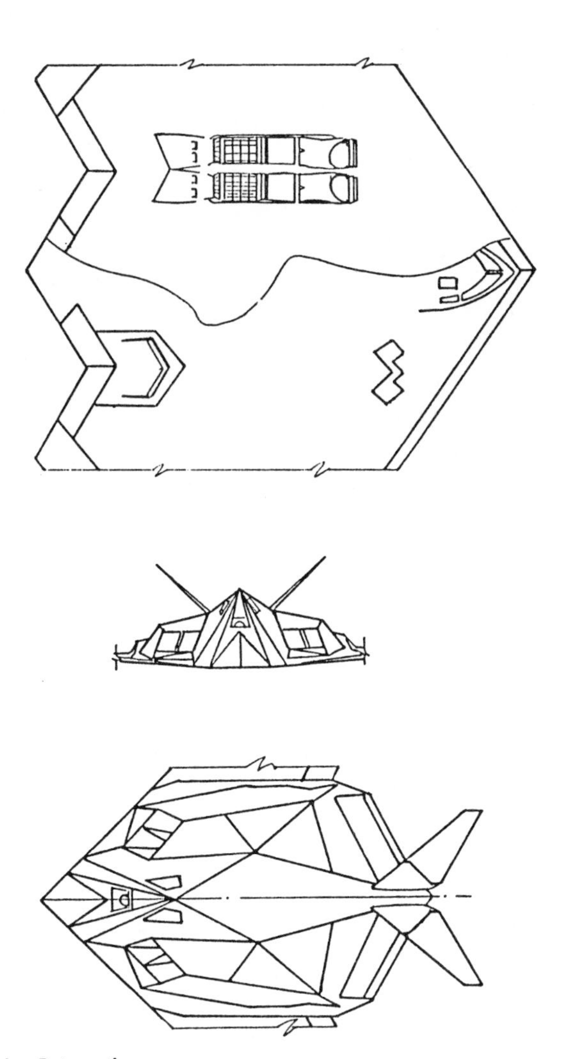

Fig. VI.1. **Wing-Engine Integration.**

Thus, the central questions before us are as follows: Does this limit constitute a particular design limit, or an authentic technology limit? And, secondly, if, indeed, this limit is an authentic technology limit, do the newer, multiaxis ETV designs, suffer, in principle, from the same limitation? I.e., can the addition of yaw-pitch, post-exit, pedal deflectors, by-pass, or surpass this limit?

In trying to answer these questions one may first note the following assertion:

The closer the multiaxis, ETV-flow passage, approaches the geometry of a shrouded, deflected, internal-flow nozzle (cf. Fig. 2), the less it should be affected by adverse, external aerodynamic effects during RaNPAS-PST maneuvers.

As an analogy to this complicated flow-field situation, but only as a highly-remote analogy, one may first examine the flow-field in and around shrouded vs. unshrouded fans at very-high AoA. The differences are obvious; thrust-vectoring, in general, is characterized by the reasonable boundary-layer behavior of expanding flows, while the air-compression stage inside fans is dominated by boundary-layer separation tendencies.

However, during PST-flights, the post-exit pedals (Fig. VI-2) encounter good boundary-layer behavior on one side, and highly separating flow regimes, on the other side. But what does it mean for the expected thrust-vectoring efficiency, and for the expected forces/moments of multiaxis-ETV during RaNPAS-PST maneuvers?

In trying to answer this question one may first examine the highly-instructive, and significant, static-test results of Berrier and Mason from NASA (209). Secondly, one may examine the design, and the flight-testing programs of the X-31. Thirdly, one may flight-test ETV–RPVs vs. ITV–RPVs during comparison maneuvers.

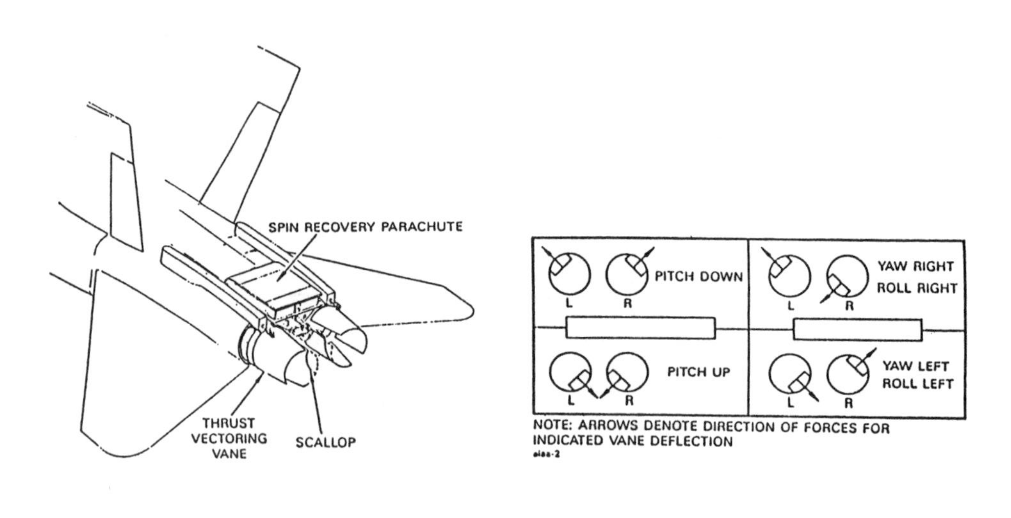

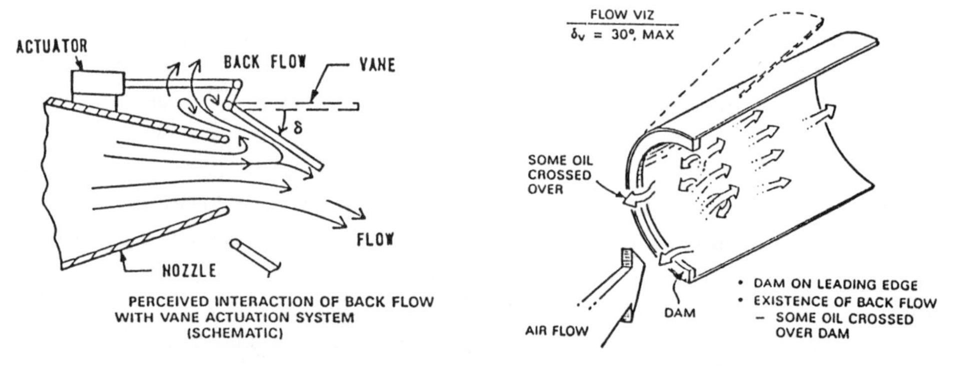

Fig. VI-2. **ETV (Pedal Type) as advanced by Berrier and Mason of NASA (208, 209).**

2.1 A FEW ASSERTIONS ABOUT EXTERNAL THRUST-VECTORING

ETV, or post-nozzle, thrust vectoring, is accomplished by single, or multiaxis, post-exit "pedals" (cf. the X-31, Fig. 7, Int.), which provide yaw-pitch, or yaw-pitch-roll controllability (by deflecting the free jet emerging from an axisymmetric (X-31) nozzle. This methodology is associated with relatively simple, readily-available, pedal/flap external devices, on one hand, and with the absence of (high-aspect-nozzle-ratio) supercirculation lift-gains (X-31); external-flow-dependent, jet-deflection propulsion/flight control laws/reliability; relatively high RCS/IR signatures (especially with circular nozzles), and longer over-all propulsion-system length, on the other hand.

Both methodologies have been proposed recently for the (partially) vectored, superagile F-16 and F-18 fighter aircraft. Hence, it is imperative to compare the two different maneuverability/controllability/propulsion potentials and limitations associated with them.

However, we must first pay our great respect to the British-American designers of the Harrier.

There is no doubt, that, somewhat similar to the historical British-American evolution of the Whittle gas turbine/jet-engine, the Harrier technology has significantly advanced the field of internal vectored propulsion, with some well-proven combat effectiveness.

However, it had been *a priori* faced with some inherent technology limitations, e.g., the absence of the powerful yaw thrust vectoring and supercirculation lift gains, and the introduction of high nozzle-drag, and heavy engine components, which reduce the thrust-to-weight-ratio of both engine and aircraft (179). The pros and cons of this technology, as well as its newly-proposed improvements in terms of advanced STOVL concepts, are available elsewhere (179), and in Appendix B.

2.2 THE PROPULSION SYSTEM OF THE VECTORED X-31

According to Herbst (188), there is no reason to use a rectangular nozzle configuration when there is no preferred direction of jet deflection. While we share Herbst's skepticism as to the priority that has been given to pitch/reversal over yaw-pitch vectoring, we may differ in our assessment of circular vs. rectangualr, and internal vs. post-exit yaw-pitch-roll vectoring nozzles. Accordingly, two, wing-integrated, rectangular, internally-directed, yaw-pitch-roll thrust vectoring nozzles (Fig. 7, introduction), may be superior to the post-exit, circular, yaw-pitch pedals, or to axi-TV nozzles.(Fig. 2).

Nevertheless, the X-31 constitutes one of the most important and most promising aircraft in the evolution of vectored aircraft. Its flight testing in 1990 in the USA would certainly become a significant milestone in aviation history. It would no doubt become the 1st PST–ETV manned aircraft in aviation history.

2.3 POST-EXIT PEDALS FOR FIXED 2D ENGINE NOZZLES

Another important contribution in yaw-pitch thrust vectoring was recently made in NASA-Langley Research Center (207–209). In one of the most promising designs (208), post-exit vanes were mounted on the side-walls of a nonaxisymmetric 2D-CD nozzle (208). While the resultant yaw vector jet angles in this design are always smaller than the geometric yaw vector angle, the widest post-exist vanes produce the largest degree of jet turning. In fact this type of propulsion system combines some feathers of 2D nozzles, e.g., lower drag and RCS/IR signatures, with the simplicity of external thrust vectoring pedals.

In fact, both internal and external, single, or multiaxis, thrust-vectoring systems have recently been proposed for the upgraded versions of (partially) vectored F-18, F-16 and F-15. However, ETV in these studies is now considered only as a means to investigate high-AoA flights.

2.4 CAN THRUST-FLAP VECTORING BE EMPLOYED?

Fig. VI-1 describes a possibility to add external-flap thrust vectoring to, say, the Advanced Tactical Bomber (ATB), the B-2. To start with, one must notice the following concepts:

1) – The fixed nozzles of all four engines are located atop the wing.

2) – The side-by-side, nozzles of each couple of engines mix their emerging hot jets with cold secondary air. The cold secondary air is venturi-pumped by the 2D nozzles from the secondary inlet slot just below the main inlet. The two streams are mixed inside a semi-confined, wing-recessed duct. This duct is open only on one side, namely; in the upward direction. Hence, the IR signature of the mixed, semi-hot jets, is drastically reduced in both the downward and sidewise directions. (However, the semi-hot jets just behind the wing trailing edge are still radiating.)

3) – Two oblique thrust-vectoring flaps may be added downstream of the semi-confined, wing-recessed ducts.

4) – Together, the two flaps form a Vee in the trailing edge of the wing. Now, both flaps may be rotated together, or separately. (At the present design they appear to be fixed.)

5) – An upward rotation of both flaps may supply nose-up pitch moment, while their combined downward rotation can turn the jets downward, until a limiting angle, by the Coanda effect. Consequently, downward rotation also provides thrust-vectoring pitching moment, i.e., in this case, a nose-down moment. This downward jet deflection also enhances lift, both by supercirculation effects (cf. Lecture IV), and by the direct lift force of the down-deflected jets. Consequently, STOL and lower approach speed payoffs may result, or TOGW may be increased.

6) – Thrust-vectoring roll may be carried out by rotating each of the two couples of

thrust-vectoring flaps in opposite directions to each other, while pitch thrust-vectoring propulsion/flight control is maintained by rotating all four thrust-vectoring flaps in the same direction.

7) – Yaw thrust-vectoring control may be affected by rotating, in each engine exhaust vee-bay, only one of the two thrust-vectoring flaps, say, only the right-hand flap.

Note also that the existing outer, vee-shaped, aerodynamic-control surfaces, are designed and controlled according to some similar reasonings, but without thrust-flap vectoring, i.e., at very low speeds and high-AoA they become ineffective, while the proposed thrust-flaps remain effective.

8) – Simultaneous yaw-pitch-roll thrust vectoring may become feasible by a properly designed IFPC system. Moreover, it remains effective even during very low speeds and high AoA, approach, runway and ground handling. Consequently, STOL and/or payload needs-specifications may be considerably revised by adding rotation to these two trailing flaps.

2.5 INTERNAL THRUST-VECTORING FOR THE (Now Cancelled) VECTORED X-29A

2.5.1 Previous Program Goals and Participants

Grumman's forward swept wing X-29A Program is a joint DARPA/NASA/USAF project. The X-29 was to be flight-tested at up to 40 degrees AoA, but without thrust vectoring.

Highly instructive simulations of thrust-vectored X-29A have, nevertheless, been reported recently by Klafin of Grumman (220).

Grumman's goal was to evaluate the controllability of the X-29A with multi-axis thrust vectoring for AoA up to +90 degrees. The 2-axis thrust vectoring was to be integrated with conventional aerodynamic flight-control surfaces. Hence, the vectored X-29A was to become a partially-vectored aircraft.

2.5.2 The Thrust-Vectoring Simulations

Klafin reports about piloted thrust-vectored simulation at 0.2 Mach, 15,000 feet and 33 degrees AoA. Without LTV the X-29A has adequate longitudinal control power for flight up to approximately +60 deg. With LTV this value increases to +90 deg. (The simulation was artificially stopped at +90 deg. AoA.) Thrust-vectored, lateral-directional control was simulated up to +90 deg. AOA. Thrust vectored augmentation was also found to reduce pilot workload with highly improved handling quality ratings. Flight control envelope was expanded by thrust vectoring to lower speed, lower dynamic pressures, and higher AoA.

Lateral-directional maneuvers are not possible at high AoA on the unvectored

X-29A (even though ailerons remain effective to 90 deg.), due to fuselage masking of the rudder. With Directional Thrust Vectoring (DTV), there was adequate lateral-directional control power for all longitudinal maneuvers and ambitious lateral-directional maneuvers, including 360 deg. rolls. Other benefits obtained with thrust vectoring are enumerated below.

2.5.3 Aerodynamics, Stability and Flight Control

The X-29A canard aerodynamic design is integrated with the wing so that the canard flow field interacts favorably with the wing and the wing flow field interacts favorably with the canard.

This close-coupled, variable incidence canard design is advantageous because the canard carries lift and also contributes to the aircraft being near neutral stability at supersonic cruise for minimum drag. However, this also causes the aircraft to be up to 35% statically unstable subsonically. This relaxed static stability is controlled with full-authority, digital FBW flight control system.

The normal flight control was employed during the simulations. Primary X-29A longitudinal control was provided by blending the canards, inboard and outboard wing flaperons, and strake flaps collectively as longitudinal control surfaces.

Aerodynamic surface control power decreases as dynamic pressure is reduced; this is compounded by decreasing control surface effectiveness as AoA increases. As expected, the addition of longitudinal and directional thrust vectoring has greatly enhanced high-AoA, PST, control power.

2.5.4 The thrust-Vectoring Nozzles

The thrust-vectoring effectiveness of the ADEN and the GE gimballed nozzles (see Appendix B), have been evaluated and simulated, using a variable-geometry inlet and the F404-GE-400 engine. Articulated sidewalls were included in the simulation of the ADEN to provide DTV (Directional Thrust Vector) in addition to its LTV (Longitudinal Thrust Vector). (The gimballed nozzle provides DTV and LTV by its very design concept – see Appendix B.)

During these studies, the gimballed nozzle was found a better choice than the ADEN. Its characteristics include: lighter weight, minimum thrust magnitude effect with deflection, thrust angle nearly equivalent to nozzle deflection angle, near-constant thrust moment application point, and pitch/yaw angle application point coincidence. Compared to the ADEN, its flow field is circular or eliptical, while the ADEN flow field is rectangular.

2.5.5 LTV Benefits

Instantaneous pitch acceleration was increased significantly with LTV. Using only dry thrust vectoring, the pitch acceleration of 0.5 rad/sec^2 was expanded from a range of

–9 to +5 degrees AoA (with only aerodynamic flight control), to +90/–50 degrees (where the aerodynamic flight control power is practically ineffective).

The pitch acceleration increment due to LTV is essentially constant with airspeed at fixed thrust levels. Moreover, LTV control moment surpasses that of the canards-flaps-strakes combined (100%) below 124 knots, for nose-down, and below 109 knots, for nose-up.

Furthermore, significant takeoff benefits were obtained with LTV: Liftoff speed was reduced 17%, distance reduced 37%, and time reduced by 20%.

Using only aerodynamic control surfaces, the loop gain can hardly be increased. However, when LTV was added, the loop gain was significantly increased.

The vectored-X-29A aircraft was found "perfectly capable of being flown to +90 degrees AoA (and higher) in the longitudinal axis, discarding any adverse inlet/engine effects, which were not simulated" (220).

2.5.6 DTV Benefits

The baseline X-29A is unstable in roll at 0.2 Mach/15,000 feet. Moreover, with the roll mode stabilized, effective rudder control is lost above 45 degrees AoA. However, above this aerodynamic limit, DTV was found effective in providing directional control power.

Fighter aircraft typically roll about the stability-axis velocity vector to maintain AoA and to minimize sideslip/lateral acceleration. Coordinated body-axis roll and yaw rates are required to perform a stability-axis roll. For a given stability-axis roll rate, the required body-axis yaw rate increases with AoA. Since rudder power decreases with AoA, yaw thrust vectoring was found to greatly augment rudder power, significantly increasing the roll rate/AoA capability as the aircraft enters into the PST domain.

2.5.7 Flight-Control Changes Due to Thrust Vectoring

The application of thrust-vectoring requires new control/coupling rules and substantial flight-control system design changes in such categories as:

- DTV to ailron cross-feed to correct DTV coupling into roll.
- Lateral-directional cross-feed paths to provide stability-axis rolls with high AoA.
- LTV gains vs. the longitudinal system loop.

2.5.8 Piloted Simulations

Thrust vectored, simulated, piloted maneuvers were started with aircraft trimmed in straight and level flight at 33 deg. AoA, with 34 lb/ft^2 dynamic pressure.

Longitudinal maneuvers included full aft control and hold, 80 deg. AoA capture and hold, and specific pitch attitude capture sequences.

Lateral-directional maneuvers included roll attitude captures and 360 deg. rolls.

These piloted simulations demonstrated that LTV improved pitch acceleration, dynamic stability, and aircraft maneuverability. Significant handling quality improvements were noted in pitch attitude and AoA tracking tasks.

2.5.9 Pilot Comments On Thrust Vectoring

Kalfin provides some pilot comments on thrust-vectoring simulated sessions, among them: "Thrust vectoring helps to control AoA. Holding 60 deg. is possible with LTV, and impossible with pure aerodynamic controls."

3.1 POWER-LIFT AIRCRAFT

According to USAF/MAC (204), powered lift is a leading technology contender for low visibility, speed, and the ability to fly low and survive in an extreme *EW* environment.

Thus, the US Air Force has been studying three different demonstrators: a generic STOVL (cf. Appendix B); an assault transport; and a tactical transport aircraft (204).

Four propulsion concepts are being investigated under the US/UK ASTOVL (Advanced Short Take-off/Vertical Landing) MOU [signed by the US DoD, NASA, and the UK MoD in 1986]: Vectored thrust; ejector augmentation; tandem fan; and the remote augmented lift system (cf. Fig. 8, Appendix B)

Upper Surface Blowing (USB) (Fig. 1, Appendix A), is being studied by Japan's National Aerospace Laboratory and NASA Ames Research Center. This involves subsonic STOL transport data, conceptual designs, and flight-test results from the Asuke STOL research transport, based on the Kawasaki C-1 (204).

An ejector-lift STOVL fighter model (cf. Fig. 8, Appendix B) had been undergoing windtunnel tests in General Dynamics during 1988 (204).

Franklin, Hynes, Hardy, Martin and Innis, of NASA Ames Research Center, have recently flight-tested a power-lift, STOL, research aircraft during approach and landing in a touchdown zone as small as 200×50 ft. (193). Using digital FBW flight control system, a head-up display, a color head-down display and nonlinear, model-following methods, the authors evaluated the influence of highly augmented control modes on the ability of pilots to execute precision instrument flight operations in the terminal area.

3.2 UNPOWERED MODELS FOR POST-STALL MANEUVERABILITY: A GERMAN PROGRAM

Ransom, of MBB/VFW, has recently described the pre-development considerations of a small, subsonic, delta-canard, highly-maneuverable, research aircraft as determined from the results of low-speed wind tunnel testing of a 1/7-scale model in MBB/

VFW's facility in Bremen, and of a 1/13-scale model of a twin-engined variant at the DFVLR, Braunschweig (194). The wind tunnel models were tested at a Mach number of approximately 0.2, a Reynolds number of approximately 10^6, and AoA from −5 to +70 deg (VFW), and from −5 to +40 deg (DFVLR), and at angles of yaw between −10 and +10 deg. The ultimate aim of this program is flight demonstration of direct-lift with vectorable engine nozzles, direct side force control, and fuselage aiming modes and post-stall modes, i.e., flight at AoA greater than that at which maximum lift is obtained. (CF. the X-31 program.)

APPENDIX A

A BRIEF HISTORICAL SURVEY ON INDIVIDUAL AND COLLECTIVE CONTRIBUTORS

A-1 The Definitions and Acceleration of US Thrust-Vectored Aircraft Programs

The US programs during the period 1973–1980 are well-covered by the combined comprehensive review of Richey (from the USAF), Berrier (from NASA) and Palcza (from the Navy) [73]. These authors stress that the contributions of USAF, NASA, and Navy and industry have demonstrated that advanced, nonaxisymmetric nozzles, integrated with tactical fighter aircraft configuration, have the potential for;

- increasing maneuverability and agility at high lift and/or low dynamic pressure conditions by thrust vectoring with supercirculation,
- reducing subsonic/transonic cruise drag compared to close coupled axisymmetric nozzles through better nozzle/airframe integration,
- improving longitudinal agility for air combat, and increased accuracy and survivability in air-to-ground weapons delivery due to steeper dive angles and higher weapon release altitude by incorporating an in-flight thrust reverser,
- reducing *IR* and *RCS* signatures due to nozzle configuration influence; also, the *IR* and *RCS* observables tend to be highly directional, thereby having the potential of greatly increased aircraft survivability against seeker missiles by maneuvers and application of thrust vectoring,
- improving take-off and landing performance and ground handling of high-thrust aircraft,
- reduced life cycle costs through projected lower fabrication costs (fewer moving parts), reduced fuel costs to perform the same mission (due to lower cruise and acceleration drag),
- reduced fleet size, as a result of improved survivability and combat effectiveness.

These authors also stress that the proper utilization of this emerging technology in future aircraft systems depends on;

- a coordinated attack on nonaxisymmetric nozzle technology to improve the data base on the disciplines of propulsion, aerodynamics, structures and vehicle *integration*,
- definition and implementation of a *meaningful, timely, flight research program to increase confidence in the technology for transition to systems applications*, to define and demonstrate advanced fighter flight characteristics, to utilize design criteria for nonaxisymmetric nozzles with thrust vectoring features, and to define full-scale performance and survivability characteristics.

Accordingly, the authors describe a series of R&D programs, which, beginning in 1977, are designed to substantiate the need for flight research programs, define their specific objectives, evaluate candidate test aircraft against these objectives, and acquire data to find the cost-effective approach to a *2D* nozzle flight research program.

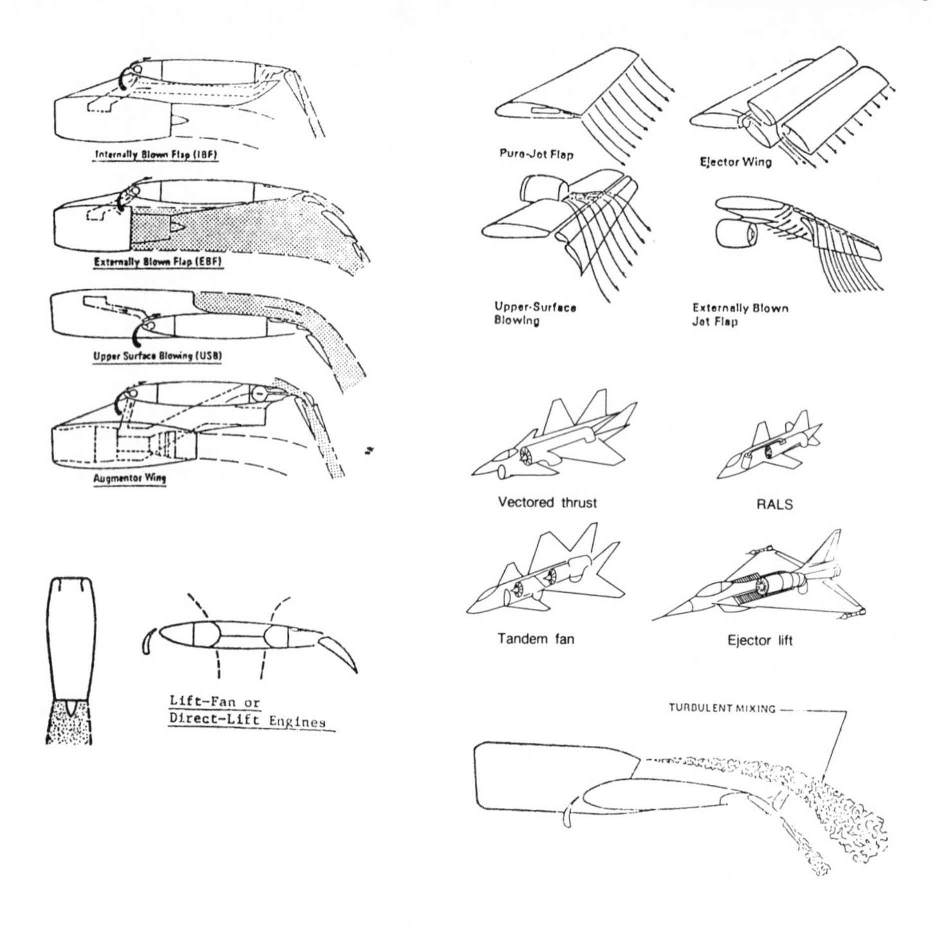

Fig. 1. **The evolution of a few power lift concepts, including some early vectored thrust control designs.** For STOVL concepts see Figs. 7 and 8 in this Appendix.

A-2 Reduction in Interference Transonic and Supersonic Wave Drag

The authors of this timely overview stress a number of additional points:

1) Exhaust nozzles have traditionally been circular in cross section to facilitate good integration with the core engine, thereby neglecting the performance penalties associated with the airframe.
2) A large part of "rounded" nozzle installation penalty for twin-engine fighters is due to the poor integration of "round" nozzles into a "rectangular" afterbody, and the very neglection of supercirculation potentials.
3) These "rounded" configurations generally have boattailed "gutter" interfairings, or base regions on the afterbody, which result in interference drag at transonic flow conditions, and wave drag at supersonic operaton. Thus, the installation of *2D* nozzles reduces external flow nozzle penalties by eliminating the separated interfairing and base regions.

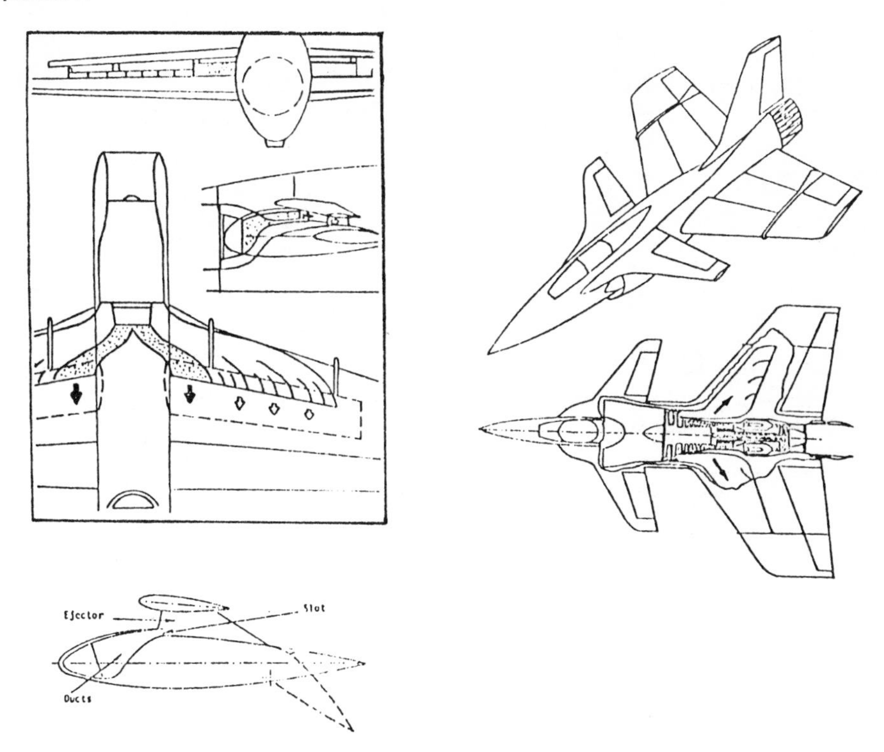

Fig. 2. **The evolution of a few "jetwing" design concepts.**
upper left:
Dividing engine cold and hot streams for separate lift enhancement sectors involving cold and hot materials, respectively.
upper right:
Core (hot) exhaust gas remains unchanged, but fan (cold) air is used to enhance lift and maneuverability.
lower left:
Internal ducts are used for inboard and outboard blowing, utilizing the *Coanda Effect* [156].

A-3 Individual Contributions

Many names are associated with the subject matter of this book. Their collective and individual contributions are well-reflected in the references given at the end of this book. What has been described in the text is, therefore, only an introduction from this author's point of view. For a full appreciation of the individual efforts one must turn to the original references.

By closely examining these references one can distinguish the collective contributions of such bodies as USAF (and especially that of Bowers, Glidewell, Laughrey, Richey and Surber at the Flight Dynamics Laboratory of WPAFB), NASA (and especially that of Banks, Berrier, Capone, Leavitt, Mason, and Pendergraft), the US Navy, industry (especially that of Herbst and McAtee) and the professional leadership demonstrated by other individuals.

Among the individuals who have significantly widened the scope and aims of this technology, one may pay a particular attention to the valuable contributions of:
Banks [29, 148], Berndt [63, 142, 198, 205], Berrier [1, 6, 18, 44, 52, 55, 73, 208, 209, 210], Bowers

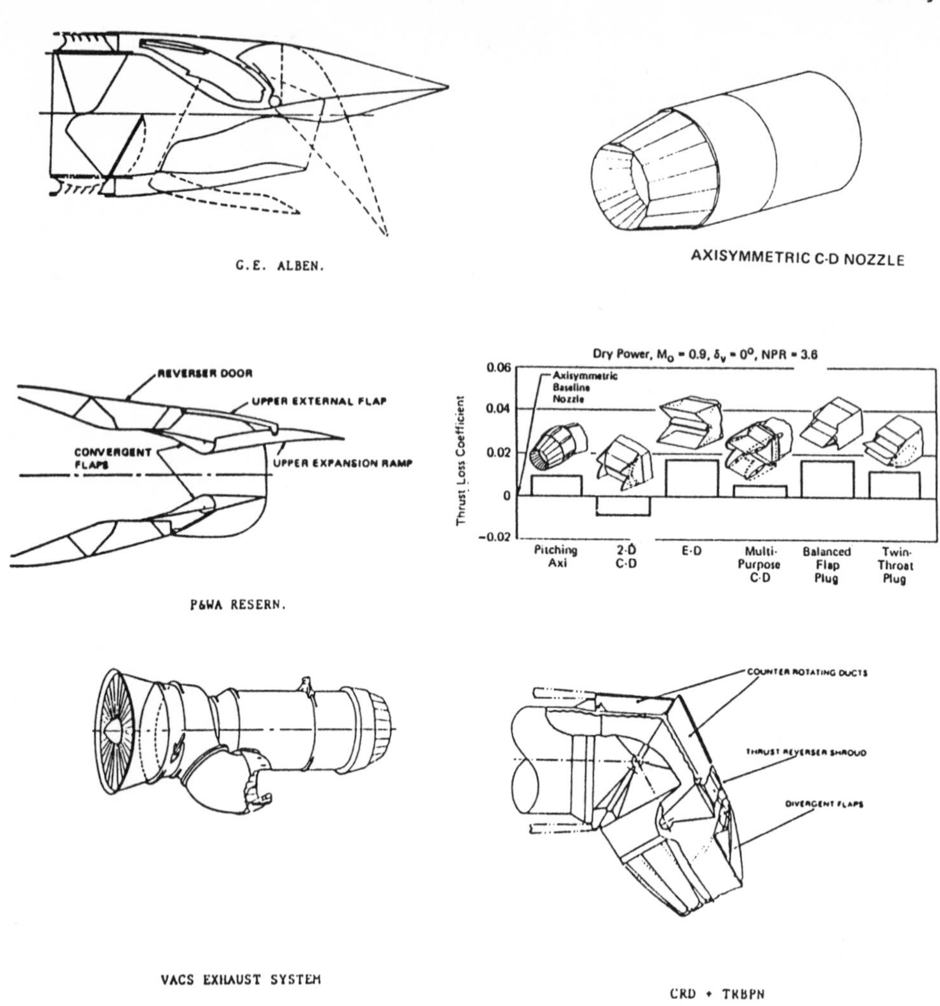

Fig. 3. **The evolution of a few early concepts in thrust-vectoring nozzles.** (The "thrust loss coefficient" data were taken from Hiley, Wallace and Booz, Ref. 72. The VACS and CRD+TRBPN nozzle figures are from Berndt, Glidewell and Burns, Ref. 142.)

[4, 11, 13, 24], Burley [33, 114, 162], Callahan [158], Capone [10, 17, 18, 20, 21, 22, 28, 37, 42, 54, 56, 49, 60, 64, 67, 68, 69, 121], Chu [90, 92, 100], Costes [182], Glidewell [5, 14, 142, 198, 205], Goetz [41, 77], Herbst [154, 188], Hiley [11, 61, 66, 72, 78, 139], Joshi [124, 140, 161], Klafin [220], Kotansky [128], Kucher [63, 138], Laughrey [4, 13], Leavitt [58, 112, 162], Mason [38, 39, 40, 52, 57, 208, 209, 215], Maiden [32, 34, 36, 40], McAtee [196], Mello [128], Miller [8, 9, 47], Nash [45, 46], Palcza [6, 41, 43, 46, 73, 75], Paulson [15, 16, 27, 29, 148, 155], Pendergraft [62, 234], Petit [40, 42, 68, 77], Re [28], Richey [6, 73, 210], Schneider [218], Sedgwick [164], Stevens [49, 70], Surber [210], Tamrat [185, 213], Thomas [95, 96], Watt [218] and Well [211].

A substantial portion of this book is actually based on the highly significant contributions of these individuals.

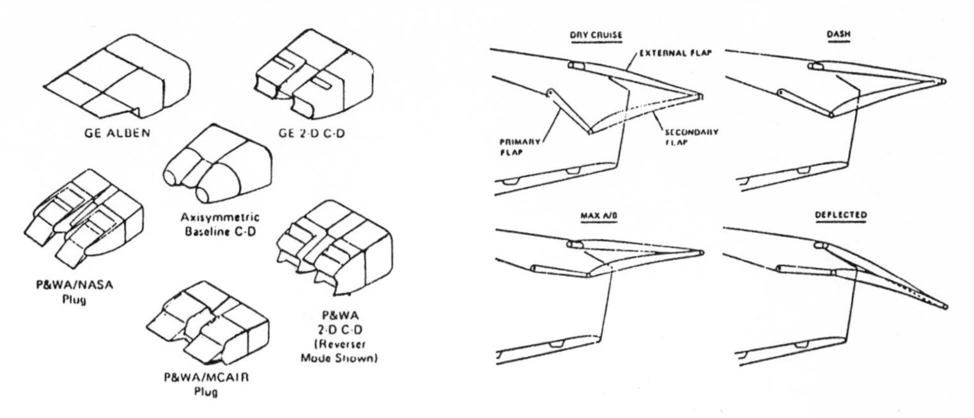

Fig. 4. **A few early nozzle competitores in the thrust vectoring "race".** The GE *2D–CD* nozzle on the upper-right is essentially the one shown in Figures I-1 and I-7. The ADEN nozzle is shown on the right during various operating conditions (cf. Fig. 5).

ATA and ATFN (See also Appendix B)

Beyond the year 2000 the Navy is likely to replace the F-14 with a version of the USAF's Advanced Tactical Fighter, known as the ATFN. Under a memorandum of understanding signed in early 1986, both the (ATF) Northrop/McDonnell Douglas YF-23 and the Lockheed/Boeing/GD YF-22 are being designed so that a navalised version can be developed without excessive modification and the Navy will monitor the USAF program.

On the attack side, the Navy is developing the Advanced Tactical Aircraft (ATA), or A-12.

ATA appears to be running about two years ahead of ATF, so it may be operational in the mid-90s. It is believed to be a subsonic aircraft powered by two uprated, nonafterburning GE-F404 engine derivatives. It will be a two seater designed for low-level, all-weather, deep-strike missions. ATA has been defined as a long-range, heavy-pay-load aircraft, using stealth technology.

The McDD/General Dynamics team has recently won the ATA/A-12 contract.

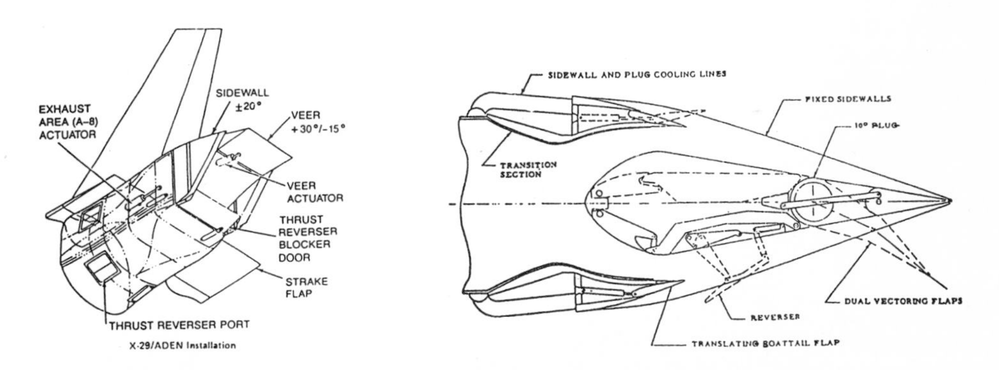

Fig. 5. **The ADEN and PWA Plug nozzles.** (A particular analysis of ADEN/X-29A integration is given in §VI-2.5.4. Its use on a STOVL aircraft is depicted in Fig. 8, Appendix B). Cf. Fig. 4.

APPENDIX B

POWERPLANT TECHNOLOGY LIMITS

(A portion of this Appendix is based on an updated version of a review published in 1984 in the *Intern. J. Turbo and Jet Engines*, Ref. 119)

"Perfection of means and confusion of goals seem – in my opinion – to characterize our age."

Albert Einstein

"Who but the Harrier pilot can thrust vector in forward flight and confuse an attacker by killing forward speed, so that the enemy overshoots into a lethal position?"

The Editor (JWRT)
Jane's All The World's Aircraft
1987–88 p. 43 [175]

B-1 Beyond the year 2000

R&D groups around the world are now shifting attention to a new generation of engine and materials technologies to provide quantum leaps in performance, reliability and survivability beyond those already on the drawing boards.

Until recently, much of this effort has been aimed at fighter engines, such as those for the *ATF, ATA and EFA*. However, since the design definition of these engines has now been completed, and since the various technology limits had been identified and set, advanced R&D and design goals are now beginning to be redefined in terms of the next generation beyond.

The payoffs of these new powerplants are expected to come from a number of new technologies. These include:

1) *RCC* Materials (see below).
2) *Vectored Propulsion Systems.*
3) *Advanced Engine Cores.*
4) *IFPC Technology.*

Current-technology engines for the ATF, ATB and ATA are described below. *The technology limits of these engines are typified by a thrust-to-weight ratio of around* 10:1 *to* 12:1, while those of the new-generation, beyond-the-year-2000, by a ratio of 20:1. These engines will include high through-flow compression systems, to reduce the number of compressor stages, segmented combustor liners and T_{T4} beyond 4000 F. Substitution of fiber-reinforced rings for compressor discs is expected to reduce engine rotor weight by a factor of 3 or 4. A test compressor of this type may be available in the 1990s for use in the Post-ATF engines. Ceramic bearings that can function with reduced quantities of cooling loads may also be available by that time (see below).

The newer engines are thus expected to be made of much lighter, but higher-temperature materials. By itself, this will result in a substantial reduction in overall *aircraft* weight, and/or sizeable increases in range and payload, and/or increased thrust-to-weight ratios beyond, say, 1.3:1.

Engine part counts may also be substantially reduced, say, to 2000, in comparison with 15,000–20,000 on current-technology fighter engines. These, in turn, have up to 50% fewer parts, 40% less supersonic fuel consumption and 20–30% reduction in life-cycle costs than the F100 fighter engine.

B-2 Uncooled Turbines/Nozzles at Up to 4000 F

Major gains in SFC and design simplicity are expected to emerge from materials such as RCC and other ceramic/ceramic composite systems. These can operate uncooled at up to 4000 F (cf. Figs. 1 and 2).

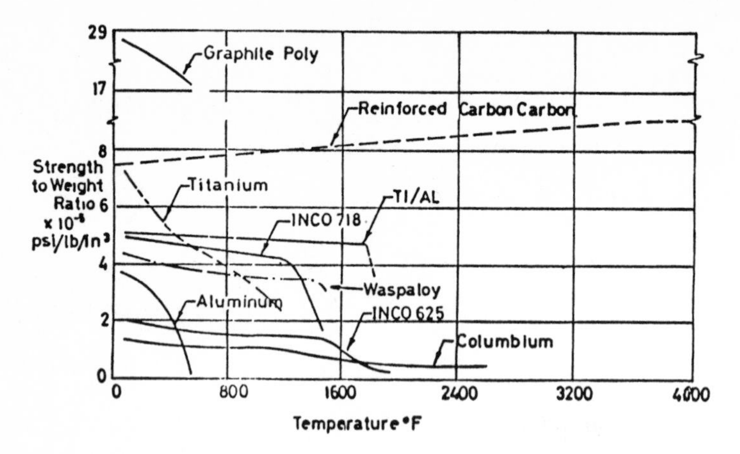

Fig. 1. **Future fighter engines will be made from higher proportions of RCC (Reinforced Carbon–Carbon Composites).**

Well-impregnated with oxidation-resistant ceramics, such as boron and silicon carbides, the new materials demonstrate excellent performance at skin-temperatures in excess of 4000 F.

Eventually not only the thrust-vectored nozzles and the transition/augmentor's ducts will be manufactured from RCC, but the hot section of the engine as well.

A major difficulty in applying this technology today to all advanced fighter engines is the lack of efficient manufacturing processes amenable to impregnating large-scale production parts (cf. Fig. 2) [Bowers, 13, Gal-Or, 146].

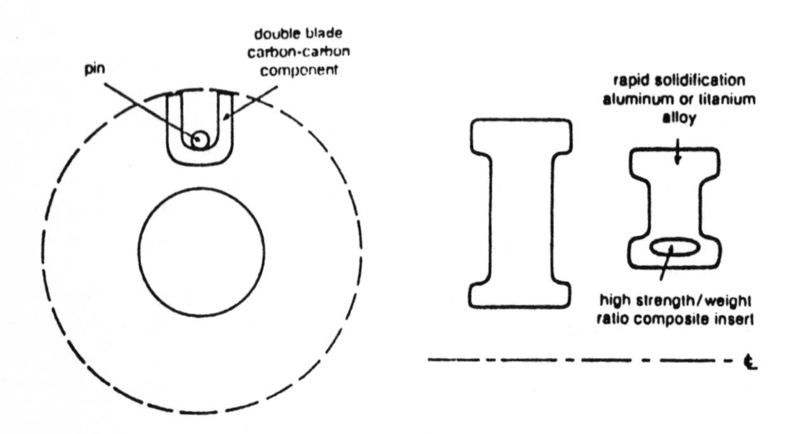

Fig. 2. **A Reinforced-Carbon–Carbon (RCC) Turbine.**

A wrap-around blade segment of RCC attached by a pin of an RCC disk may characterize future engines.

The bladed ring, or "bling", shown here is much squatter than conventional designs. It is reinforced by a composite insert to withstand the stresses of high rotational speeds.

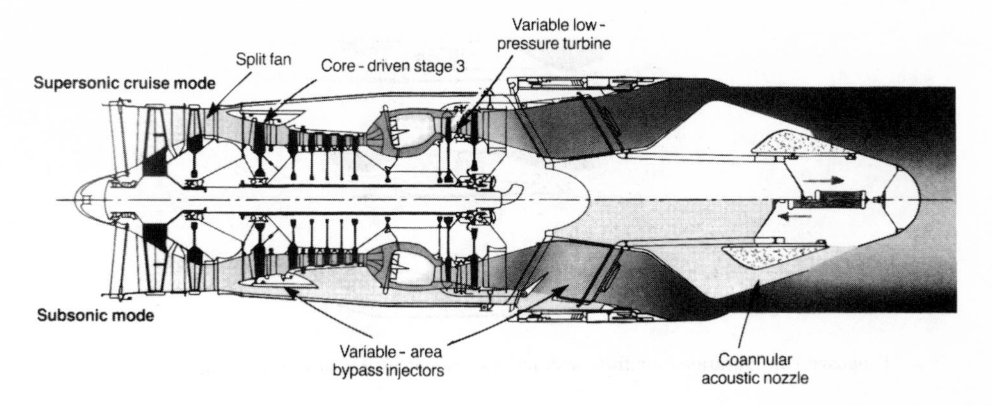

Fig. 3. **The variable-cycle engine proposed by General Electric (cf. § B.12).**

In recognition of the key role materials will play in future power-plant technology, the so-called *Integrated Technology Plan for the 1990's*, has been evolving from *USAF Forecast II Analysis* [as adapted by joint Aero Propulsion Laboratory and the Materials Laboratory Projects at USAF WPAFB, and in the US aero-engine industry]. These projects can lead to 20:1 *T/W* ratio in future fighter engines with the same life expectancy as that of the ATF engine, (e.g., increasing inspection times with respect to around 1800 cycles for the ATF engine hot section parts.)

Current emphasis is on damage tolerant designs, using materials that are configured to limit propagation of cracks.

The new integrated design approach to turbine blades and disks, balances between what the aerodynamicist may say he needs, what the thermal requirements are, and what the durability/maintainability designer may figure as a reasonable interval for inspection.

Contra-rotating spools and single-stage high and low pressure turbines may characterize the new engines. Both GE and PWA currently use single crystal superalloys in the turbine, improved titanium alloys in the compressor system, and RCC as external skins for the thrust-vectoring nozzles. For more details on the ATF engine see § B.11.

B-3 RCC Thrust Vectoring Propulsion

RCC composite materials offer the best potential for drastic improvements in hot-section propulsion system performance, especially in optimal integration of wing and engine, and in reducing about 50% in the combined weight of a typical high-performance fighter engine.

Substantial increases in maximum thrust can be materialized, since practically no cooling flow is required to protect RCC composites. Indeed, Figure 1 shows a dramatic reduction in strength for "metals" at skin-temperatures in excess of 1600°F, while RCC composites maintain their strength at temperatures in excess of 4000°F.

This fact, combined with almost negligible thermal expansion, in comparison to "metals", also results in increased durability and reliability.

However, to prevent oxidation of the carbon fibers and of the inner matrix at elevated temperatures, one must improve the current-technology methods to *impregnate* the matrix porosity with carbides, etc., using specially-developed new processes [146].

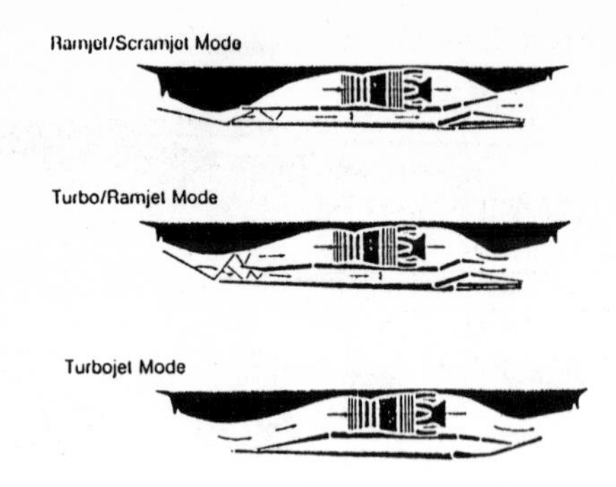

Fig. 4, **Proposed configurations for turbojet/ramjet/scramjet modes of a future engine.**

It should be stressed that no coating can offer the oxidation protection and durability available by suitably-impregnated Carbon–Carbon composites.

Another important parameter of RCC is its electromagnetic capacity to absorb radar waves. (This subject will be covered in more detail in Volume II.)

Hence, most future propulsion systems will contain higher proportions of RCC in their various, highly-loaded, hot sections, and in their auxiliary systems.

B-4 USMC/STOVL Aircraft.

In the last decade, the US Marine Corps has taken a lead in the development of STOVL aircraft. The AV-8B Harrier II, developed by McDonnell Douglas for the USMC, has already demonstrated an advance in performance and safety over the AV-8A.

In the long run, the USMC may want to replace all of its current combat aircraft-AV-8Bs, F-18s, and A-6s – with a single STOVL Advanced Combat Aircraft (ACA). The "desired operational capabilities" include a 300n.m. intercept mission radius, with a speed of Mach 1.4 sustained on the outbound leg, and a 450n.m. offensive counter-air mission radius.

The USMC may consider the STOVL as the key to control airpower, free from the dependence on a Navy ship.

References 149, 150 and 197–201 provide comprehensive reviews on the new-type of powerplants proposed for these aircraft. Fig. 7 shows the location of the engines and nozzles, while Fig. 8 shows some additional details.

B-5 Supersonic STOVL Aircraft: The NASA Programs.

Wind tunnel tests using a full-scale model of the General Dynamics E-7 supersonic STOVL aircraft design, fitted with a de Havilland ejector system, have been conducted in 1988 in the NASA–Ames 40×80-ft. wind tunnel. The nonflying E-7 model is equipped with a Rolls-Royce Spey engine [cf. Figs. 6, 7, and 8].

In an arrangement unique to the E-7, engine fan flow is collected and fed through a plenum to either a forward or aft duct. The forward duct directs the flow to an ejector augmentor in the aircraft's wing, while flow directed to the aft duct is expelled through a thrust nozzle located in the rear of the aircraft. Butterfly valves control the direction of the flow.

After the wind tunnel work at NASA–Ames is completed, full-scale ejector system components will be tested on the NASA–Lewis powered lift facility. The facility was opened in late 1986 to support vertical lift propulsion tests that will be required under the joint US/UK advanced STOVL program. The US/UK program is assessing several candidate propulsion technologies and airframe concepts that could lead to the development of a supersonic STOVL demonstrator aircraft in the mid-1990s.

The powered lift facility includes a triangular frame with 30-ft. long sides. The sides are supported 15 ft. off the ground by load cells that provide a six-component force measuring system. The facility can measure vertical forces up to 20,000 lb., axial forces up to 30,000 lb. and lateral forces up to 5,000 lb. in positive and negative directions.

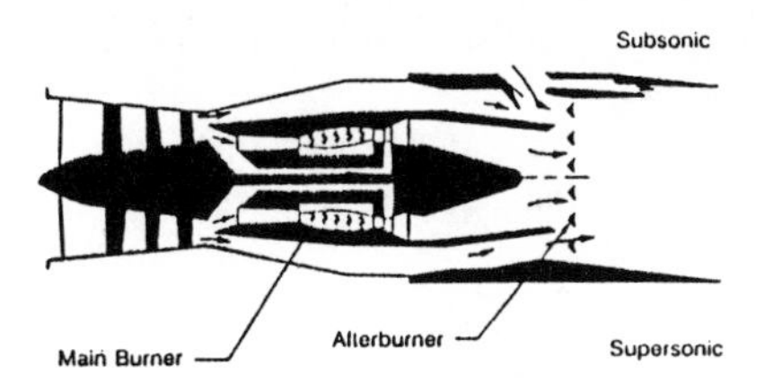

Fig. 5. **A variable stream control engine.**

B-6 Ejector Tests: The NASA Programs.

Additional tests conducted at the NASA–Lewis powered lift facility were supportive of the ejector approach. Designed to evaluate the thrust performance of the ejector system, these tests indicated that the design requirements for a STOVL aircraft would be exceeded by the de Havilland configuration.

The model is to be modified to incorporate a two dimensional converging/diverging aft nozzle and a ventral nozzle that will provide vertical thrust.

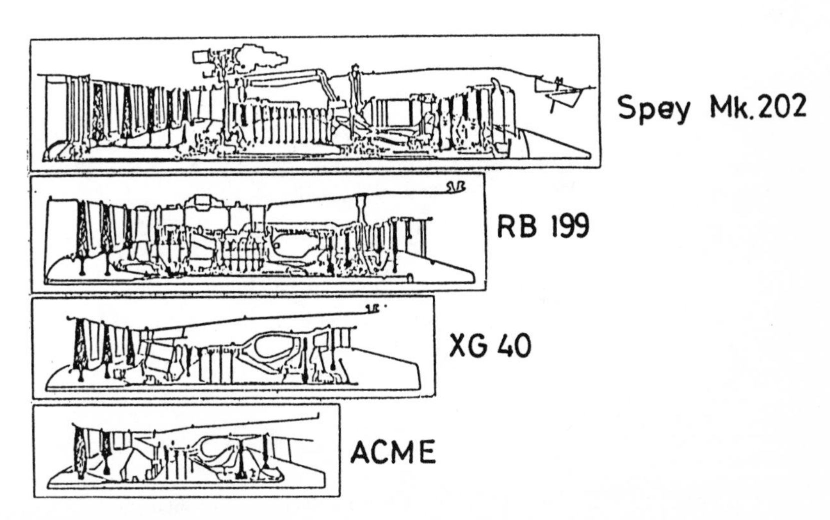

Fig. 6. **Trends in British Powerplants.**

The US and the United Kingdom are also studying airframes and powerplants based on the remote augmentator fan, vectored thrust, ejector augmentor and tandem fan propulsion concepts (Fig. 8). Related to this venture is a NASA and US Defense Dept. study of lift/cruise systems.

Proper propulsion design is clearly the key to a supersonic STOVL aircraft, and much of the work performed under various agreements and contracts is aimed at developing such advanced propulsion technologies.

Significant supersonic STOVL propulsion technology issues can be grouped into several areas that include [cf. Fig. 8]:

- Development of high thrust-to-weight ratio engine cores with sufficient bleed capacity for supersonic STOVL aircraft attitude-control systems.
- Reduction/avoidance of hot gas ingestion (Fig. 1, Appendix D).
- Development of integrated flight/propulsion controls to ease pilot workload during takeoff, transition and landing (Appendix F).
- Development of supersonic inlets that can be used in high angle-of-attack attitudes and low-speed situations; thrust vectoring nozzles, and efficient low-loss ducts, valves and collectors (Appendix F).

Under investigation are low-loss fan collectors, valves and ducting; hot gas ingestion avoidance; short diffusers and supersonic inlets with high angle-of-attack capability, and integrated flight/propulsion controls. Programs to explore thrust-deflecting and vectoring nozzles and thrust augmentation are also in the planning stages (Lecture V).

B-7 Thrust Vectoring Control of Slow-Moving Airships

Design concepts for thrust vectoring of very slow-moving airships may serve as a simplified simulation for vectored-RPVs, or for VTOL-vectored aircraft at vertical takeoffs. Nagabhushan and Faiss (192) have used a simulation using two ducted fans mounted one on either side of the airship, each having the capability of tilting in pitch and roll to give vertical and lateral thrust for control. An auxiliary thruster at the bow or stern of the airship, which augments its directional control, was investigated.

It has been found that the tiltable ducted fans provide the airship with greater operational flex-

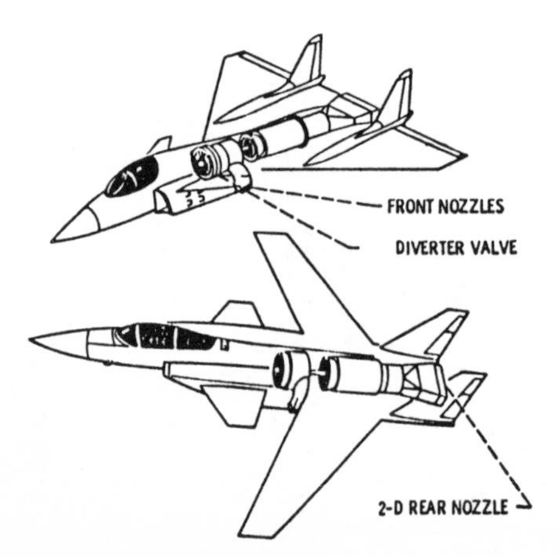

Fig. 7. **The evolution of concepts of a STOVL fighter aircraft – (general lay-out of a delta-Canard and a forward swept wing aircraft with canard [149, 150]).**

ibility, especially during takeoff and landing. Thrust vectoring to provide roll control was found to be effective while ground handling. The bow/stern thruster was found to give excellent directional control, which significantly improved lateral maneuverability. Thrust reversal and vectoring rates were found to be important design parameters.

B-8 Advanced Compressors

Compressor stall margins (cf. Fig. F-6) are expected to increase substantially to afford high-α-β maneuvers without engine stall/flame out. This increased margin will be combined with *advanced* IFPC systems (Fig. F-7), new PST-inlets and new thrust vectoring systems as described in the main text. Advanced vectored engines may also incorporate rectangular cross-section afterburner [cf., eg., Fig. I-13, 14].

These new compressors incorporate tip clearance control (cf. Table 1) and low-aspect ratio blades to gain beneficial compressor margins and characteristics without adding to the compressor weight.

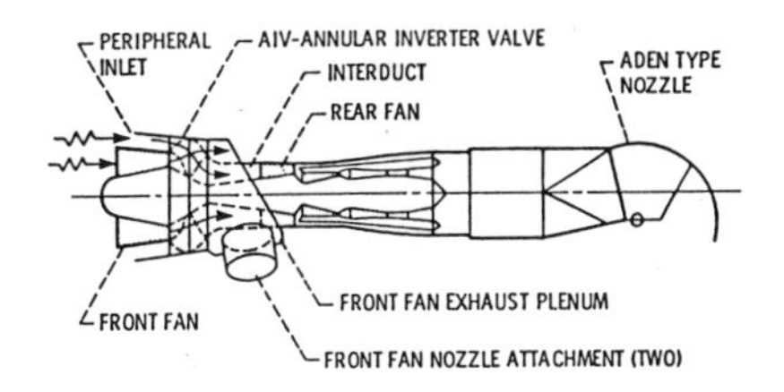

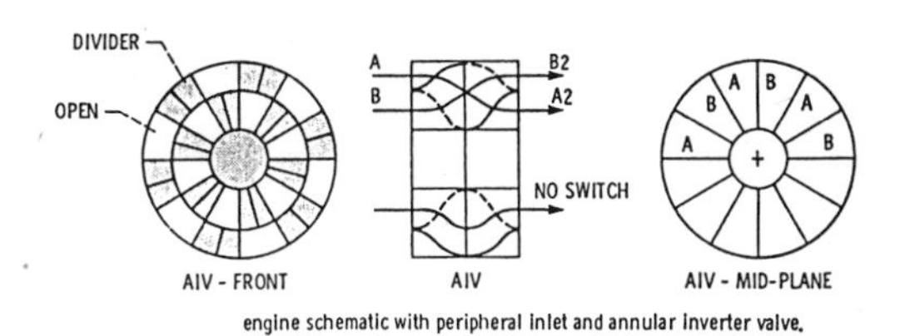

engine schematic with peripheral inlet and annular inverter valve.

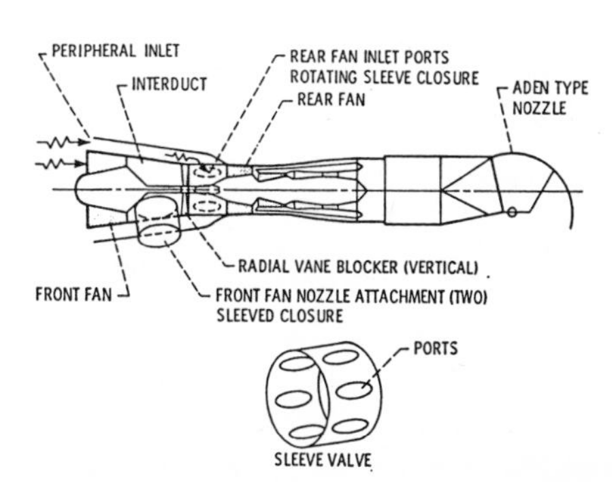

●Fig. 8 **The evolution of propulsion system concepts for STOVL fighter [149, 150, 197–201].** For the geometry of the ADEN see Fig. 5, Appendix A.

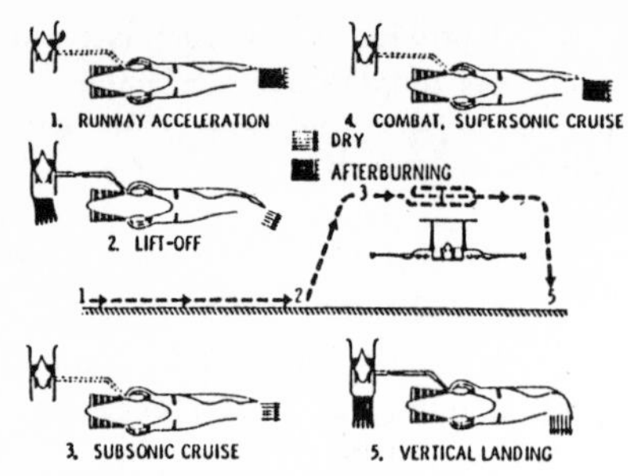

TBE/turbo-compressor lift engine propulsion modes for supersonic STOVL.

HIGH THROTTLE
BYPASS DUCT
LOW THROTTLE

AFTERBURNING TURBO-COMPRESSOR
AFTERBURNING TBE/(TURBO-COMPRESSOR - STOVL ONLY)
DUCT
MANIFOLD
RALS
AFTERBURNING TURBOFAN/RALS
TURBO-COMPRESSOR
AFTERBURNING TURBO-COMPRESSOR
DRY TBE/(TURBO-COMPRESSOR - STOVL AND CRUISE)

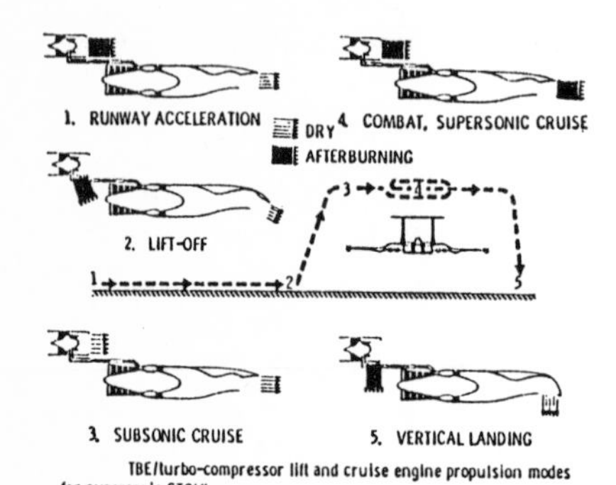

TBE/turbo-compressor lift and cruise engine propulsion modes for supersonic STOVL.

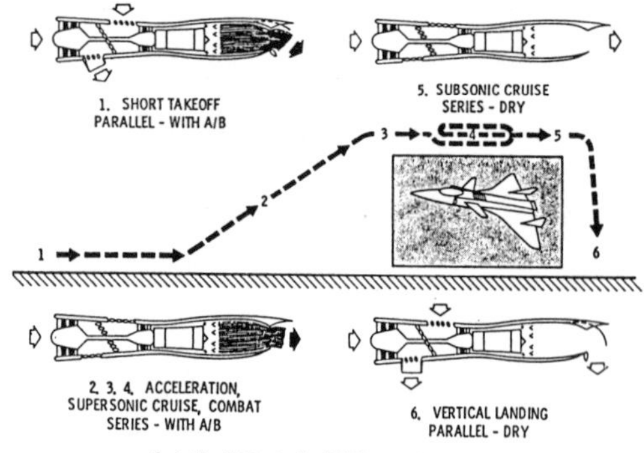

Series/Parallel Tandem Fan (SPTF) propulsion for supersonic STOVL.

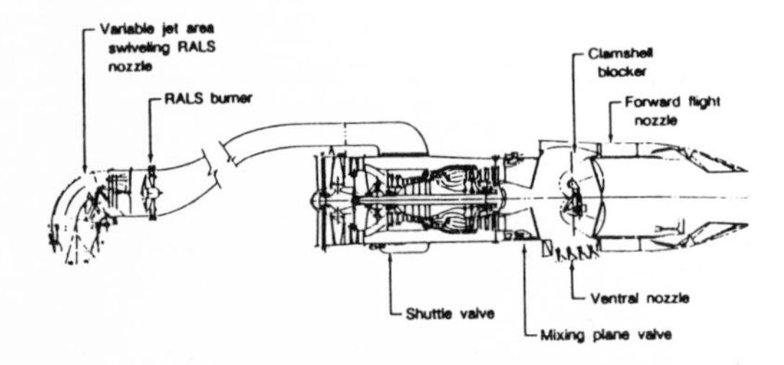

RALS Example

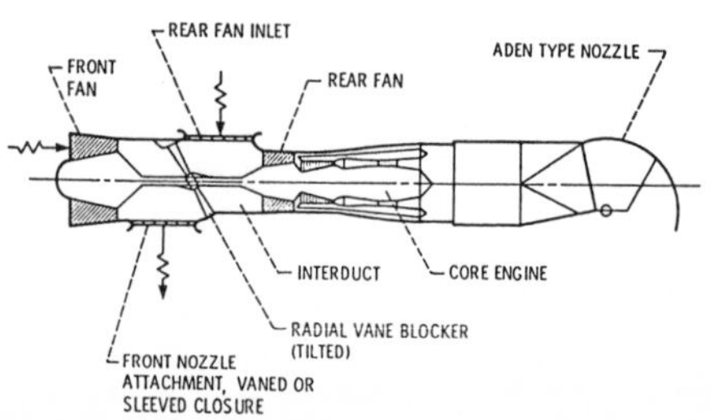

Series/Parallel Tandem Fan (SPTF) in top inlet configuration.

Fig. 8 (continued)

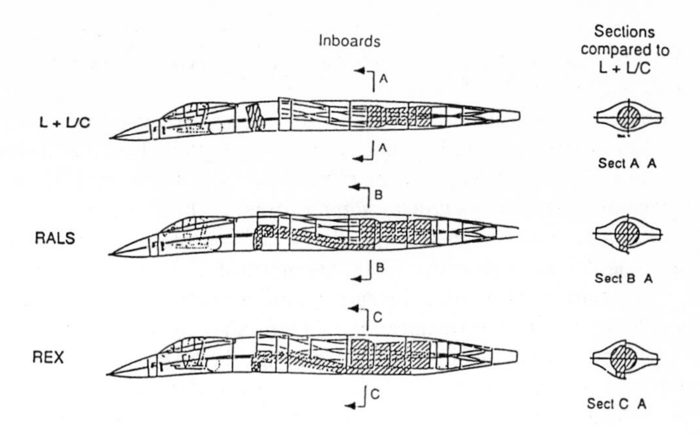

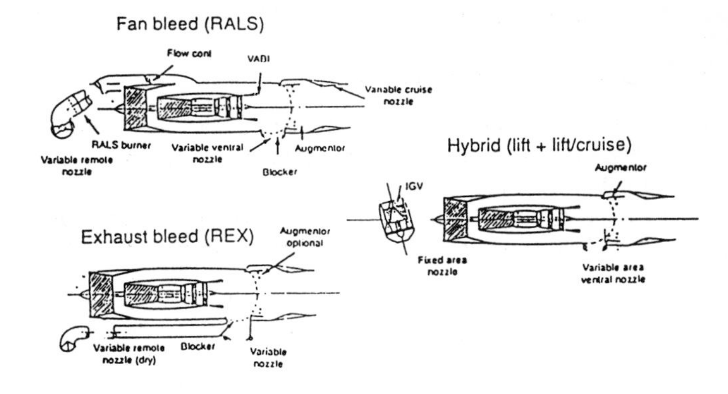

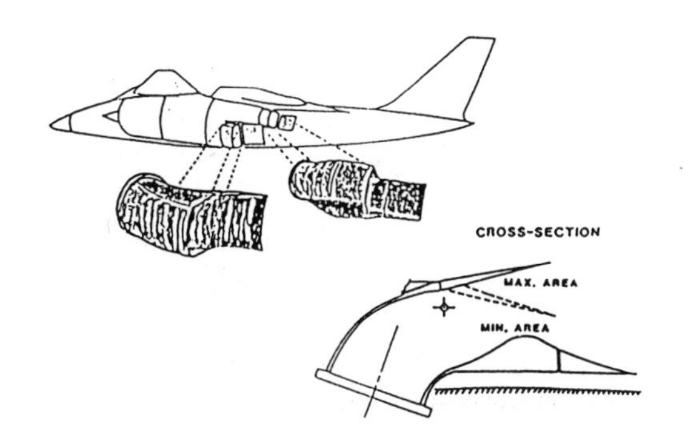

Fig. 8 (continued)

B-9 Current-Technology Limits and Trends

Much has been written on the advantages of the vectoring/reversing nozzles over current-production nozzles. References 1 to 14 give *general information and reviews* concerning the *design and propulsion integration* of various thrust vectoring schemes in fighter applications.

General *aerodynamic* considerations, including the effects of deflected thrust on the aerodynamic characteristics of *canard configurations* and *vertical tail loads*, in single and twin engine aircraft are covered in references 15 to 31.

Performance characteristics of various 2-*D* nozzles are dealt with in references 32 to 42, with reference 42 devoted to the expected performance of the F-18 fighter. The performance of various "ADEN" and "SERN" nozzles is given in references 43 to 47, with 47 devoted to the expected performance of a *forward swept-wing fighter* utilizing thrust vectoring.

Altitude performance of turbojet engines equipped with such nozzles is described in references 48 to 50, while reference 51 deals with the F-100 engine, and 52 with yaw vectoring.

A comprehensive comparison between different types of nozzles may be found in references 53 to 69, while *engine integration* with advanced fighter aircraft is covered in detail in references 70 to 80. Thus, F-15 integration studies are available, among other places, in references 70 to 71.

The *mechanics of flight*, as well as *flight and engine control* considerations, are dealt with in references 81 to 87, while *exhaust plume thermodynamic effects, infrared characteristics and signatures* are covered in references 88 to 92.

Computer programs and *mathematical models* associated with 2-*D* vectoring/reversing nozzles, especially with flow characteristics, are to be found in references 93 to 101.

B-9.1 Current-Technology Cooling Requirements

A major problem in the production of current nozzles is the *cooling requirements* of the converging-diverging flaps and sidewalls of optimal nozzle configurations (cf. Lecture I). While reference 102 describes the effects of shocks on film cooling of a full-scale turbojet exhaust nozzle having an external expansion surface, reference 103 is dealing with the *temperature fields* to be found in such a nozzle, in subsonic and supersonic performance (cf. Lecture III).

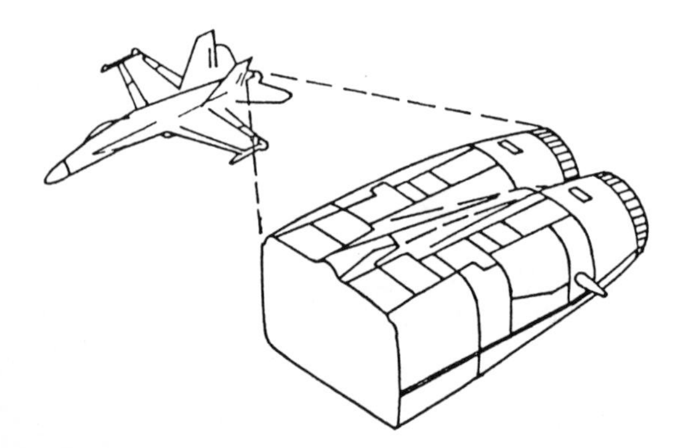

Fig. 9. **F-18 fighter with current-production nozzles.**

Figs. 9 and 10 show the F-18 fighter with current and pitch-only vectoring nozzles, respectively.

Being a non-symmetric flow channel, one expects the 2-*D* nozzle to develop non-uniformities of the temperature field in mid flap, as compared to the nozzle corners. This is clearly indicated in our recent test results (Lecture III). Moreover, *C–D* angles must be optimized to minimize flow-separation phenomena (which cause temperature and thrust distortion).

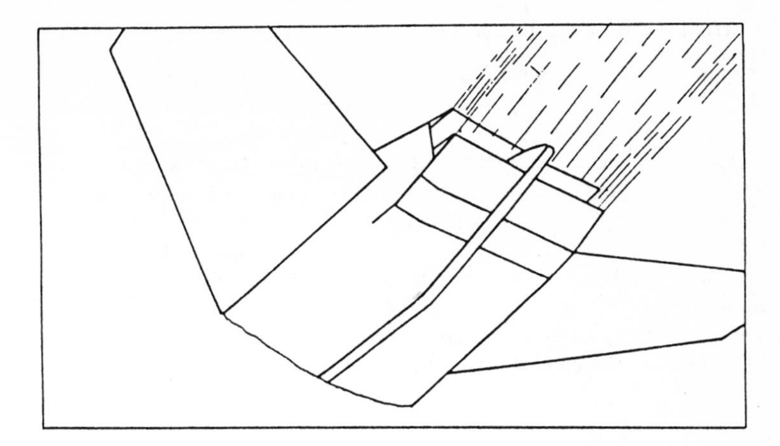

Fig. 10. **Pitch-only, 2-*D*, *C–D* nozzle, as planned for the F-18 fighter (Capone and Berrier, 56).**

B-9.2 Current-Technology Engine Trends

In Table 1, a comparison is made between F-100, F-404, F-110, PW-1128, F-118, and the YF-120, and YF-119 of the new ATF. A number of long-standing trends in engine R&D and design methodologies can be drawn from this Table. These are enumerated below. Further details are given in §B.11.

Thrust to Weight Ratio Trends

According to Table 1, the maximum thrust to weight ratios is expected to increase from 9 to 20 and beyond.

By-Pass Ratio Trends

By-pass ratios in fighter engines tend to be reduced, thereby transforming fighter turbofans into "leaky turbo-jets" [Turbofans may be less suited to supersonic flight than turbo-jets, and, with relatively large by-pass ratios, say 0.7:1, are more susceptible to the well-known "stagnation-stall" phenomena, especially during *A/B* relight at high altitude, low Mach numbers and high rates of turn]. However, V.C. engines bypass the bypass problem (Fig. 3).

Pressure Ratio and Turbine Temperature Trends in Current-Technology Fighter Engines.

Engine pressure ratios remain almost constant, while Turbine Inlet Temperatures (TIT) keep increasing. [Air pressure and temperature at inlet end increase with flight Mach number. Hence, when multiplied by the compressor, they rapidly approach engine pressure-temperature limits, whereby fuel flow must be cut back to prevent excessive TIT. This fuel cut-back limits fighter performance speed and the thrust to weight ratio of a given engine design, unless, of course, its hot part sustains higher TIT values. Higher thermodynamic efficiencies, combined with these considerations, would, therefore, dictate high TIT values. The maximum TIT target for the ATF engine may, therefore, be a few hundreds degrees F higher than that in the current engines, thereby requiring the use of further improved blades and disks – see below.]

B-9.3 Current-Technology Integrated FADEC-Aircraft Controls

Fighter engines continue to change from hydromechanical control systems to various *FADEC systems (Full Authority Digital Electronic Control).* Unlike the hydromechanical control systems these schemes may be fully integrated with aircraft flight controls.

This trend is strongly affected by the increased number of engine variables to be controlled, (especially in the new types of geometrically-variable engines) and by the increased capability of FADEC systems to provide faster (stall-free) accelerations, lower SFC values, safer operation, and higher degrees of maintainability and adaptability to engine failure or degeneration.

Fully-integrated FADEC-cockpit computer systems provide the best solution for advanced fighters equipped with vectoring/reversing nozzles. They free the future fighter pilot from various engine/flight working loads associated with current fighters during take-offs, cruise, combat and landing. (For future trends see Appendix F: IFPC).

TABLE 1. Current-Technology Fighter Engine Trends [119]. (see also § B.11)[+]

	Current production engines				Engines undergoing tests	
Engine Designation	F-100* (PW 220–229)	F-404**	F-110** F-118**	PW1120		Vectored Engines [GE37; YF-120, PW5000; YF-119]
Engine Manufacturer	PWA	GE	GE	PWA	PWA	GE & PWA
Major applications	F-15 F-16	F-18 F-20 117A	F-16 B-2 F-15	F-4?		ATF, ATFN
Year of introduction	1972	1982	1985/8	1987	1990s	1990s
Type	T-F	T-F	T-F	"leaky" T-J	T-F	2-*D* nozzles. V.C. (GE)
T/W maximum (thrust to weight ratio) [year]	7.8	7.5	7.3	7.23	8.5	
Bypass ratio	0.69	0.35	0.78	0.15–0.19	–	
SFC (specific fuel consumption; intermediate (dry))	0.72		0.67	0.8	0.85	
SFC at Max A/B	2.17		2.08	1.86	2.00	
TIT(Maximum turbine inlet temperature), F° [year]	2595	2530	2682	>2550	>2600	<3230 [4000/2000]
Turbine blade type [year]	DS	SC	SC	DS		SC ([RSR/RCC/2000])
PR (pressure ratio	24	25	29	27	27	
Thrust (lb.); intermediate T–0, S/L (dry)	14700	11000	19000	13570	16720	35000
Max A/B, T–0, S/L	23830–29000	16000+ 19000	27500–29080	20600	~32000	
Nominal Airflow, lb/sec	225	140	254	178		
Weight (lb.), dry	3040	2136	3830	2848	3225	
Control System	EEC/ FADEC	H-M[+] Elect.	ACC & H–M/E	ACC & DEEC	ACC & DEEC	ACC & FADEC & IFPC
Includes a vectoring nozzle	No***	No***	No***	No***		yes
Max. diameter, in	46.5	34.8	46.5	40.2	47	
Length	191.2	158.8	181.9	162	209	

[+] All data presented have been accumulated from unclassified US public releases.

* Current production engines are given here for a comparison with the new fighter engine. F-100-PW-229 is an improved version of the F100 engine.

** RM12 is a derivative of the F404. It powers the Swedish Saab JAS 39 Grippen fighter. F110-GE-129 is the improved successor to the F110-GE-100 used on the F-16 fighters. It is flight-tested on an F-15E. F-118 powers the B-2. It is based on the F101/110 core (cf. B-11). F404–F5D2 is to power the A-12 (ATA).

*** Existing axisymmetric exhaust nozzles can, nevertheless, be replaced by Axi-TV nozzles, or by advanced 2-*D* vectoring/reversing nozzles as stressed in the text.

ACC = Active Clearance Control.
DEEC = Digital Engine Electronic Control
FADEC = Full Authority Digital Engine Control
DS = Directionally Solidified
SC = Single Crystal.
RSR = Rapid Solidification Rate
RCC = Reinforced Carbon–Carbon

DEEC and FADEC abilities to diagnose engine problems are now being demonstrated in actual service.

[For instance, a DEEC or a FADEC system may optimize engine performance with a failure in engine fuel pump, or, in the nozzle, and allow the pilot to bring the aircraft safely to base.]

Turbine Blade Pyrometer Systems

Most engines use a set of thermocouples in the low-pressure turbine section to measure gas temperatures. Then, using various schemes, the "calculated TIT" value is indicated and controlled.

General Electric, for instance, has recently introduced into the F-110 engine an optical pyrometer which directly monitors high-pressure turbine blade metal temperatures. However, the blade metal temperature is monitored in the F-110 only for exceedance, i.e., it is not governed as a scheduled value.

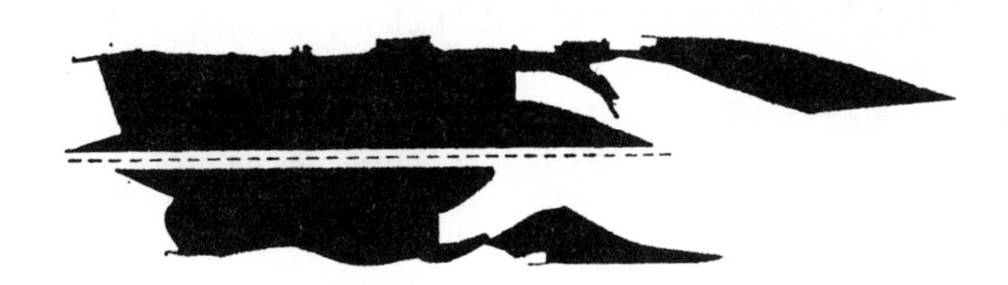

Fig. 11. **Above: The ATF Engine. Below: The Post-ATF Engine is expected to double the *T/W* ratio (to be more than 20:1) and to be much shorter than the ATF engine (as depicted in this comparative-size scheme).**

15 billion US dollars are expected to cause this change. Called Integrated High-Performance Turbine Engine Technology (IHPTET), the program's time-table proceeds till the year 2003.

Note the increasing size of the nozzle in comparison with current-technology core-engines (cf. Figs. I-1, 4, 8, 12, 13 and 14, Table 1, and § B.11).

Afterburner Light-Off Detectors (LOD)

An advanced *A/B* fuel management system, light-off detector, and a dual ignition system, have been introduced by PWA into their PW-1128 engine. GE "LOD system", as used on its F-110 engine, consists of an ultra-violet flame sensor, which views the innermost (flame-holder) "V" ring in the afterburner duct.

To prevent hard *A/B* light-offs at high fuel flows during *A/B* ignition (which may cause fan stall), the F-110 initially allows only a minimal flow rate of the *A/B* innermost "V" ring zone. Only after the flame propagates around the innermost ring, does the LOD system allow higher fuel flows to the afterburner. Hence, irrespective of the pilot's initial *A/B* throttle position, the LOD system provides smooth *A/B* thrust increase.

Bypass Air Combustion Delays

The afterburner of the F-110, for intance, includes a scaled version of the F-101 mixed flow type, with convoluted flow mixer. Here, bypass air, and hot gas flows emerging from the turbines, mix in the plane of the flame-holder, and 90% of the core flow is completely burned, before fuelling of any bypass air is initiated. Such design methodologies provide improved altitude performance, especially during *A/B* relight and acceleration.

Commonality

An increased degree of commonality with current-production engines is stressed by GE and PWA.

For instance, the PW-1128 maintains about 80% parts commonality with the 1985 F-100 production configuration.

Single Crystal Turbine Blades

Increased turbine durability, as well as improved cycle performance, are achieved by employing single crystal turbine airfoil materials (superalloys), with advanced internal cooling methodologies.

Much has been written on these advanced cooling methodologies, and the most recent reviews are available in No. 1 and No. 2 issues of the 1984 Volume of the *Intern. J. of Turbo and Jet Engines.*

Advanced Fracture Mechanics and Bearing Design

Gas turbine bearings operate at relatively high temperatures and rotational speeds. The new trends in advanced aircraft engines require higher bearing speeds, which towards the next decade or so would probably reach 3×10^6 DN [bearing bore in mm multiplied by the shaft speed in RPM].

Current production jet engine speeds are limited to $1.5\text{–}2.3 \times 10^6$ DN (T–700, F-100, JT-9D, F-101, F-404, CF6, etc.). [Jet lubrication is normally used to lubricate and cool the bearing material. Here, the placement, direction and number of nozzles, jet velocity, lubricant type as well as flow and removal rates into and out of the bearing and case, are all critical for obtaining safe operation and long bearing life. Improved bearing structure, materials, coating, and hardness, are also critical in increasing bearing reliability and load capacity. Even trace elements, nonmetallic inclusions, entrapped gases, and especially the processing method, affect bearing lifespans and reliability.]

Much research work is now being directed towards the improvement of fracture mechanics methodology (see also a review by B.L. Koff of PWA in the aforementioned Journal volume on materials and processes as well as in the pumping capacity of the bearing [which includes the design of many radial holes in the race to cool and lubricate the bearing]).

Durability and Survivability

Most technical design decisions associated with the development of advanced fighter engines are based on the following categories:

1. Performance-including survivability, operability, observability and emissions.
2. Reliability, durability, maintainability, and safety.
3. Weight.
4. Cost (of both acquisition and ownership).

For instance, the durability development cycles in turbo engine design include:

1. Mission requirements definitions.
2. Duty cycle definitions (time vs. "Max", "Cruise" and "Idle" variations).
3. "Best" configuration selection (usually after many iterations and reviews involving critical components).
4. Analysis and design.
5. Hardware manufacture.
6. Instrumented laboratory tests verifying the design intent.

7. Flight test evaluations.
8. Accelerated Mission Testing (AMT) and qualification.
9. Field data service evaluations.

The durability cycles are usually combined with such subcycles as *Component Improved Programs (CIP)*, in which field data and proposals emerging from all formal (international) users are combined to improve durability/maintainability/reliability/operability levels, as well as to reduce ownership cost.

The trade-off between cost-effectiveness and durability/reliability/operability is certainly to influence the development of the new ATF, and a whole new generation of ATF derivative engines.

European Fighter Engines

A new (unvectored) *European Fighter Aircraft (EFA)* is now being planned by five European nations (U.K., W. Germany, France, Italy and Spain). It may be propelled by the EJ200.

Early prototypes of the EFA will be propelled by the RB-199 engine. During these early phases of the program the new EFA engine would be developed and tested. The new EFA engine may also serve for re-engining the Tornado ADV fighter. *However, the lack of TV-maneuverability may be a major cause to make the EFA obsolete.*

Meanwhile France's Snecma is further developing its M-88 engine as a candidate future fighter engine. [The M-88 has evolved from experience gained with the M-53 which propels the Mirage 2000 fighter.] As the EJ200, it is unvectored.

The unvectored EJ200 turbofan is based on RR-XG-40 demonstrator (Fig. 6), and is aiming at *T/W* ratio of 10, a compression ratio of 25, and a bypass ratio of approximately 0.4. In contrast to the relatively-complex, 3-shaft RB199, the EJ200 is a 2-shaft engine with fewer parts and a modular design for easier maintenance. It will include single-crystal turbine blades and wide-chord fan blades. The powerplant will be the product of a multinational joint venture known as Eurojet. It brings together RR, MTU, Fiat Aviazione and Sener.

B-10 Whole-System Engineering Approach [119]

More than ever before the new fighter engine design would require the "whole-system-engineering" approach in engine and component R&D and production. It is required because of the increasing need to integrate advanced engines with wing's design, aircraft flight control, *2D* nozzles and variable geometries, and, of course, with the aircraft structure.

Nevertheless, the methodology required for such a practice has not yet been fully developed. Engine designers and manufacturers still tend to leave R&D of such systms to specialized bodies and individuals. As a result, things tend to "fall between the chairs", with poor integration and a resulting performance far below the optimal required.

Designing an engine, or an engine component, in isolation, ignoring these systems and their drastic effects on aircraft and component design, performance/reliability/durability, is no longer accepted around the world. Yet, the effects of these needs on component R&D and production have not yet been fully understood.

Included in this approach are optimal and rapid integration of new results obtained in the various disciplines of engineering, and in materials science and computers. Hence, as TIT values are increased beyond 3000F, new materials, new processes and new cooling methods are required. It is, reasonable therefore, to expect continual government support in the development of critical components such as yaw-pitch vectoring nozzles, FADEC-IFPC systems, RCC, and RSR (rapid solidification rates) metallurgy. In fact, the list of subjects is unlimited. So there is no need to specify here too closely how broad, or how narrow, this list should be at any given time in the evolution of this technology. Instead, one may stress here only the need to develop the awareness and the correct sensitivity to constantly re-educate ourselves towards increasing adaptations of such practice.

B.11 New Fighter Engines

The ATF Engine

As they currently stand, the design criteria for the thrust-vectoring ATF engine include a thrust rating of 35,000 lb, and 40% fewer parts than current-generation engines. The engine features pitch-only, or pitch/reversal, partially-vectoring capabilities (Figs. I-1, 4, 5, 8 and 8 in the Introduction). It is to sustain flight at Mach 1.5, without recourse to afterburners. Nozzle materials and cooling loads would thus become less demanding in comparison with current, *A/B* engines. Moreover, if incorporated in later design versions of this engine, internal yaw-thrust-vectoring vanes would have to sustain lower temperatures and use less cooling air than current *A/B* nozzles. In addition, the development of RCC materials for nozzle and *A/B* systems may remove most, or all of the current cooling penalties (§ B.2).

The GE competing engine for the ATF is known as YF 120-GE-100 (GE37). It began tests in May 1987, and is scheduled to start flight testing in 1990. Thrust vectoring will be added later in the flight program (as 2*D*-TV nozzles with low observables, or as Axi-TV nozzles). According to GE the engine is of variable-cycle design, i.e., the bypass ratio may be set [relatively] high, for subsonic, or low, for supersonic flights (cf. Table 1 and Fig. 3).

The PW competing engine is known as YF 119-PW-100 (PW5000). It began tests in October 1986 (cf. Table 1). The unvectored version will be flight tested in 1990.

The USAF expects first-production engines to be available by the mid-1990s. Post-ATF Propulsion systems are, most likely, entirely different (see the main text).

The F-117A, B-2 and ATA (A-12) Engines

Four, unreheated, F118-GE-100 turbofan engines propel the subsonic, stealth, "flying-wing", Advanced Technology Bomber (cf. Fig. VI-1). These engines are believed to be unreheated derivatives of the GE F101-F110 engine equipped with 2*D* nozzles.

The USAF F-117A is believed to be powered by two General Electric F404 engines.

The other new (USN) fighter program, the subsonic, twin-engine, shipborne, attack aircraft, the ATA, now known as A-12, has already been sewn up by GE. The firm will supply a derivative version of its versatile F404 fighter engine. Another growth version of this engine, the RM12, is being developed by GE in association with Volvo Flygmotor to power Sweden's unvectored J-39 Gripen fighter.

The Up-Rated F-16/F-15 Engines

GE's F110-129, increased-performance fighter engine, has been recently flight-tested in a USAF F-16C. Rated at 29,000 lb. thrust, the F110-129 is the successor to GE's F110-100/F-16 engine, and is competing to power the F-16Cs, and F-15Es, scheduled for delivery to the USAF in 1991.

PW's F100-PWA229, increased-performance fighter engine, has been recently tested in a USAF F-15. Together with GE-F110-129 it competes to power the F-16Cs, and the F-15Es, etc.

The Engines for the USM-AV-8B Harrier and the Rafale

The British *RR* Pegasus 11–61 (pitch-only) vectored engines, are to propel the USM-AV-8B Harrier II, while the new, unvectored, French Snecma M88 engines are designed for a family of common-core engines with thrust ratings of 7.5 to 10.5 tons, with overall compression ratio of 25 and bypass ratio ranges from 0.25 to 0.60. The first unvectored Rafale-*D* fighter aircraft prototype is scheduled to be equipped with M88 engines. It is expected to be flight-tested in 1991. (The present prototypes are propelled by the versatile F404-GE-400 engines.)

The MiG 29 and Su-27 Propulsion Systems

During full afterburning each of the two Tumansky *R*-33*D* produces 18,300 lb. of static thrust. The engine nozzles are of contemporary, axisymmetric, variable, converging-diverging design.

Normal MiG-29 takeoff weight is 33,000 lb. ($T/W > 1.0$) while the maximum value is 39,000 lb. ($T/W < 1.0$).

The engine inlets have a multisegment ramp system that includes a hinged door, which closes during takeoff, landing and ground operations to protect the engines from encountering FOD. Simultaneously, overwing louvered vanes are opened to provide air to the engines. However, additional air still enters the engines through a small gap just above each inlet upper lip. As the aircraft reaches rotation speed, these doors/ramps are opened, i.e., rotate downward and the overwing vanes are closed.

The Su-27 is powered by two Lyulka AL-31F turbofan engines, rated at 27,500 lbs each. Thrust vectoring is currently added to these engines. (For a discussion of the "Pougachev's Cobra" see Fig. 4 in the Introduction.)

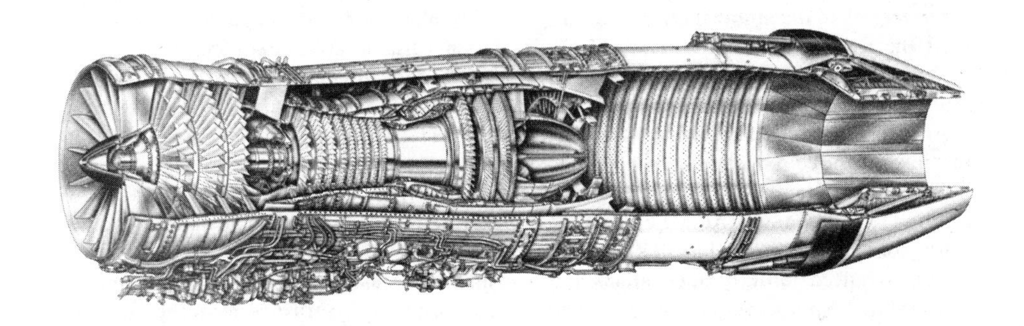

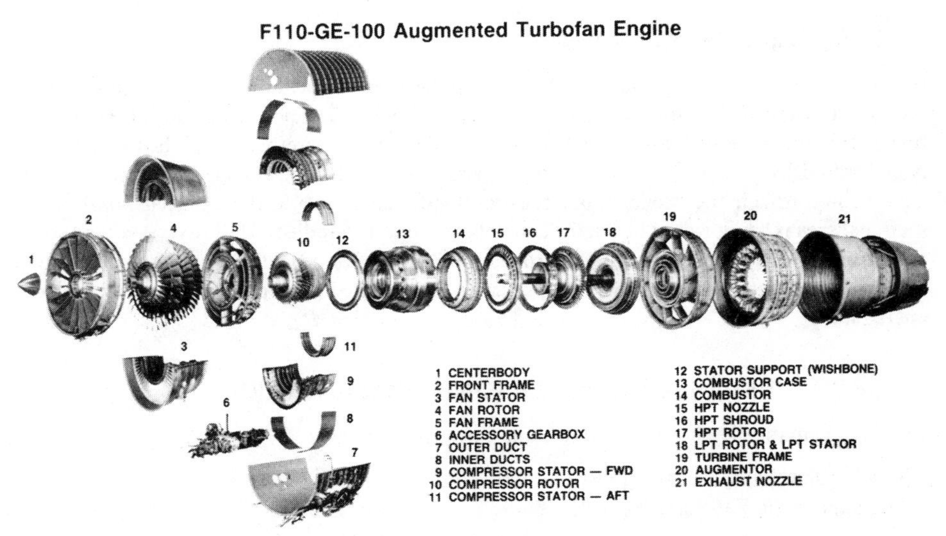

Fig. 11a. **An example of current-technology, unvectored, fighter engine.** See also § B.11 and Table 1.

B.12 THE NEW GE VARIABLE CYCLE ENGINE [See also the ATF engine in parg. B-11]

Very high exhaust jet velocities are required to maintain supersonic flight. For instance, the exhaust jet velocity generated by the Concorde's Olympus engine is about 1800 ft/sec. However, only about 1200 ft/sec may be recommended to reduce exhaust jet noise during takeoffs in populated areas.

The design approach to meet both high and low jet velocities requirements, as well as to improve aeropropulsive efficiences and expand flight envelopes, is the Variable Cycle Engine (VCE) which functions as a **turbofan** during **subsonic** mode of operation, and as a **turbojet** during **supersonic** operation.

There are a number of methods to design VCE. A new GE-VCE design is shown in Fig. 3. It is based on variable-area bypass injectors and a 'Coannular Acoustic Nozzle' [This nozzle may be combined/replaced with a proper thrust vectoring nozzle for advanced, non-augmented, fighter engines]. This GE-VCE functions as follows;

1. During the Subsonic Mode

In this mode the variable-area bypass injectors are open and fan cold-air bypasses the core engine and, then, at the aft turbine section, injected into the core flow portion of the exhaust nozzle, emerging to the ambient air through an **inner variable nozzle** whose exposed geometry redirects the flow in the axial direction [Fig. 3]. It is a variable area nozzle controlled by the side-motion generated by the actuators in the far right of the drawing [inside the central nozzle bluff-body]. The bypass ratio during this subsonic/turbofan mode may vary, say from 2 to 2.5.

An interesting phenomenon has been employed here: Out-directing the hot core gas flow and in-directing the cold bypass air reduces noise [for the hot, external, less-dense, stream mixes better with the internal cold air than the other way around]. The inner variable nozzle is therefore referred to as **'Coannular Acoustic Nozzle'** [Fig. 3]. Moreover, the internal cold air flow cools the inner nozzle parts and reduces IR signatures.

For advanced **military applications** the **'Coannular Acoustic Nozzle'** may be combined/ replaced with a proper [pitch-only, or pitch-yaw,or pitch-yaw-roll] **thrust-vectoring nozzle. In any design the new nozzles acount for about 50% of the overall length of the engine.**

2 During Supersonic Cruise Mode

In this mode the [cold-air] bypass ducts are closed [see the vanes at the lower and upper aft turbine section] and all fan air is forced through the core engine. The powerplant now functions as a turbojet engine. Bypass ratio during this operational mode is, therefore, zero. The turbojet mode is performed by closing the variable-area bypass injector 'doors'. Under these conditions the 'cold' flow through the **'Coannular Acoustic Nozzle'** stops, while the external nozzle cross-sectional area is being varied by engine controls to properly handle the hot gas efflux for optimal conditions [Fig. 3]. Consequently, improved aeropropulsive efficiencies and expanded flight envelopes characterizing **turbojet engines are substantiated during this supersonic mode of operation.**

TURBINE BYPASS AND SUPERSONIC FAN ENGINES

Fig. 12 schematically shows two additional powerplants:

- The Turbine Bypass Engine [TBE].
- The Supersonic Fan Engine [SFE].

In the first type, during full-power conditions, a compressor discharge air flow is bypassed

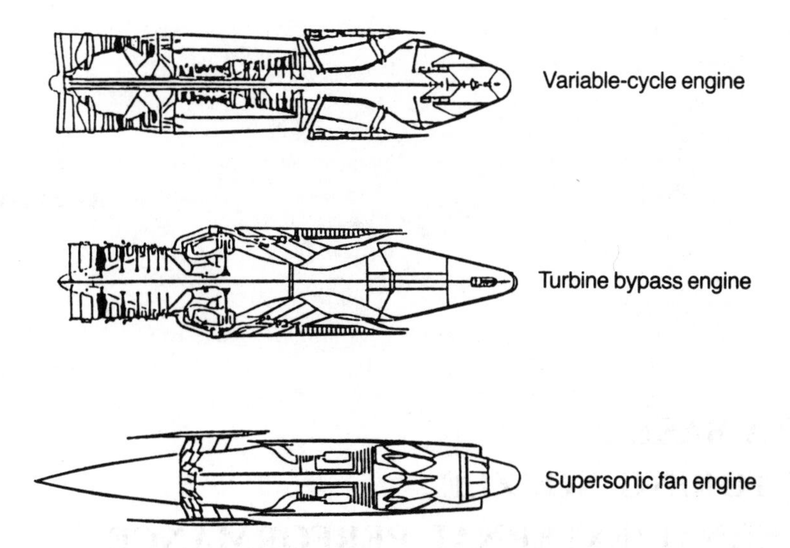

Fig. 12. **VCE, Turbine Bypass and Supersonec Fan Engines.**

around the high-pressure turbine and reintroduced so that it passes through the low-pressure turbine. Thus, the HP turbine performs at near optimum speed throughout the throttle range, which may significantly reduce SFC. This design concept was conceived by Boeing in the late 1970s and subsequently improved by PWA.

In the second type the long and heavy inlet system may be replaced/reduced by a supersonic throughflow fan which, unlike conventional fans, can efficiently handle air at **supersonic internal flow regimes**. This design may result in 25% lower engine weight and about 22% lower SFC values. The potential breakthrough of this design concept lies in the development of a reliable, low-cost, supersonic throughflow fan and its integration with the core engine.

APPENDIX C

DATA BASE-1: VECTORING NOZZLE INTERNAL/EXTERNAL PERFORMANCE

●C-1 Vectoring Nozzle Internal Performance: Low NAR

Re and Leavit (28, 58, 112, 162) have used the NASA Langley static test facility to evaluate the effects of geometric design parameters on *2D–CD* nozzles at NPR up to 12 (see Figs. 1 to 7).

The (machined) nozzles had thrust deflection angles varying from 0° to 20.26°, throat aspect ratios varying from 2.012 to 7.612, throat radii from sharp corners to 2.738 cm., expansion ratios from 1.089 to 1.797, and various sidewall lengths (cf., e.g., Fig. 1).

The results of this investigation (cf. Figs. 2 to 8) indicate that two-dimensional, low NAR convergent-divergent nozzles have static internal performance comparable to axisymmetric nozzles with similar expansion ratios. Nozzle expansion flap curvature (radius) at the throat had

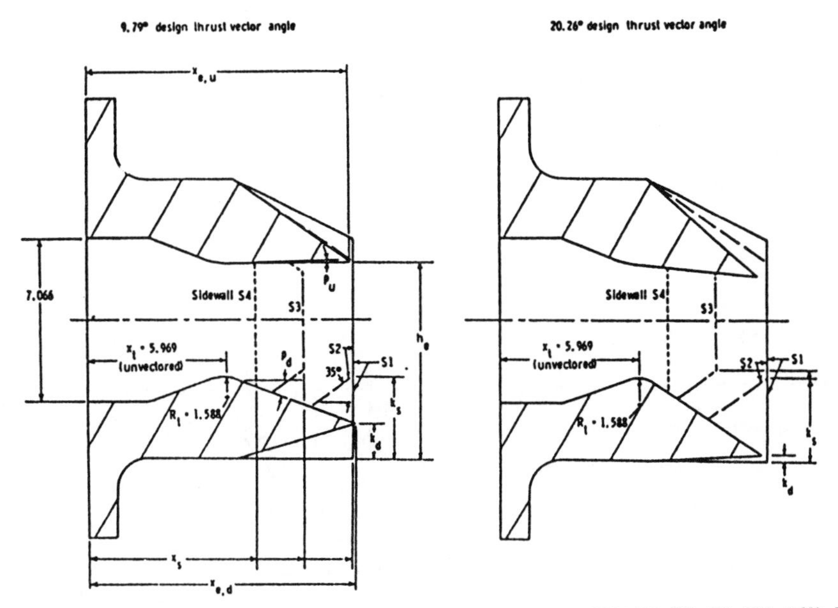

●Fig. 1 **Vectored-nozzle geometry as investigated by Re and Leavit of NASA (28, 58, 112, 162).** Their results are shown also in Figs. 2 to 7. All dimensions are in centimeters unless otherwise indicated.

AR = 3.696 and 7.612.

Configuration	Unvectored A_e/A_t	δ_v, deg	Sidewall	Sidewall Length Unvectored $\frac{x_s - x_t}{x_e - x_t}$	x_s, cm	k_s, cm	$x_{e,u}$, cm	$x_{e,d}$, cm	k_d, cm	h_e, cm	ρ_u, deg	ρ_d, deg
A1V10	1.300	9.79	S1	1.000	11.557	—	11.405	11.608	1.600	8.547	-.08	19.50
A2V10	↓	↓	S2	↓	↓	3.493	↓	↓	↓	↓	↓	↓
A3V10	↓	↓	S3	.614	9.398	3.937	↓	↓	↓	↓	↓	↓
A4V10	↓	↓	S4	.227	7.239	—	↓	↓	↓	↓	↓	↓
A1V13	1.166	13.22	S1	1.000	11.557	—	11.176	↓	↓	7.912	-6.93	↓
A2V13	↓	↓	S2	↓	↓	3.493	↓	↓	↓	↓	↓	↓
A3V13	↓	↓	S3	.614	9.398	3.937	↓	↓	↓	↓	↓	↓
A4V13	↓	↓	S4	.227	7.239	—	↓	↓	↓	↓	↓	↓
A1V20	1.300	20.26	S1	1.000	11.557	—	↓	11.328	.274	↓	↓	33.58
A2V20	↓	↓	S2	↓	↓	3.493	↓	↓	↓	↓	↓	↓
A3V20	↓	↓	S3	.614	9.398	3.937	↓	↓	↓	↓	↓	↓
A4V20	↓	↓	S4	.227	7.239	—	↓	↓	↓	↓	↓	↓

AR = 2.012.

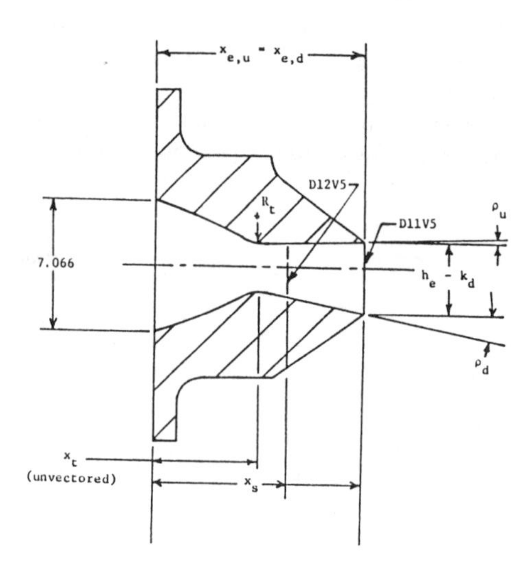

Configuration	AR	Unvectored A_e/A_t	δ_v, deg	Sidewall Length Unvectored $\frac{x_s - x_t}{x_e - x_t}$	x_t, cm	x_s, cm	$x_{e,u}$ and $x_{e,d}$, cm	$h_e - k_d$, cm	ρ_u, deg	ρ_d, deg
D11V5	3.696	1.443	4.82	1.000	5.779	11.557	11.557	3.967	1.28	10.92
D12V5	↓	↓	↓	.253	↓	7.239	↓	↓	↓	↓
F5V5	7.612	↓	4.84	1.000	8.890	11.557	↓	1.926	1.33	11.00

AR = 3.696 and 7.612.

Fig. 1 cont.

little effect on the thrust coefficient, but the discharge coefficient, C_{D8}, decreased by as much as 3.5 percent when the radius was reduced to zero (sharp throat). Nozzle throat aspect ratio, NAR, had little effect on thrust coefficient over the range of nozzle pressure ratio tested. A nozzle geometrically vectored at angles up to 20.26° turned the flow at least as much as the design vector angle, once nozzle pressure ratio was high enough to eliminate separation on the lower expansion surface.

●C-2 The Limitations of Wing Maneuver Devices and Vectoring In Supersonic Fighter Aircraft

Thrust vectoring technology may require incorporation of auxiliary trimming devices such as canard surfaces, strakes, or nose jets (cf. Figs. II-1a to 1d).

Aerodynamic improvements may be achieved through the use of wing maneuver devices such as drooped leading-edge and trailing-edge flaps.

Vortex flow control may also be considered in the design of advanced fighter aircraft.

However, the interaction between these various technologies may *negate* individual performance gains.

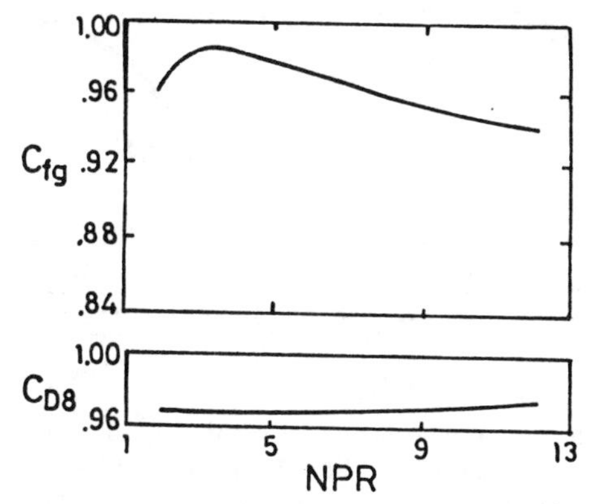

Fig. 2. **Re and Leavit's test results show variation of nozzle thrust and discharge coefficients with nozzle pressure ratio for $AR = 5.8$ nozzle with $A_e/A_t = 1.09$.** (Cf. Fig. III-11a.)

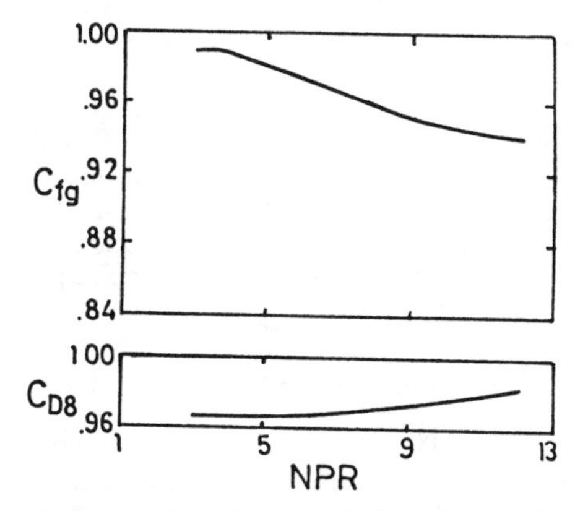

● Fig. 3. **Variation of nozzle thrust and discharge coefficient with nozzle pressure ratio for $AR = 7.6$ nozzle with $A_e/A_t = 1.09$ (Re and Leavit, NASA).**

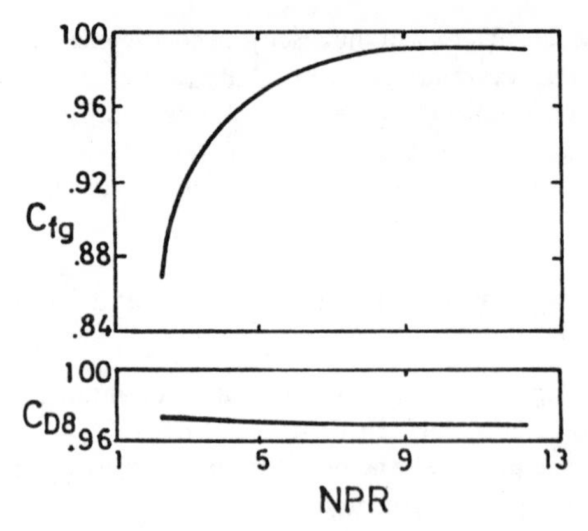

●Fig. 4. **Variation of nozzle thrust and discharge coefficients with nozzle pressure ratio for $AR = 5.0$ nozzle with $A_e/A_t = 1.8$ (Re and Leavit, NASA).**

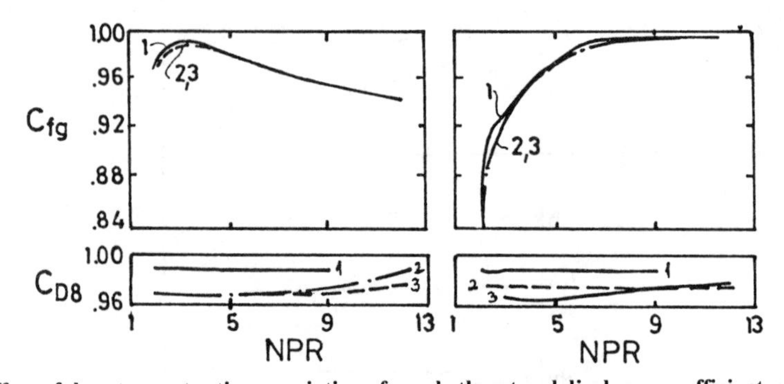

●Fig. 5. **Effect of throat aspect ratio on variation of nozzle thrust and discharge coefficients with nozzle pressure ratio for low and high expansion (cf. Fig. III-11a, Re and Leavit, NASA).**
Key: Left Figure: $A_e/A_t = 1.09$. Right Figure: $A_e/A_t = 1.8$. 1) $AR = 3.7$; 2) $AR = 5.8$; 3) $AR = 7.6$

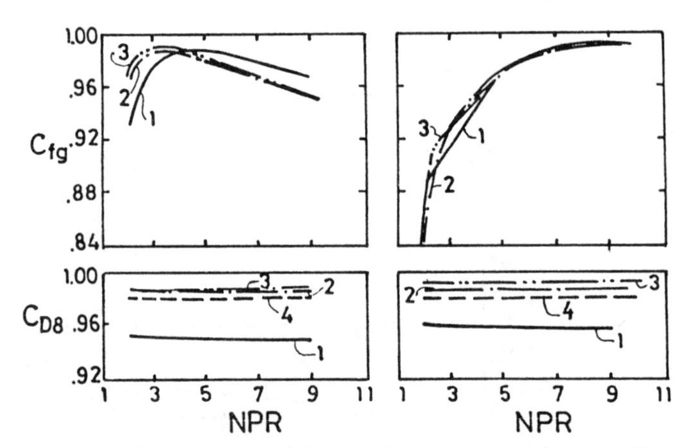

●Fig. 6. **Effect of throat radius on the variation of nozzle thrust and discharge coefficients for low and high expansion ratio nozzles.** [$A_e/A_t = 1.09$ (left figure) and $A_e/A_t = 1.8$ (right figure), $AR = 3.7$].
Key: 1) $R_t = 0$; 2) $R_t = 0.68$ cm; 3) $R_t = 1.59$ cm; 4) $R_t = 2.74$ cm. (Re and Leavit, NASA).

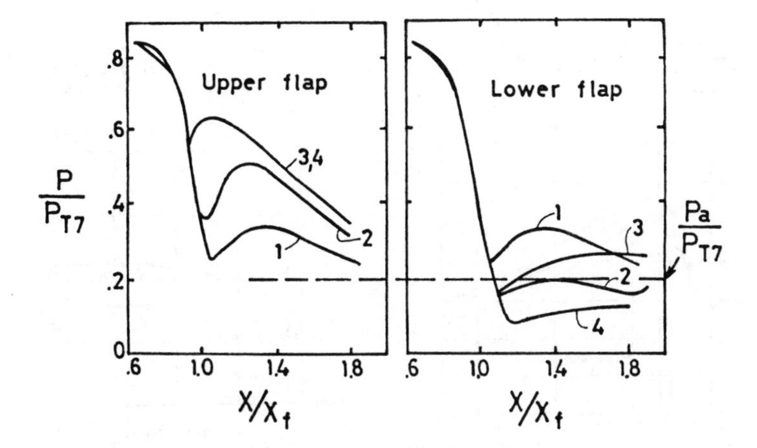

Fig. 7. **Effect of nozzle flap angle on flap centerline pressure distribution with full-length sidewalls (Re and Leavit, NASA).**

Key:

X/X_f-dimensionless distance along flap walls.

1) Pressure distribution on flaps with no vectoring ($\delta_v = 0$).

2) $\delta_v = 9.8°$, $A_e/A_t = 1.3$, NPR = 5.

3) $\delta_v = 13.2°$, $A_e/A_t = 1.17$, NPR = 5.

4) $\delta_v = 20.3°$, $A_e/A_t = 1.3$, NPR = 5.

(cf. Fig. I-6).

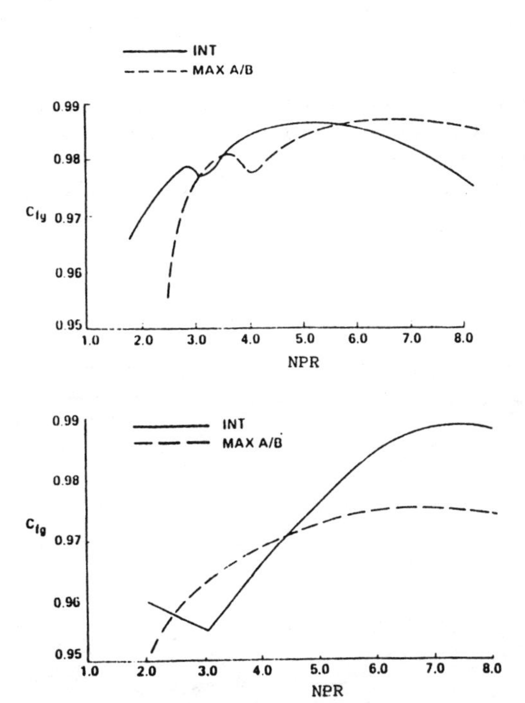

●Fig. 8. **Variation of the GE F-404 axisymmetric nozzle thrust coefficient C_{fg} with NPR for static, S.L. conditions, at "Intermediate" (INT) and maximum afterburning (Max A/B) power settings.** The lower graph represents similar tests but with the *2D ADEN* nozzle shown in Fig. A-4.

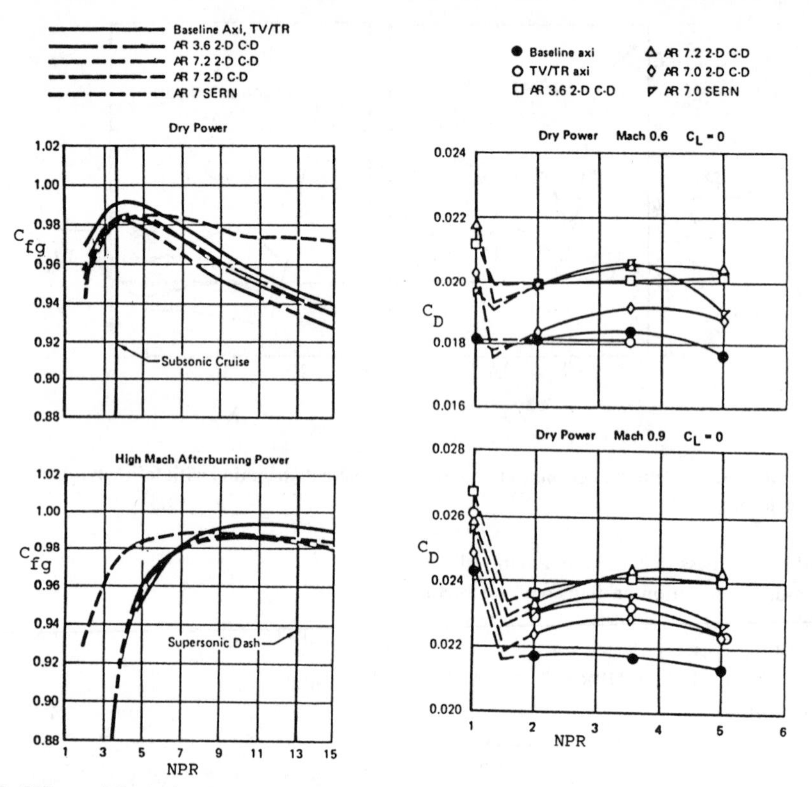

●Fig. 9. **Hiley and Bower's test results show the effect of *AR* on C_{fg} for an *AR* range between 3.6 and 7.2 [11].** The experimental results shown on the right-hand side are for a *total configuration installed drag for unvectored flight with various vectoring nozzles.* The results depicted show that the drag values for *2D–CD* nozzles are less than those of current-technology nozzles. However, this is probably not a general rule.

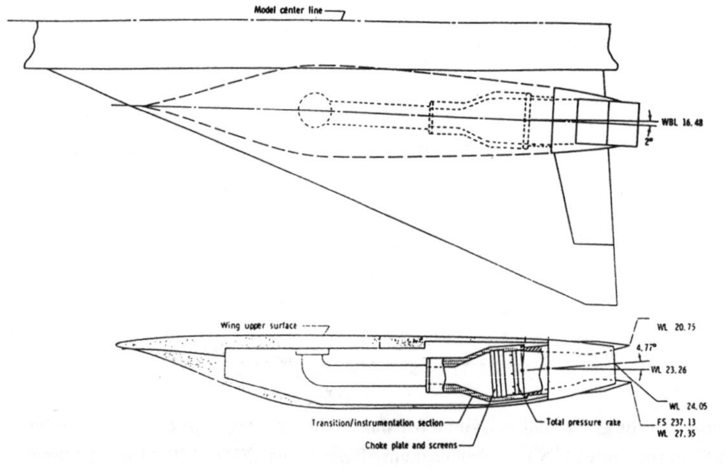

●Fig. 10. **NASA nacelle/nozzle installation.** All dimensions are in centimeters [Capone, 1983].

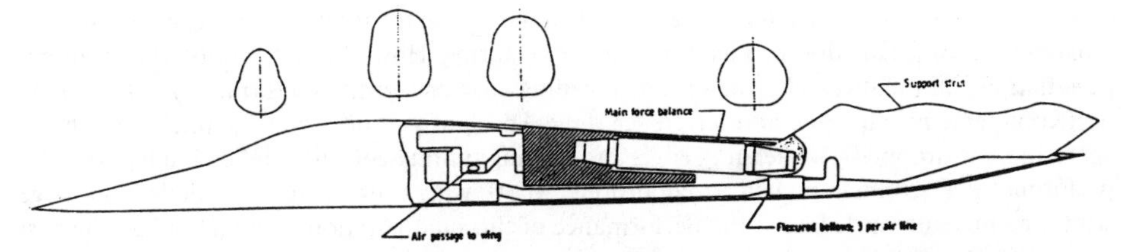

●Fig. 11. **Body arrangement and internal flow hardware in the NASA tests [Capone, 1983].**

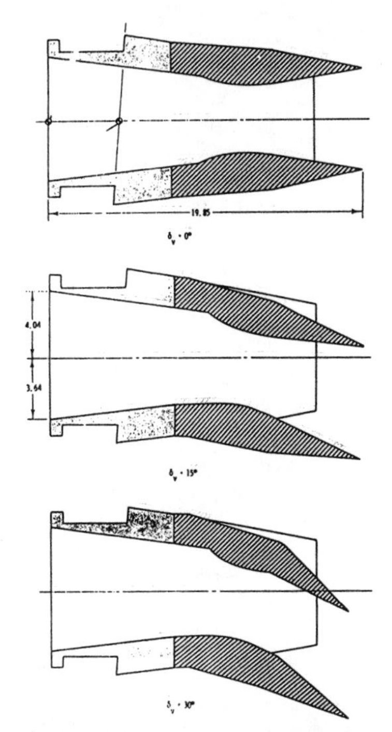

Fig. 12. **The *2D–CD* nozzle configurations in the NASA tests [Capone, 1983].**

To evaluate such interactions Capone, of the NASA Langley Research Center, has conducted the following instructive study [1983]:

The wing maneuver devices consisted of a drooped leading edge and a trailing-edge flap. Thrust vectoring was accomplished with *2D–CD* nozzles, located below the wing in two single-engine podded nacelles. A canard was utilized for trim pitch-only. Thrust vector angles of 0°, 15°, and 30° were evaluated in combination with a drooped wing leading edge and with wing trailing-edge flap deflections up to 30°. The investigation was conducted at Mach numbers from 0.60 to 1.20, at angles of attack from 0° to 20°, and at nozzle pressure ratios from about 1 (jet off) to 10. Reynolds number based on mean aerodynamic chord varied from 9.24×10^6 to 10.56×10^6.

The mutual interference effect of deployment of the drooped leading edge in conjunction with thrust vectoring was beneficial to untrimmed drag-minus-thrust polars, because an additional drag reduction was obtained that was greater than the sum of the individual drag reductions due separately to either the drooped leading edge or vectoring. However, deflection of the trailing-edge flap in combination with the drooped leading edge and thrust vectoring caused an unexpected increase in incremental interference drag! The configuration with 15° pitch-only thrust vectoring, the drooped wing leading edge, and 7° wing trailing-edge flap deflection had *the best* performance at trimmed maneuver conditions. At 30° wing trailing-edge flap deflection, large trim drag increments degraded the performance of this configutation, although it had the best untrimmed performance.

The wind-tunnel models employed by Capone are shown in Figs. 10 to 12.

APPENDIX D

DATA BASE – 2: SYNERGISTIC EFFECTS

D-1 Synergistic Effects of Thrust-Reversal

D-1.1 Thrust Reversal Effects in Landing Procedures

Some aircraft make a constant AoA approach until touch-down, while others have to flare to avoid an undercarriage disaster.

Thus, high lift, deceleration devices, and undercarriage strength, must be designed and adjusted to the landing requirements.

Deceleration devices include thrust-reversal and aerodynamic braking. However, during the ground roll the aerodynamic brakes are ineffective and at low landing speeds they become unsafe.

Consequently, optimum use of thrust reversal may become an important design criterion in the development of STOL aircraft, especially when short field landing techniques are required at the low approach speeds and steep glidepath which characterize vectored aircraft. (See, however, our reservations in Figs. 12 and 21, Introduction.)

The synergistic effects of thrust-reversal also include high vs. low pilot's work loads, and hot-exhaust-gas reingestion as depicted in Fig. 1. This figure demonstrates the interaction between forward speed and the ground-reflected hot exhaust gases streaming from the lower part of the thrust-reverser port to the ground.

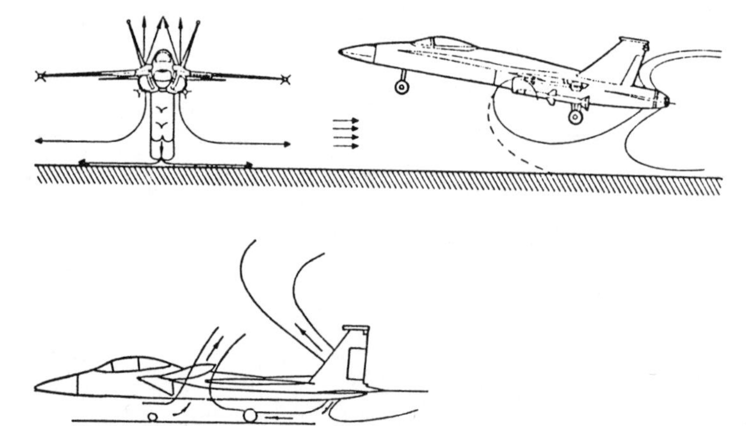

Fig. 1. **Hot-gas reingestion through engine inlets is caused by ground effects in improper use of *TR* during landing.** Hence, one must shut-down *TR* below a given ground speed. Other asymmetric effects caused by the upper and lower hot jets are discussed in the text and depicted in Figs. 10 to12.

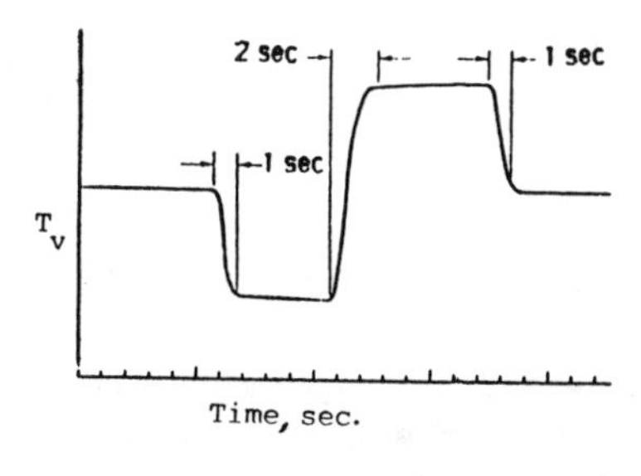

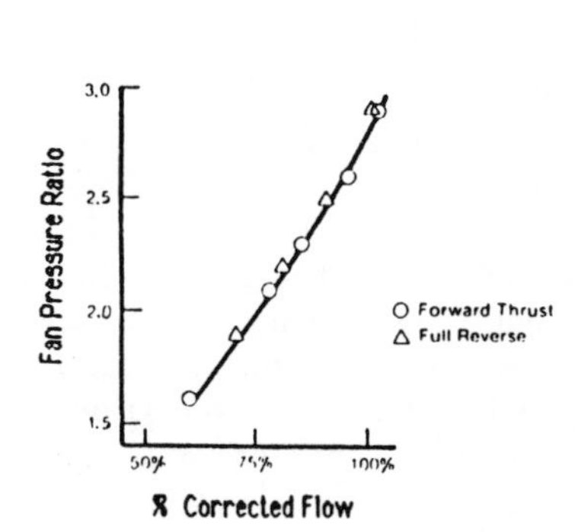

Fig. 2. **Typical pitch vectoring response times (51).**

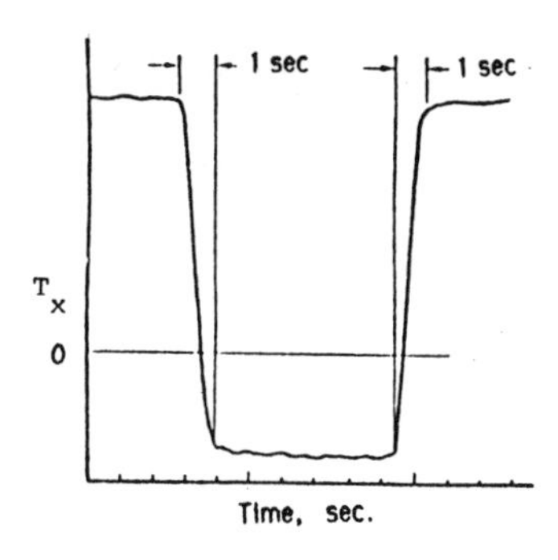

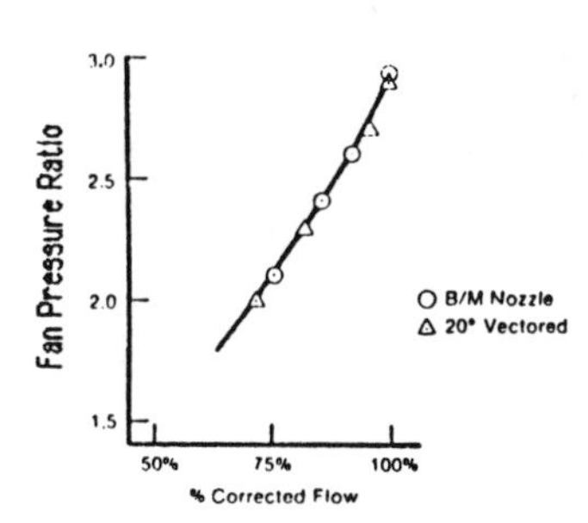

Fig. 3. **(left) Typical reversed thrust response times** (minus values of T_x represent TR) (51).
●Fig. 4 . **(right) Reversed thrust (top) and vectored thrust (below) stabilities are high as shown by the depicted performance with bellmouth inlets and T_x, T_y full-transient variations [51].**

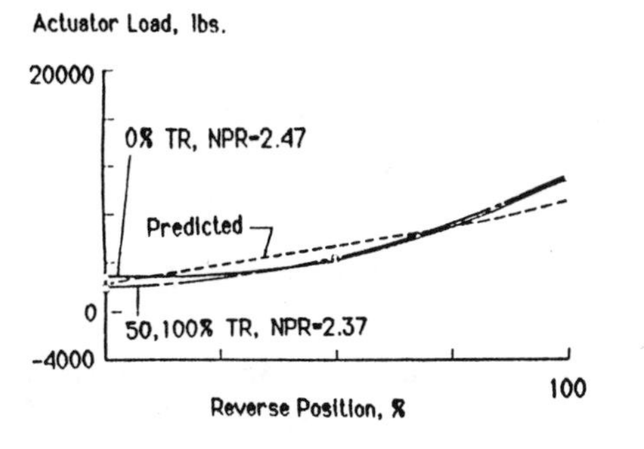

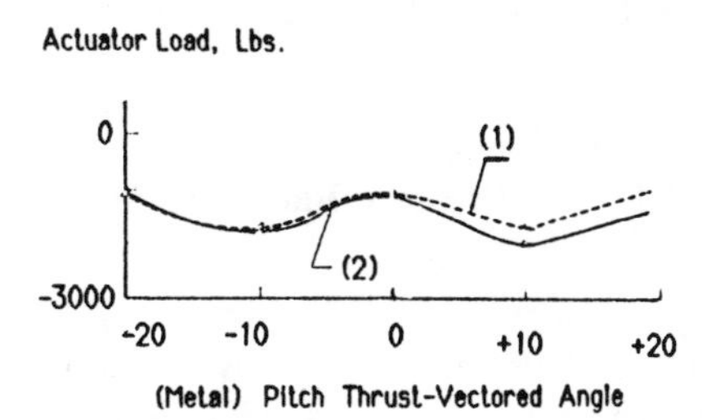

●Fig. 5. **The variations in the force loads on *2D–CD* actuators during pitch vectoring (right) and thrust reversing (left) are shown to be quite moderate.** (1) Fixed load calculated from wall pressures. (2) Average actuators load (51).

As far as the aircraft ground roll speed V_g is subject to the condition

$$V_g > \bar{V}_{\text{reflected forward jet}}, \tag{1}$$

the *TR* device may remain operative.

However, when

$$V_g \leqq V_{\text{reflected forward jet}}, \tag{2}$$

the ground-reflected hot gases may enter the engine intake and cause engine stall, instability, or flame-out. This situation must be avoided, and, consequently, *TR* must be shut-down before the condition described by eq. (2) is reached. Moreover, the stagnation point generated just below the aircraft inlets may cause foreign objects (and rain water) to be air-suspended from the ground and hit engine compressor blades, causing "foreign-objects" damage.

D-1.2 Hot Gas Ingestion

A one-tenth scale, vectored thrust McDonnell Douglas 279-3 supersonic STOVL model was used in a wind tunnel in 1987 to investigate the effect of ground proximity and airframe-mounted flow diverters on hot-gas ingestion.

To permit a wider range of hot-gas-reingestion investigations, NASA-Lewis has prepard a new system which allows variations in model height, angle of attack, pitch, roll and yaw. The operating temperatures in this system may be as high as 1000F.

Replacing vertical TR jets by horizontal (yaw) TR jets has been proposed recently (179) for the elimination of the ingestion problem.

Another design factor is the deployment time during thrust vectoring and reversing (see below).

D-2 Deployment Times Limitations During Thrust Vectoring and Reversing.

1-second maximum deployment time for full-range pitch thrust vectoring, and for thrust-reversing, has emerged as the minimal requirement. However, shorter deployment times may be demonstrated by a few designs (179).

Fig. 2 shows typical pitch vectoring response times. Thus, the vertical force component T_v may be changed from 0 to −20 degrees in 1 second time interval, and from −20 degrees to +20 degrees in 2 seconds (51).

Fig. 3 shows typical reversed thrust response (minus values of T_x represent thrust-reversal). Thus, the transition from full forward thrust to full thrust-reversing should not exceed 1 second. Note that, due to inherent space limitations, the thrust reversal efficiency is normally less than 67%.

D-2.1 Fan Pressure Ratio and Actuator Loads Variations During Thrust Vectoring.

Fig. 4 shows that thrust vectoring variations, and thrust-reversing, may have negligible effects on fan pressure ratio (51). This is demonstrated by the negligible fan pressure ratio change for the thrust-vectoring variations and thrust-reversing shown in Figs. 2 and 3. Here fan-pressure stability is demonstrated by a performance comparison between bellmouth (unvectored) and 20 degrees vectoring deployment operations.

An example of the variations expected in the force loads on 2D–CD actuators is depicted in Fig. 5.

D-3 Nozzle Efficiency During Thrust-Reversal.

Fig. 6 demonstrates decreasing thrust-reversing efficiencies with increasing NPR values. This phenomenon is almost diametrically opposed to forward-thrust-nozzle efficiency changes with NPR, up to a critical value of NPR (cf. Fig. 13, Introduction).

These characteristics depend on NAR and on other nozzle variables. However, they clearly demonstrate internal nozzle stagnation-point effects associated with thrust-reversal turning of the flow (normally about 135 degrees).

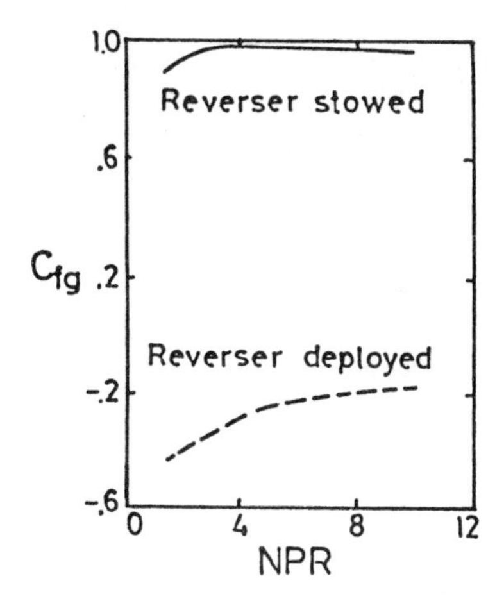

Fig. 6. **Thrust-reversing efficiency variations (NASA, Capone and Berrier, Ref. 56).** Note the decrease in thrust-reversing efficiency with increasing NPR values. (These findings should be compared with those shown in Figs. 7a and 7b.)

Unlike the decrease of C_{fgR} with increasing values of NPR, and the low efficiency values of thrust-reversal reported by Ref. 56 (Fig. 6), Ref. 11 reports much higher C_{fgR} with increasing NPR (cf. Fig. 7a). Thus, at NPR = 3.5, $C_{fgR} = -0.7$, and is not less than -0.6 during the entire range of subsonic-supersonic mode of operation. Obviously, the efficiency depends on nozzle geometry and internal design factors, such as NAR.

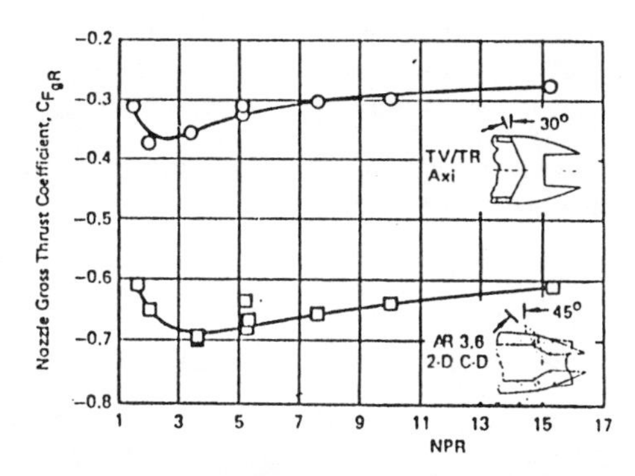

Fig. 7a. **Unlike the drastic decrease of C_{fgR} with NPR, and the low efficiency values of thrust-reversal reported by Ref. 56 (Fig. 6), Hiley and Bowers [11] report much higher efficiency values, and only a moderate decrease of C_{fgR} with increasing NPR values [11].** (see also Fig. 7b.)

It should be noted, however, that the comparison made in Fig. 7 between axisymmetric nozzle and an $AR = 3.6$ *2D–CD* nozzle, is slightly misleading, for the former deflection angle is limited to 30 degrees, while that of the latter to 45 degrees. Further experimental work may therefore be required to finally assess this problem.

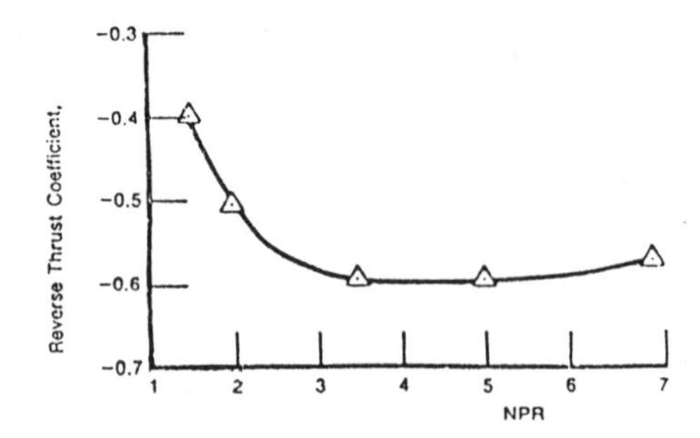

Fig. 7b. **The experimental data of this reference are more in line with those of Hiley and Bowers [11] (Fig. 7a) than with those reported by Ref. 56 (Fig. 6).** The data are for a *2D–CD* vectoring/reversing nozzle [Stevens, Thayer and Fullerton, 49]. (See, however, the text.)

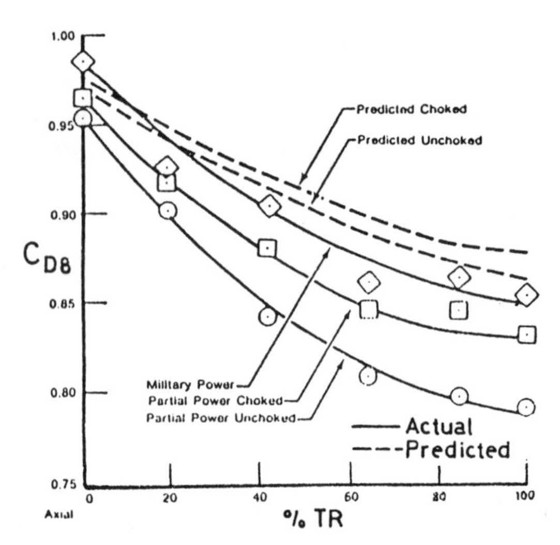

●Fig. 8. **Variations of the discharge-coefficient during transition from axial thrust to full thrust-reversing [Callahan, 158], at different power levels.**

●D-4 C_{D8}-Variation With Temperature and Thrust-Reversing

The definition of C_{D8} is given by equation 7 in Lecture III. Various experimental values of C_{D8} during forward thrust with low *AR*, *2D–CD* nozzles (and different A_e/A_t ratios), are shown in Appendix C–9, including the effect of δ_v-vectoring on C_{D8} values.

As expected, the values of C_{D8} should be lower during thrust-reversing, mainly because of boundary-layer thickening effects near, and around, the internal stagnation point. This is dem-

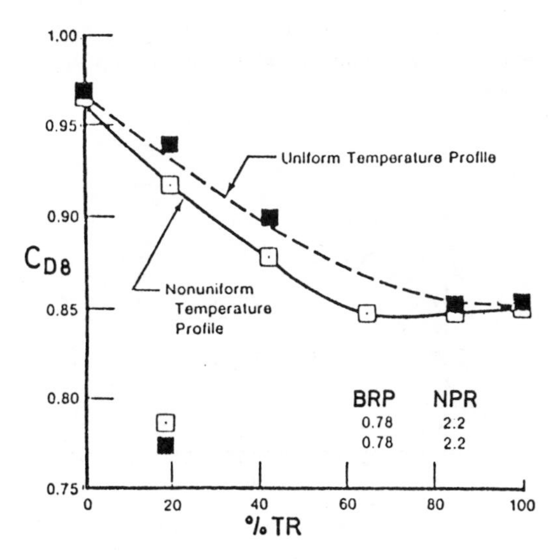

●Fig. 9. **Effect of temperature on the discharge coefficient [Stevens, Thayer and Fullerton, 49].** Non-uniform temperature profiles reduce the values of the discharge coefficient during various deployments of thrust-reversing at a constant NPR value.

onstrated in Figs. 6 and 7. Figs. 8 and 9 show the effects of thrust-reversing on C_{D8}, including the actual behavior in non-uniform exhaust-gas temperature distributions.

These figures show that the values of C_{D8} decrease with increasing NPR values (higher percentage of "full reverse").

●D-5 Asymmetric, Aircraft Tail/Thrust-Reversal Effects

Northrop low-speed wind tunnel facilities had been employed by Glezer, Hughes and Hunt (157), to evaluate thrust-reverser effects on the aircraft-tail-surface aerodynamics of an F-18-type configuration at conditions representing approach phases.

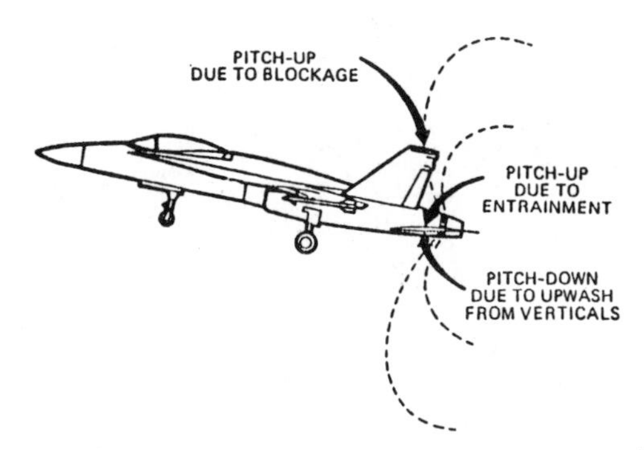

Fig. 10. **Reverser-induced, asymmetric effects contribute to changes in longitudinal stability (Miller, 8).** These asymmetric, aerodynamic forces generate pitch-up and pitch down moments on the aircraft. IFPC rules must, therefore, incorporate proper responses to these effects (see also Figs. 1, 11, 11a and 12).

A major feature of these tests was a systematic buildup of reverser jets and tail surface components to identify and understand the major aerodynamic forces which the reversers generate on the aircraft-tail surfaces.

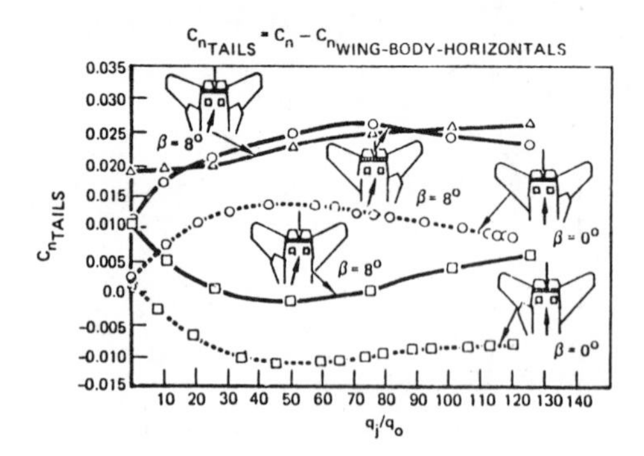

●Fig. 11. **Reverser-induced yawing moments on vertical tails for $\alpha = 0°$ [157].**

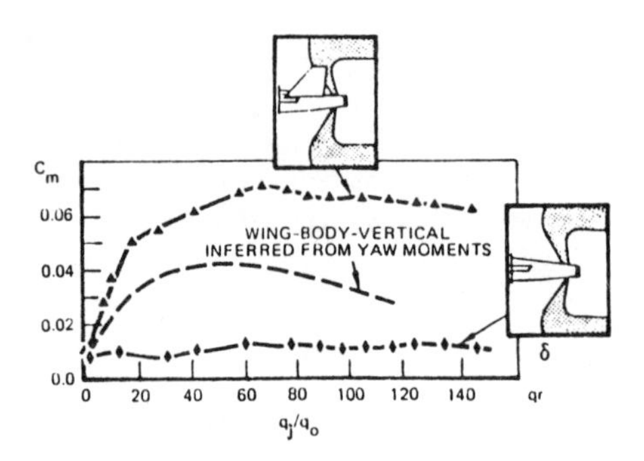

●Fig. 11a. **Asymmetric pitching moment as a function of *q*-ratio.** $\alpha = 0°$, $\beta = 0°$ [Glazer, Hughes and Hunt, 157].

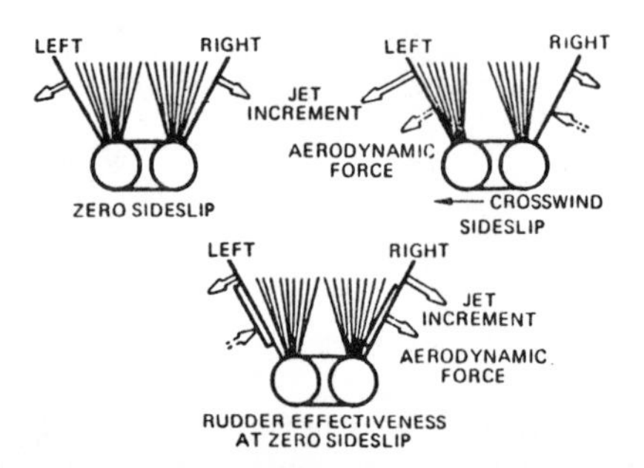

●Fig. 12. **Sketch of jet asymmetric interactions with aircraft vertical tails, rear views [Glazer, Hughes and Hunt, 157].**

It is shown that the upper jets may produce an aerodynamic "blockage" between the two *vertical tails* (Fig. 10). The induced aerodynamic effects generate pitch-up and pitch-down moments, which, in turn, change the longitudinal stability of the aircraft.

In sideslip flight, or with the rudders deflected, this "blockage" is asymmetric, and results in increases in directional stability and rudder effectiveness.

The reverser jets may induce a strong "entrainment flow" (Fig. 10) on the horizontal tails, which amplifies the tail load, resulting in either a pitch-up, or a pitch-down effect, depending on the tail setting (cf, Fig. 11).

However, installed in a twin-engined fighter, with a single vertical tail, the reversers may cause smaller effects on longitudinal stability. Nevertheless, they still produce an incremental pitching moment, changes in directional stability, an increase in horizontal tail effectiveness and a decrease in vertical tail effectiveness.

Since the vertical tails in the F-18 fighter aircraft are canted 20 degrees outwards, the *yaw moments* measured on the individual vertical tails may be associated with a *downward* force, and hence, with a *pitchup* moment (Fig. 10).

D-6 Swirling Flows in Fixed, Rectangular Engine Nozzles

An interesting experimental and numerical investigation was reported recently by Sobota and Marble [177]. In their experimental studies the authors employed the blower/flow-conditioners shown in Fig. 13, while the numerical study has been based on the computational grid shown in Fig. 17. Their interesting results are shown in Figs. 14 to 16 and 18 to 23.

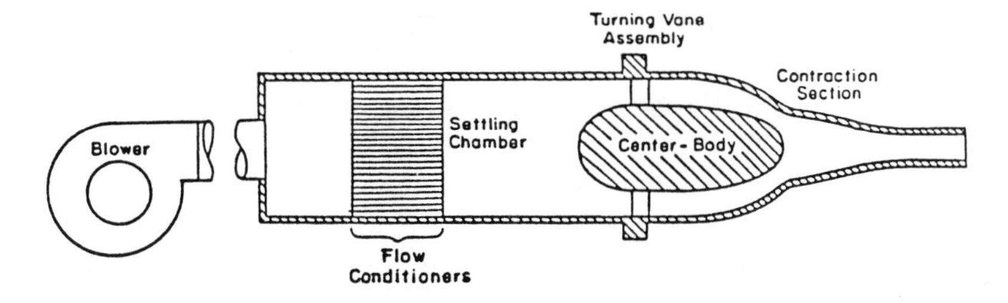

Fig. 13. **The Experimental System of Sobota & Marble [177].**

As expected, the results indicate that the flow-field between the turbine discharge, and the nozzle exit, is highly influenced by the vorticity distribution at the turbine exhaust. Such information may help the engine/nozzle designer to "tailor" the vortex distribution at the nozzle exist by a proper design of the turbine discharge and the intervening ducting and flow dividers. It may be of great interest to see the expansion of this pioneering work into the variable-vectoring-nozzle-domain.

Some of the Sobota-Marble conclusions are given below:

(i) – The use of steady-state, laminar, incompressible, Navier-Stokes equations, on a digital computer, can adequately model the qualitative features of the flowfield (under a proper set of operating conditions).

(ii) – In the low and high swirl cases, the majority of the vorticity may be introduced in the neigh-

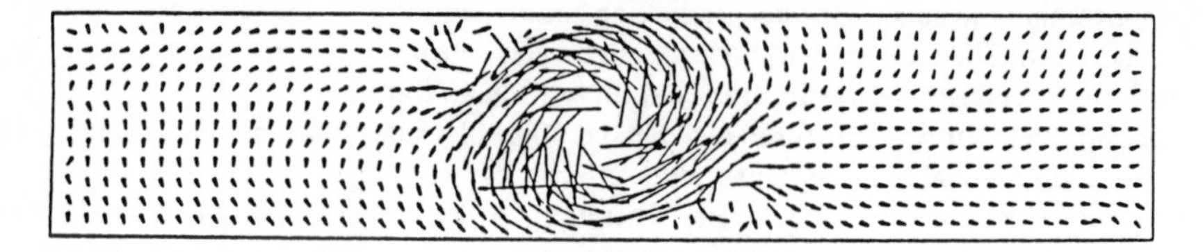

Fig. 14. **The Cross-Flow Velocity Vector Plot, 15 Degrees Blade Angle (Measured) [177].**

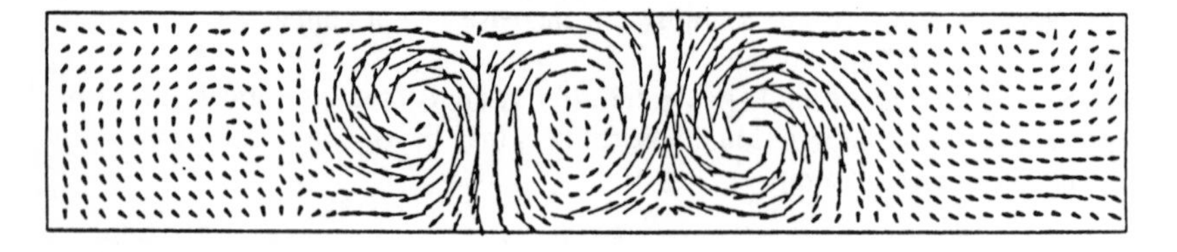

Fig. 15. **The Cross-Flow Velocity Vector Plot, 30 Degrees Blade Angle (Measured) [177].**

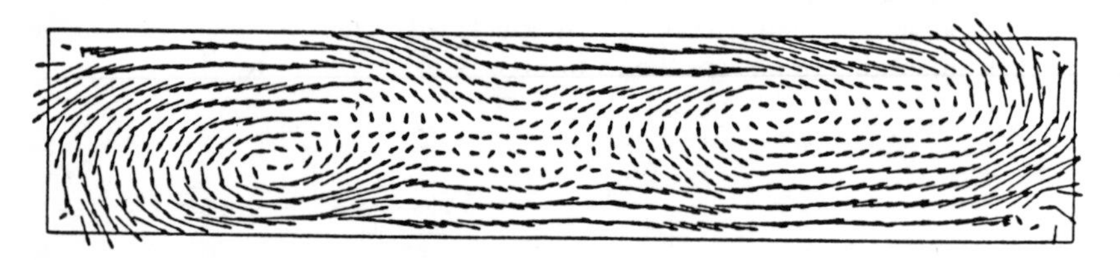

Fig. 16. **The Cross-Flow Velocity Plot, Split Blades (Measured) [177].**

borhood of the blade root and the blade tip in the boundary-layers on the centerbody and the outer wall, respectively. The axial vorticity introduced at the blade-root may be convected along the centerbody to the downstream tip, and, then, along the centerline of the transition section. This results in the formation of a strong vortex along the central axis of the duct. The pressure gradient established on the outer wall of the duct, as a result of the strong central vortex, may promote the separation of the boundary-layer from the outer-wall, into the main stream. This separation is evident in the formation of two streamwise vortices, one on each side of the central vortex, which have a sense opposite to the sense of the central vortex.

(iii) – In another test case, a set of blades that introduce swirl into only the outer-half of the inlet-annulus is used. This, in effect, introduces a cylindrical sheet of streamwise vorticity into the flow at the inlet, at a radial-position halfway between the centerbody and the outer-wall. There is also axial vorticity of the opposite sense introduced into the boundary-layer of the outer wall. In this case the separation of the boundary-layer on the outer wall does not occur to a significant extent, but is apparent in the formation of two small, streamwise vortices in a diagonally-opposed corners of the rectangular, exit-cross-section. The cylindrical vortex sheet, when convected through the annular to rectangular transiton-section, is distorted in such a way as to conform to the shape of the outer-wall of the duct. In the regions where the vortex sheet is most distorted, it tends to roll up. The result is the formation of the two axial, streamwise vortices, at the exit plane of the duct.

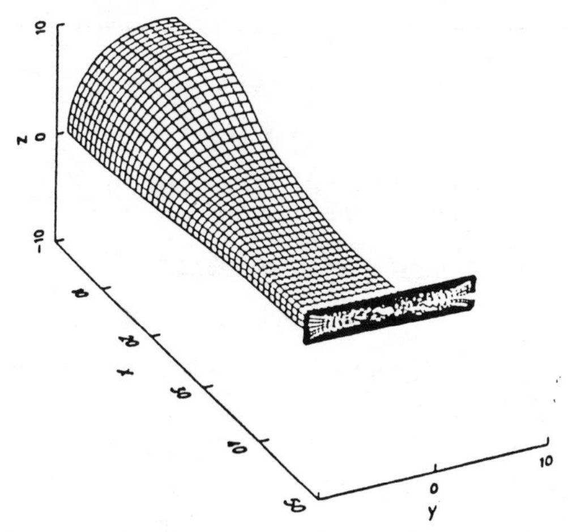

Fig. 17. **The Outer Surface and Exit Plane of the Computational Grid employed by Sobota and Marble [177].**

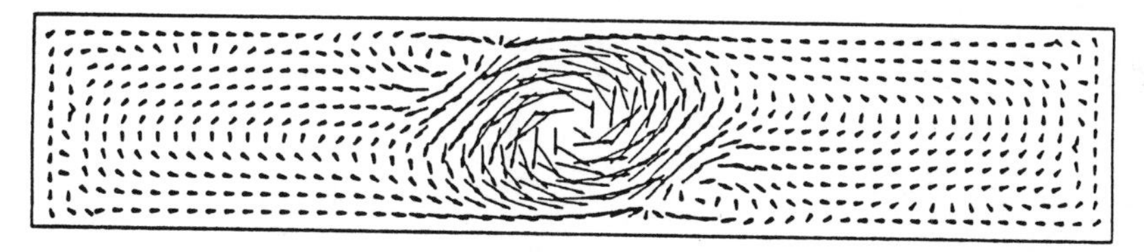

Fig. 18. **The Cross-Flow Velocity Vector Plot, 15 Degrees Blade Angle (Computed) [177].**

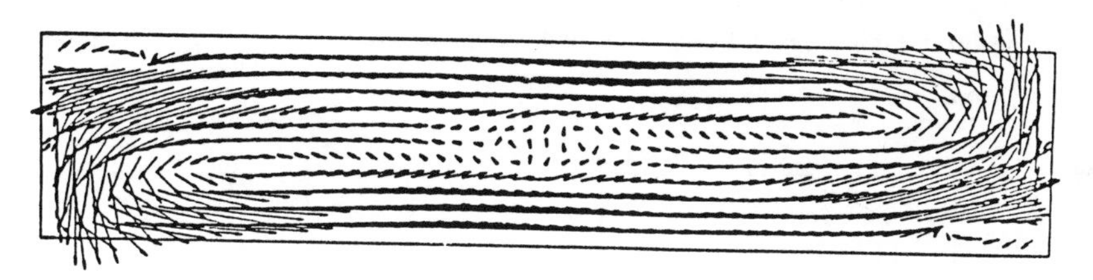

Fig. 19. **The Cross-Flow Velocity Vector Plot, Split Blades (Computed) [177].**

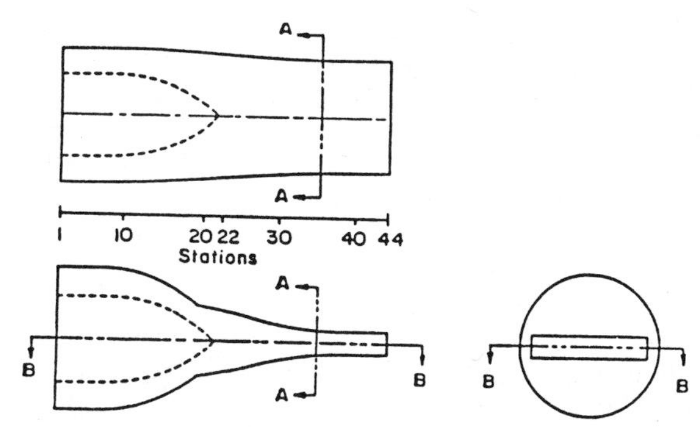

Fig. 20. **Sections and Stations in the Sobota/Marble Study (177).**

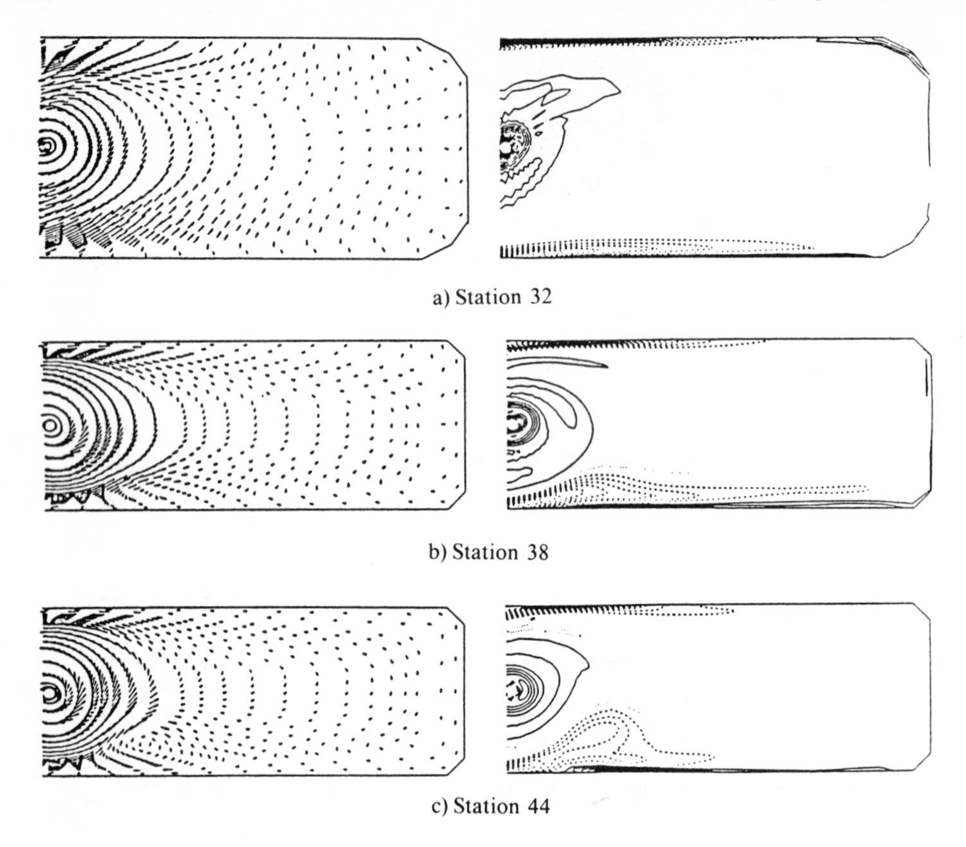

Fig. 21. **The Cross-Flow Velocity Vector Plots and Axial Vorticity Contour Plots, 15 Degrees Blade Angle [177].**

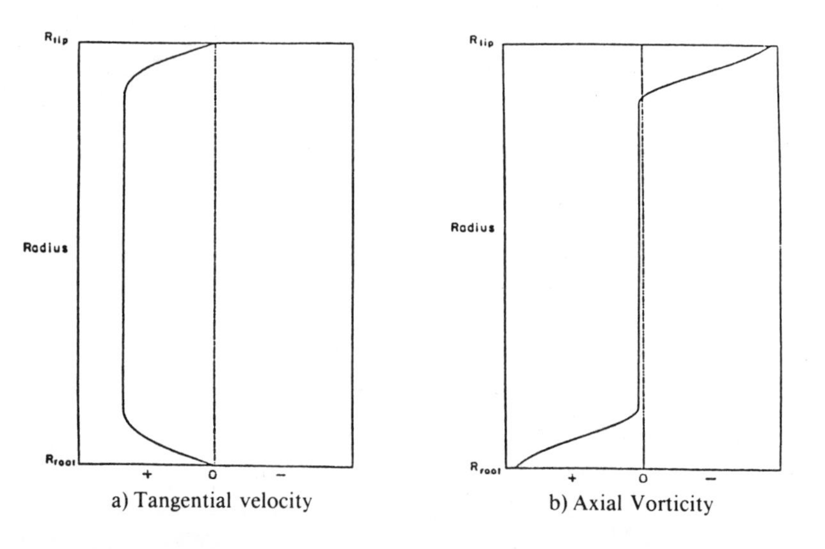

Fig. 22. **Inlet Conditions, 15 Degrees Blade Angle [177].**

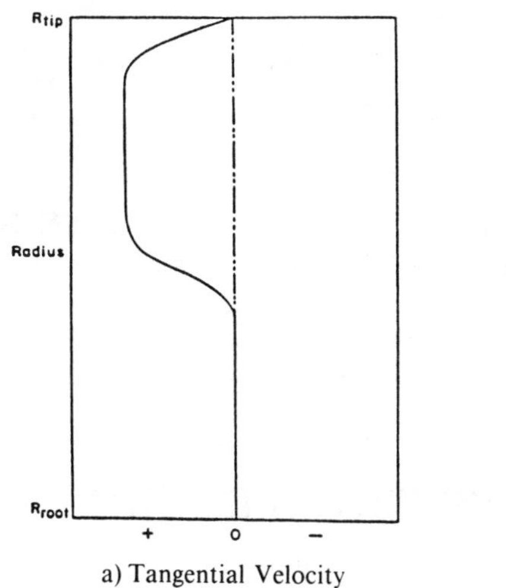

a) Tangential Velocity

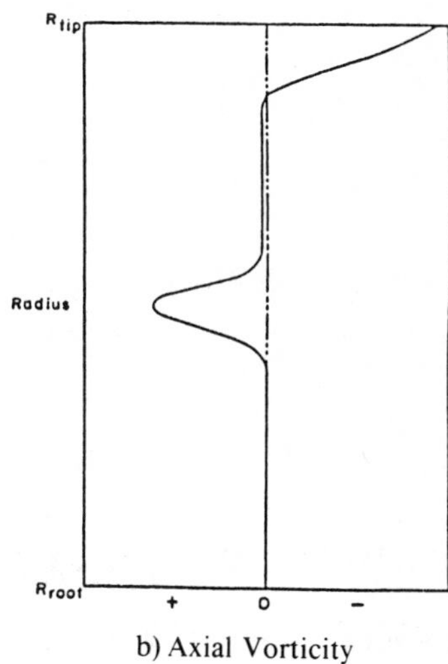

b) Axial Vorticity

Fig. 23. **Inlet Conditions, Split Blades [177].**

(iv) – The vortex patterns created in the low and high swirl cases appear to be highly conducive to the mixing of the rectangular jet, particularly in the high swirl cases. The adjacent positioning of strong vortices of opposite sense may tend to draw the cold ambient fluid between them. This large-scale mixing may rapidly reduce the size of the potential core-region of the jet (cf. Ref. 178).

APPENDIX E

TEMPERATURE DISTRIBUTIONS INSIDE VECTORING NOZZLES

This Appendix briefly describes a test facility for studying the cooling requirements for components of vectoring nozzles.

To investigate the cooling requirements of the nozzle flaps and of the sidewalls, one may design and construct the *2D–CD* nozzles shown in Lectures I to III. The first one (Fig. III-3), may be operated without cooling for the sake of comparison. Its task may include the evaluation of maximum temperature-fields in subsonic or in supersonic nozzle gas flows. Secondly, one may have to repeat these tests with a properly-cooled nozzle.

Using such a methodology, Figure III-1 shows the same nozzle with film-cooled slots. (Both models have the same size and internal-flow geometry.)

●The Test Facility

Model size is designed to match the maximum flow rate of a GTC P85-180 gas turbine available in the JPL. This turbine provides air bleeding from a two-stage centrifugal compressor, regulated at up to 1 kg/s air-mass flow rate (at up to 3.5 bar).

The test facility and the relevant instrumentation are schematically depicted in Fig. 1. Exhaust gas conditions have been simulated by means of a combustion chamber [using a combustor and an ignition system of the T56 turbo-prop engine]. Alternatively, the air-mass flow may be supplied to the combustor from a centrifugal fan (8), through a large (stainless-steel) heat exchanger (7), (with air heating up to 250°C for better fuel evaporation).

Chormel-alumel thermocouples are connected to a universal data logger (24), which, in turn, is linked to an interactive graphic computer system for test-results acquisition, computation and graphing. Remote *TV*-control of the tests is conducted from a well-isolated control room (Fig. IV-2).

The wall temperatures, calculated as T_w/T_g, *may be expressed as*

$$LTC = \frac{T_w - T_A}{T_g - T_A},$$

where T_A is the air-cooling average temperature.

Tests may be conducted with various thrust vectoring angles. (In the example shown in Figs. III-1, 2a, 2b: $\delta_v = -7, 0, 16, 17$ and 23 degrees. At subsonic flows the local Mach number was varied from 0.07 to 0.22, and at supersonic flow conditions from 1.0 to 1.19).

Conclusions: The test results indicate that in the *subsonic* flow regime the temperature distribution may be approximately two-dimensional, i.e., the main temperature variation is alongside the nozzle central axis, while in the *supersonic* domain, a three-dimensional temperature distribution may be developed. The throat section of the nozzle is the most critical region, in which the temperatures are maximized during all thrust vectoring conditions. Under

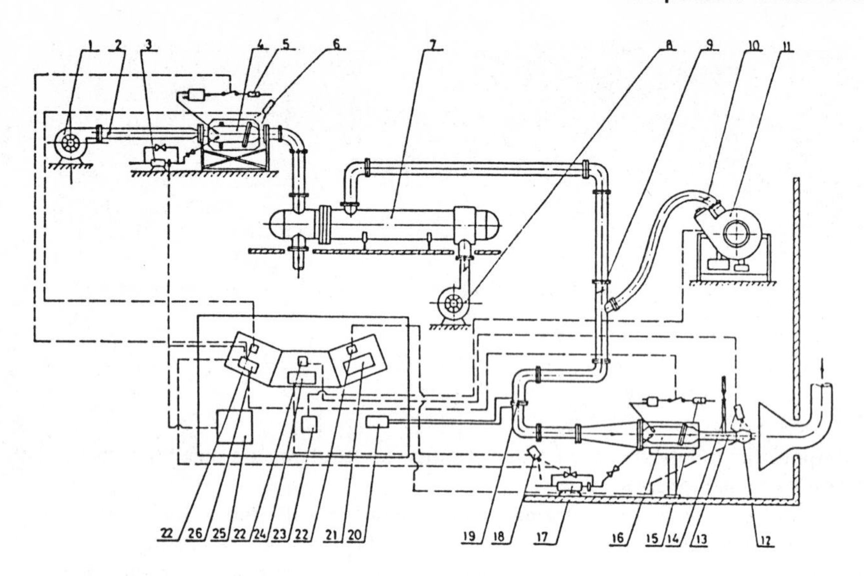

Fig. 1. **Test Rig for the Evaluaton of Temperature Profiles During Thrust Vectoring (with and without nozzle cooling).**

1, 8	Electric motor-driven centrifugal fans.
3, 17	Fuel pumps.
5, 16	Ignition systems.
7	Stainless-steel heat exchanger.
2, 9, 10	Butterfly valves.
11	GTC P85–180 gas turbine with air bleeding from a two-stage centrifugal compressor.
6, 13, 18, 22	Three *TV*-systems (closed circuit).
12	Thrust vectoring/reversing nozzle.
19	Air-mass-flow orifice panel.
23	GTC-engine control panel.
14	Water cooling system.
20, 21, 24, 25, 26	Temperature, pressure and fuel-flow data loggers & indicators.

subsonic conditions thrust-vectoring generates temperature differences between the lower and upper divergent flaps, by generating higher temperatures on the *lower* flap during *"downward" thrust-vectoring*, while under *supersonic* conditions, the thrust-vectoring causes lower temperatures on the "convex" flow-path of the divergent flap i.e., a downward thrust-vectoring causes lower temperatures on the lower divergent flap.

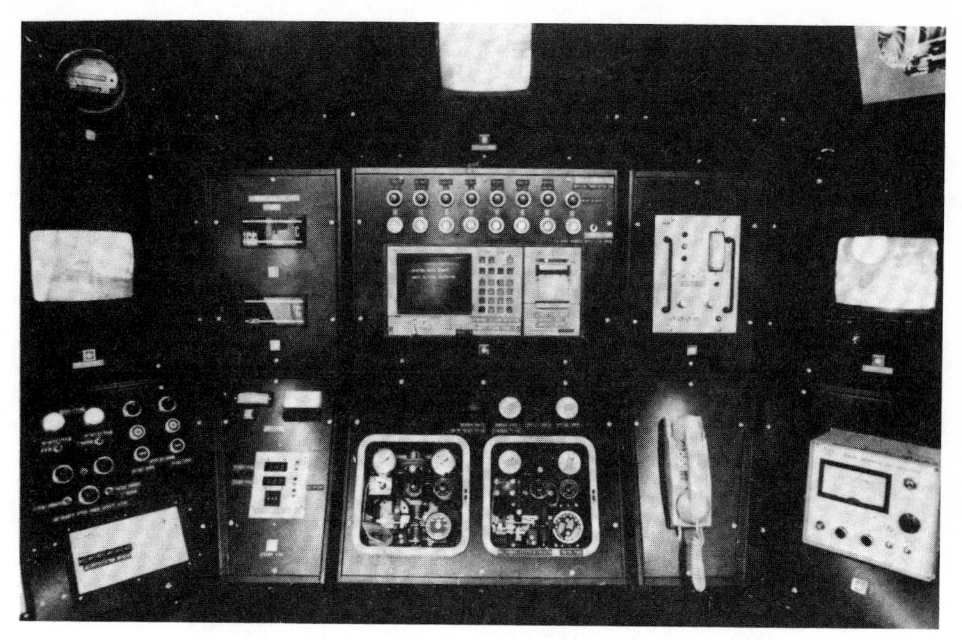

Fig. 2. **Details of the JPL data loggers, indicators and controls depicted in Fig. 1.**

APPENDIX F

LIMITING ENGINE-INLET ENVELOPES AND IFPC

F.1 The Yet Undefinable Matrix of PST-Intake Variables

Intakes of hot-jet-powered aircraft require minimum variations in airflow qualities at station 2 (compressor inlet). This means minimum variations from uniformity in composition, temperature, pressure and axial velocity distributions at station 2. (Water, rocket, or jet-engine exhaust gas characteristics, are typical distortion generators, in addition to high-α-β maneuvers, etc.) However, no real engine intake can achieve full uniformity in flight. Thus, each engine manufacturer must specify the maximum allowable distortion limits for each relevant parameter, at any engine inlet attitude throughout the flight envelope (with an *'uninstalled'*, or, preferably, with an *'installed'* power system). Since compressor stall may become a limiting factor (at high altitudes, small Mach numbers and at high-rates-of-turn), the problem of engine-intake compatibility during PST/PSM/RaNPAS becomes quite complicated by a matrix of yet unknown variables/limiting conditions (235–237).

Consequently, there is no universally-accepted method to define, *a priori*, actual distortion limits to 'installed' propulsion systems of highly agile aircraft. Moreover, there are, yet, no *bona fide* PST inlets. (Cf. Fig. 1).

On the other hand, there are various methods to avoid, or by-pass, some specific distortion problems, e.g., to activate special engine control subroutines during weapon firing, icy or stormy-cloud penetrations, or during operations in rain, or in highly-dusty environments (179).

Engine malfunctioning may first be subdivided into four categories:

- *Aircraft operating conditions* (e.g., M, Re, attitude in incidence and yaw).
- *Intake design adequacy* (e.g., fixed or variable geometric control, shape, size, bleed, relative position to the aircraft, staggering angle, simple or multiple ducting, top, bottom or side-type, etc.).
- *Engine's margin variations* (e.g., the margin between accelerating/decelerating paths and the surge line during rapid acceleratin-deceleration transients at any point on the various agility metrics flight envelopes). Cf. Fig. 6.
- *External influences* – such as weapon firing, ice, rain, dust, temperature, pressure and density variations at station 2. (Cf. Fig. 2.)
- The simplest indication of airflow departure from uniformity, under constant temperature and composition, is the variation of total pressure across the compressor inlet. These variations are subdivided into "steady" and "unsteady" (or "dynamic" and "time-variant").
- Intake spatial dynamic non-uniformity may be called "turbulence", while the "time-variant" types may cause "buzz" (internal intake shock oscillations).

Ultimately all these variables must be brought together in a total operational system, i.e., in a

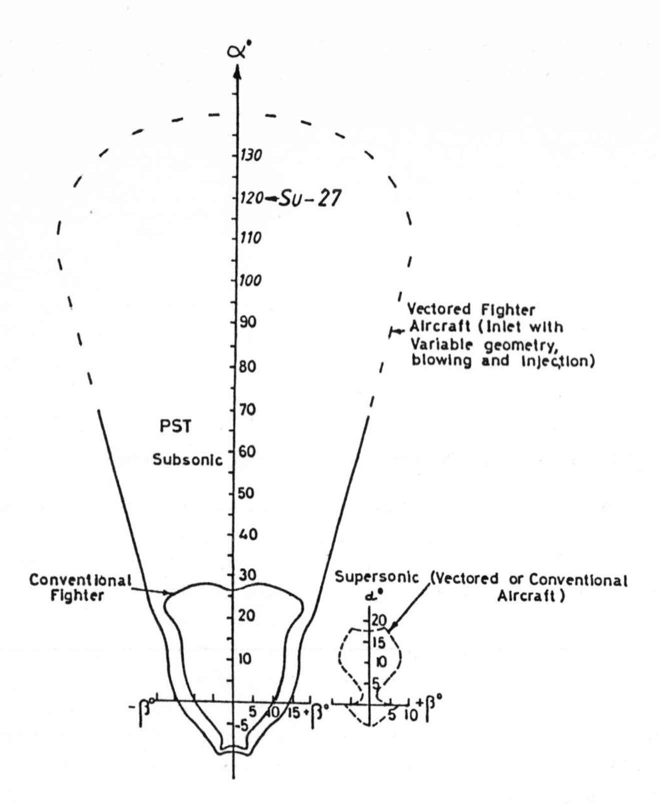

Fig. 1. **The maneuver envelopes of PST,vectored aircraft at subsonic and supersonic speeds (an example).** PST, vectored aircraft are expected to perform inside these envelopes, especially during high-α-β maneuvers at subsonic speeds (cf. Introduction and Lecture II). Our vectored F-15 and F-16 RPVs, and the PVAs, now operate well into the subsonic PST domains depicted.

well-integrated, whole-system design. Unless some compromises in cost and performance are considered, this is not an easy task.

F.1.1 Limited Maneuvers Envelopes

Vectored aircraft, especially highly-maneuverable, **strike-fighters**, have to operate close to the limiting conditions imposed by the strength of the aircraft structure, by the ability of the pilot to withstand high g-loads (caused by rapid braking and turning, as well as by strong sidewise variations — each with its particular limiting factor); by intake-engine performance at high-α-β and high-α-β time-variations, and by a few other factors to be discussed later.

Two examples of limiting maneuver-envelopes for fighters, one for PST-subsonic maneuvers, and one for supersonic speeds, are illustrated in Fig. 1.

As enumerated in the main text, these examples represent maneuverability envelopes expected for vectored aircraft, both manned and unmanned.

Let us consider for instance, *TR* braking at low subsonic speeds, which may be combined with very high drag through a rapid increase of AoA (cf. Figs. 21 in the Introduction, as well as Figs. II–2, 3, 4 and Figs. II-11 to 14).

Under these conditions an unsuitable inlet may cause engine-surge or flameout, and rob the aircraft of its ability to perform safely in line with its mission.

Consequently, PST-intakes must be developed prior to the flight-testing of manned vectored/ PST fighter aircraft. However, as we shall see, this is not going to be a simple task.

F-2 'Classical' and PST Intake Designs

The design of advanced engine intakes for vectored/PST/Stealth aircraft cannot follow the classical, "disciplinary" approach. Instead, it must follow a new, whole-system, highly-integrated methodology (§ B.10).

Such a new methodology may extract relevant tools from classical (different) "disciplines" such as conventional propulsion, aerodynamics, flight-control, materials, radar-waves-absorption-reflection theories, computers, and wind-tunnel/propulsion-laboratories testing methodologies. It may then attempt to integrate them into a new field of study and practice.

Thus, unlike the situation in classical wing-aerodynamics, where flow separation is generally taken to mark the technology limit of operation, the new engine intakes for super-agile aircraft may be required to function satisfactorily under separated-flow regimes, or to continuously vary their geometry in response to variations in α, β, etc. Thus, from the practical viewpoint, a few preliminary conclusions may be considered from the very outset:

- Flight-tested/experimental-data-bases and exact-intake-flow-theories, especially for PST/ vectored aircraft during high-α-β maneuvers, are not yet available.
- Flush, variable-geometry/subsonic/supersonic, high-aspect-ratio, low RCS or PST-inlets must be developed now (cf., for comparison, the B-2 inlets and the integrated methodology discussed in the Preface and the Introduction).
- In wind-tunnel-model-testing of vectored aircraft, the vectored engine may not be well-represented. (The internal flow undergoes only a loss of total pressure corresponding to the pressure recovery assessment. This, in effect, is an "internal drag" of the model, and it must be assessed as such in order to reduce the relevant external drag from measurements of total drag forces on the model. Cold air-suction and injection further complicates the dynamic simulation of vectored models in wind-tunelling.)
- The generally accepted definition of engine thrust leads to the inclusion of an intake drag

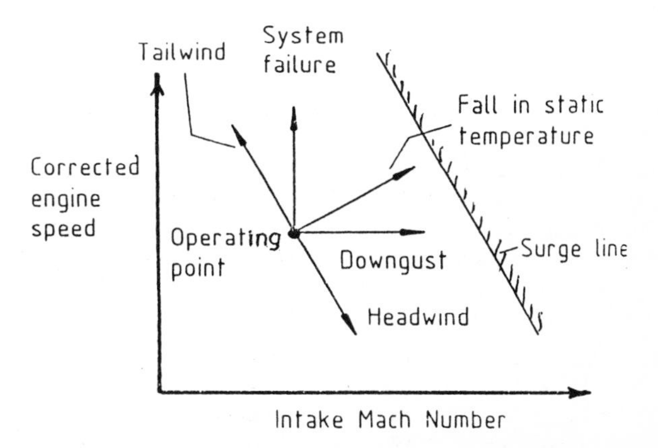

Fig. 2. **An example of uncontrolled responses of the propulsion system to the transients indicated.** Engine's surge-stall margin is the distance between the operating point and the surge line (Fig. 6). The surge line varies, however, with altitude, engine deterioration rate and history, trim, and IFPC characteristics [179, 133]. See also Lecture III.

known as "pre-entry drag", or "additive drag". This concept must now be re-examined and re-assessed, say, for AoA ≥ 90°.

- The presence of the wing section wetted by the inflow, or the aircraft canard ahead of the intake, must be taken into account under unorthodox flight conditions, including Post-Stall (PST) performance during supermaneuverability turns as those shown in Figs. II-2, 3 and II-11–14, and discussed in the Introduction.

(For details on conventional intakes the reader is referred to such comprehensive texts as that of Seddon and Goldsmith's *"Inlet Aerodynamics"*, Collins, London, 1985) [133].)

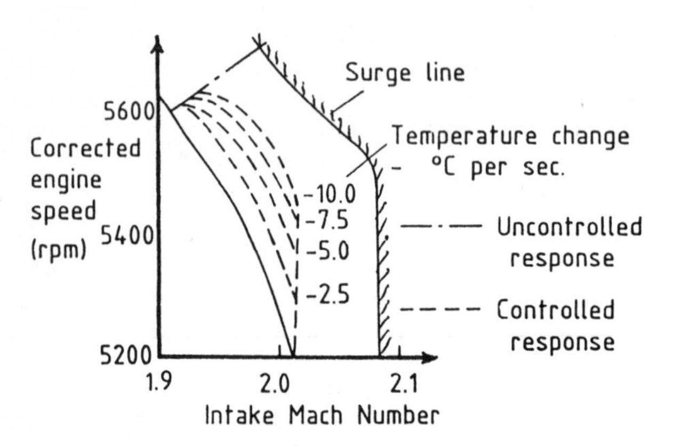

Fig. 3. **Controlled system response to temperature transients (Concorde data) [133].** Cf. Fig. 6.

F.2.1 Engine Inlet Scale-Up From Vectored/PST RPV Flight Testing.

In our current research programs, we employ a-9ft-long vectored F-15 demonstrator model, equipped with variable canards, yaw-pitch-roll vectored nozzles and on-board computer-metry systems. The aim of the on-board computer-metry systems is to evaluate inlet air distortion under flight conditions otherwise unattainable in wind-tunnels, or with RPVs powered by hot jet engines. To fly this model one must first install a couple of single-stage or multiple-stage ducted-fans, and, later, as the PST inlet design and testing is completed, to install hot vectored jet engines. Since jet engines are highly sensitive to inlet high-α-β distortion, one cannot install them on a PST/vectored RPV prior to the completion of STAGE IV; namely, the flight-testing of PST inlets (IV-4).

This methodology supplies, *under "scale-up" restrictions*, inlet flight data for new PST/ variable-inlet designs to be employed in future vectored aircraft.

F-3 IFPC Transient Definitions*

In the following we refrain from well-documented designs of such systems as rotating engine-nacelles, jet-induced-flow-over flap, and reaction-control systems. We concentrate, instead, on the emerging technology of vectored aircraft as defined in the Introduction and in Lecture I.

* This subject will be taken up again in Vol. II.

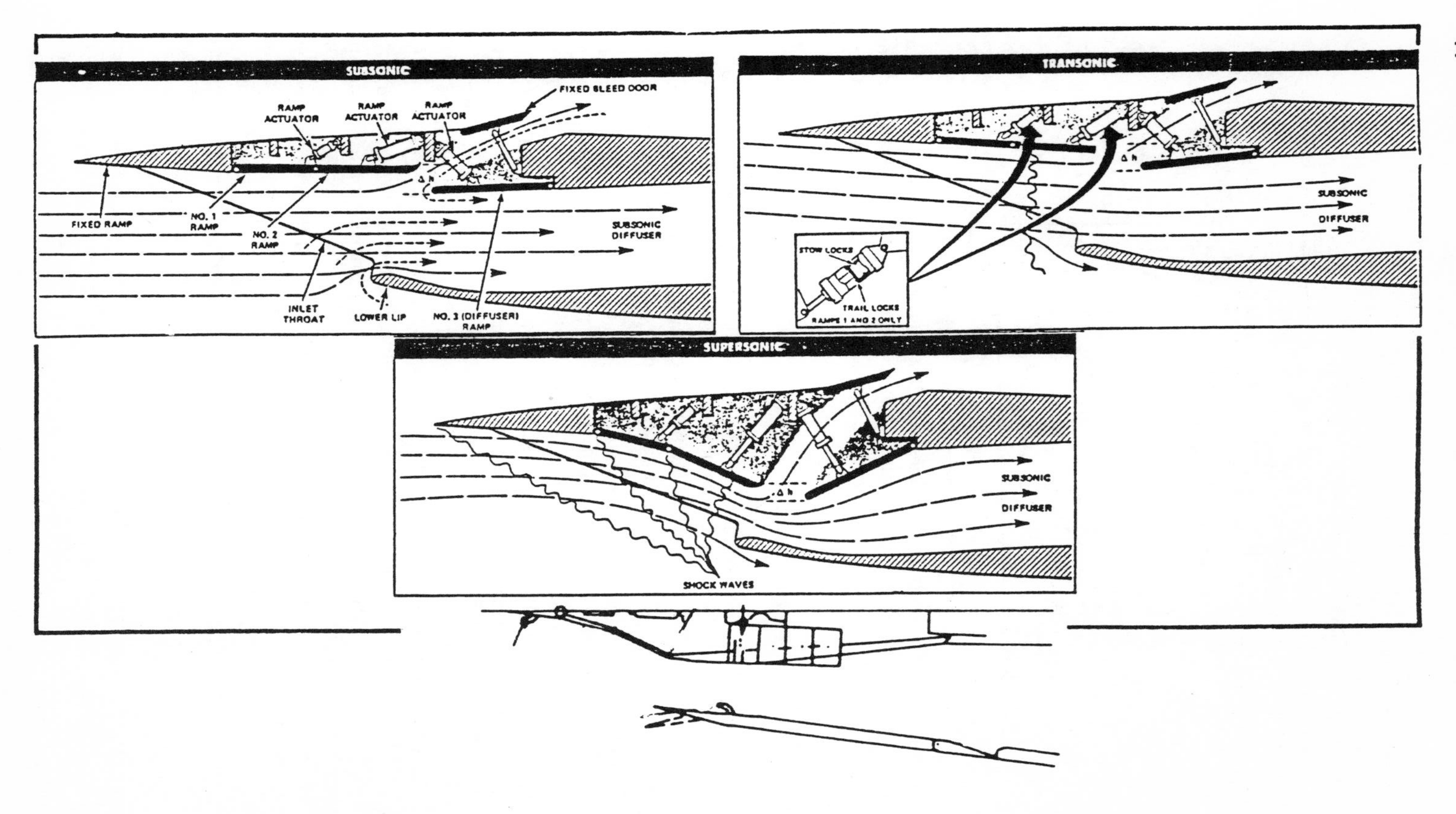

Fig. 4. **An example of conventional technology, variable-geometry inlets.** Such inlets must be modified and redesigned for PSt. (Depicted are the F-14 and the B-1 engine inlets). Note the rotatable vanes at the lower wall of the B-1 inlet, and at inlet lips.

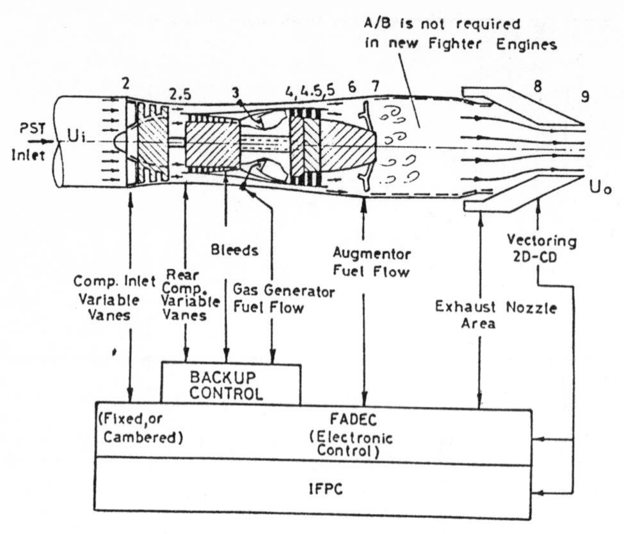

Fig. 5. **Conventional Engine Thermodynamic Stations and Main Controlled Variables.** IFPC must also include the thrust-vectoring control rules enumerated in the main text, as well as those depicted here, those associated with advanced HIDEC (cf. Appendix B), and, most important, those enumerated in Figs. 7 and 8.

To gain a better physical understanding of characteristic transients to be caused by PST/RaNPAS maneuvers, one may first establish comprehensive laboratory and flight tests aimed at the development of new variable PST-inlet geometries and control rules (§ IV-4). However, prior to this effort, one may have to assess propulsion-aircraft responses to transient effects such as:

- A rapid fall in static temperatures increases both M_2 demand and intake Mach number M_1, before reaching the surge line. (Fig. 3 shows the difference between uncontrolled and controlled responses to temperature transients.)
- A sudden headwind increases M_1, but reduces M_2 demand, and vice versa for a tail wind.
- To preserve the required margin between engine operating line and engine surge line must be the time-invariant objective of advanced PST–IFPC (Figs. 5, 6). This objective includes engine-status/history-deterioration rate, mechanical trim, IFPC trim, etc. (Lecture III).
- To shorten the acceleration time, the closer must be the engine operating transient path to the surge line (without, however, causing engine surge in the entire flight envelope) (cf. Fig. 6).
- As one moves from civil to military engines the required engine transients must be significantly shorter.
- Vectored strike/fighters may require the fastest transients. Hence, IFPC for strike/fighters may have to be the most effective (at the smallest possible engine margins – cf. Fig. 6).
- Widening engine/IFPC margins is a major goal for any new vectored propulsion design.
- Widening IFPC margins requirements for vectored aircraft are entirely different from those required for advanced conventional aircraft. This is due to a number of reasons as listed below:

(i) Most current technology military engines are designed with improved margins over their entire operational range, i.e., from start-up/idling, to full military, and up to full *AB* power. This is due, in part, to *Pilot's throttle variations/demands* during air-to-air, or air-to-ground modes of current pilot tactics. Since margins decrease with increasing altitude and

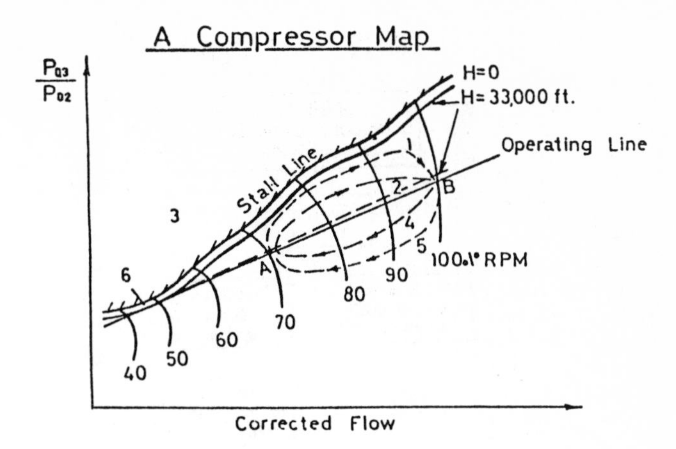

Fig. 6. **Limiting factors in acceleration–deceleration of military and civil engines are affected by many factors, but mainly by altitude, aircraft speed, engine RPM, and, engine history and inlet distortion level.**

The following definitions apply:

1) A characteristic rapid acceleration path of some military engines.
2) A moderate acceleration rate (e.g., a limiting regulation for civil engines).
3) Compressor-stall domain.
4) Same as (2) for deceleration.
5) A characteristic rapid deceleration path.
6) The problematic gap between operating line and stall line during starts and restarts in altitude.

with decreasing aircraft speed, the mission should be well-defined for any IFPC mode of operation.

(ii) Most critical air-to-air and air-to-ground pilot's modes with vectored aircraft would probably be accomplished with (fixed) full-throttle (while rapidly changing the IFPC commands (cf. Figs. II.2 and II.10).

(iii) While IFPC requirements may be relaxed as one moves down from full military power to cruise conditions and below (*AB* modes may not be required in most vectored aircraft), most PST maneuvers are expected to be carried-out with full throttle.

(iv) Variable, flush, top–bottom intakes must be developed for minimum inlet-air distortion at extremely high-α-β transients. Here, the B-1 (Fig. 4), the MiG-29 and the Su-27 inlet designs are of particular interest.

(v) The development of combined-cycle, distortion-insensitive, surge-line-insensitive, cold/hot propulsion systems must be given priority (cf. Lecture IV).

(vi) The greatest innovation efforts for vectored-aircraft intakes should concentrate on low and medium subsonic maneuverability, without relaxing transonic and supersonic demands, and without relaxing stealth benefits.

(vii) Current fly-by-wire (FBW) aircraft, such as the F-16, use three or four redundant control channels to prevent aircraft loss from a single failure. The channels are connected together at control surfaces by complex and expensive actuators, and multiple surfaces are used for aerodynamic redundancy. That redundancy must now be well-integrated with a new type of F-16/IFPC methodology.

(Traditional aerodynamic *decoupling* of control surfaces – the elevator provides only pitch, the ailerons only roll, for example – helped keep mechanical flight control systems

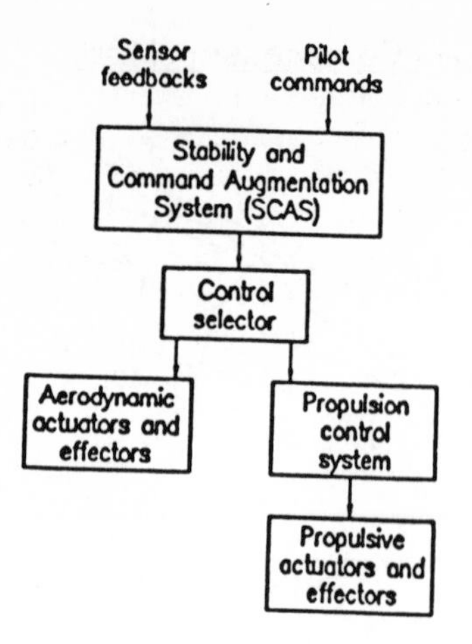

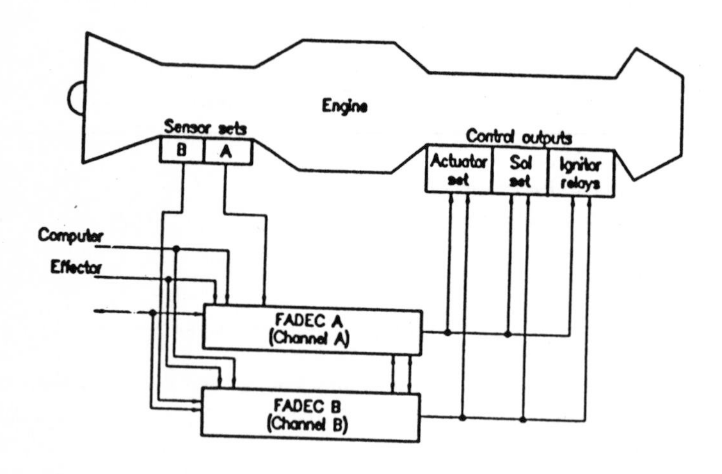

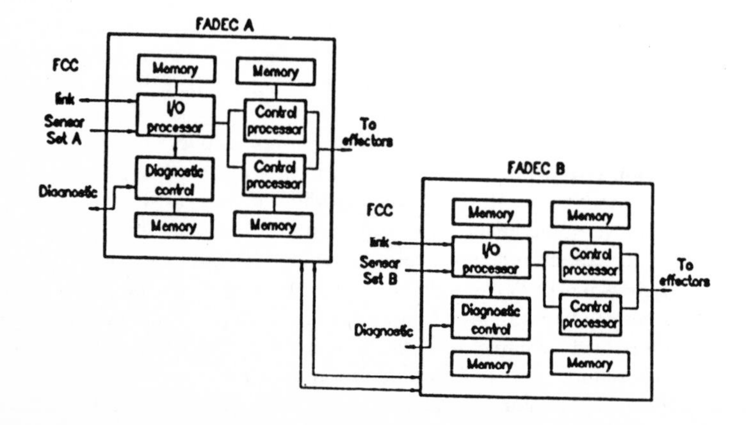

Fig. 7. **IFPC Computer Architecture** (145, 104, 110, 128, 235–239).

simple, but is no longer necessary with electronic flight controls. This gives the designer new degrees of freedom in control surface layouts.

Various self-repairing, improved control concepts are now being tested by the USAF-WPAFB Flight Dynamics Laboratory. For instance, simulations of emergency cases, such as an aileron separation from a test RPV may have to be investigatged to evaluate how the IFPC system reconfigures to employ the remaining control surfaces to cope with the emergency. These tests may be intended for a few payoffs: Improve resistance to battle damage, reduce flight control purchase and operational costs, and improve combat effectiveness.) (cf. Appendix G).

(viii) Integration of all on-board avionics is expected to facilitate the introduction of integrated cockpit displays that will provide future pilots with significantly improved tactical situation awareness. Such integrated systems will include IFPC navigation and communications systems, identification friend-or-foe systems, sensors, weapons and electronic warfare systems and provisions to provide future fighter pilots with three-dimensional visual, audio and tactile displays and cues, while permitting voice, head, and eyesight activation of aircraft systems and rapid reconfiguration of cockpit controls and displays (235–239).

(ix) The proposed integrated systems may also provide means for pilot state monitoring, so that the aircraft's automated flight/propulsion controls could, temporarily, fly the aircraft and recover from hazardous situation, in case of indications of pilot incapacitation due, for example, to a *g*-induced loss of consciousness.

●F-4 Short Inlets

Short inlet development is another activity which must be supported by analytic modelling, wind tunnel tests and flight-testing of RPVs. Since weight and balance considerations in a super-

	Enhanced Modes			
Effector \ Mode	STOL	Cruise	Combat	Conventional
Symmetrical Primary Jet Vectoring	X	X	X	
Differential Primary Jet Vectoring	X	X	X	
Symmetrical Rotating Vane	X	X	X	
L/R Differential Rotating Vane	X	X	X	
T/B Differential Rotating Vane	X	X	X	
L/R, T/B Differential Rotating Vane	X	X	X	
Symmetrical Gross Thrust	X	X	X	⊗
Differential Gross Thrust	X	X	X	⊗
Symmetrical Stabilator	X	X	X	X
Differential Stabilator	X	X	X	X
Symmetrical Canard	X	X	X	X
Differential Canard	X	X	X	
Symmetrical Aileron	X			
Differential Aileron	X	X	X	X
Symmetrical Rudder	X	X	X	X
Differential Rudder	X			
Symmetrical Flaperon	X	X	X	⊗
Differential Flaperon	X	X	X	
Speedbrake				⊗
Symmetrical Main Gear Brakes	X			⊗
Differential Main Gear Brakes	X			⊗
Nose Gear Steering	X			⊗

L/R · Left to Right T/B · Top to Bottom
⊗ Manual Operation Only

Fig. 8. **Four flight modes of the Vectored-F-15-STOL IFPC (Mello and Kotansky, 128) (cf. Fig. 7).**

sonic STOVL aircraft may position the engine farther forward in the fuselage than is currently normal, the distance between the powerplant and the engine air inlet may be reduced, shortening the diffuser length. Short diffusers typically perform poorly and current work may have to focus on this issue. The new needs of PST-inlets may cause similar problems.

Using computational methods and wind tunnel test data, NASA-Lewis researchers have recently fabricated a supersonic short diffuser model configured with several types of boundary layer controls. The model incorporates holes of varying porosity and distribution for suction, discrete jets for blowing and distributed slots for blowing control. This topic will be taken up again in Volume II.

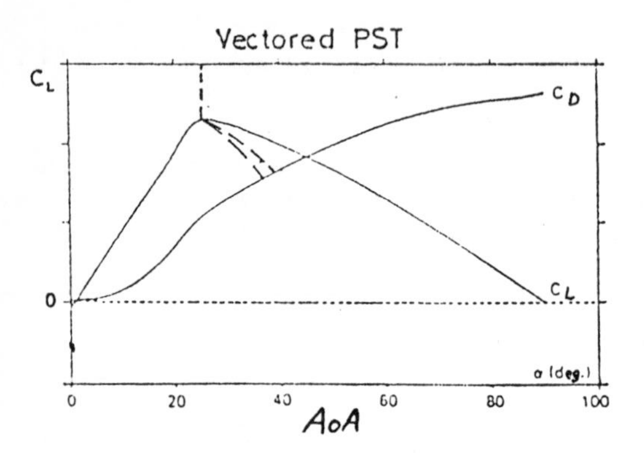

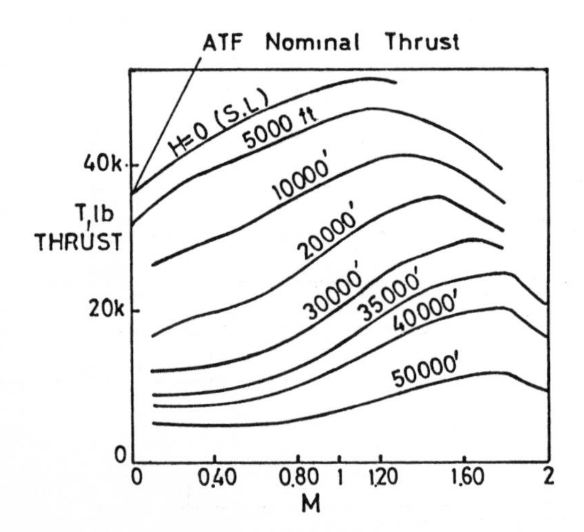

Fig. 9. **Other Inputs to the IFPC System during Vectored PST Flight (Mello and Kotansky, 128).** These include the variation of (total) thrust with aircraft Mach number and with altitude.
Note that the ATF-engine delivers 35,000 lb.(f) at $M = 0$, $H = 0$. Cf. Appendix B. Note also that the very definition of **net** thrust at PST–AoA must be reassessed at the inlet station.

APPENDIX G

ON SOME OTHER VECTORED/STEALTH AIRCRAFT CONSIDERATIONS

G.1 Combat Effectiveness of Vectored Aircraft

The following simulations may be considered in association with the ones given in the main text, or as a separate entity. They were recently performed in France by Costes, Ph. and H. T. Huynh, for a schematic model of a canard-configured, delta-wing, pitch-only, partially-vectored fighter aircraft (DRET, STPA Contracts-88/89-p. 85/ONERA's Activities – see also Costes paper: Ref. 182).

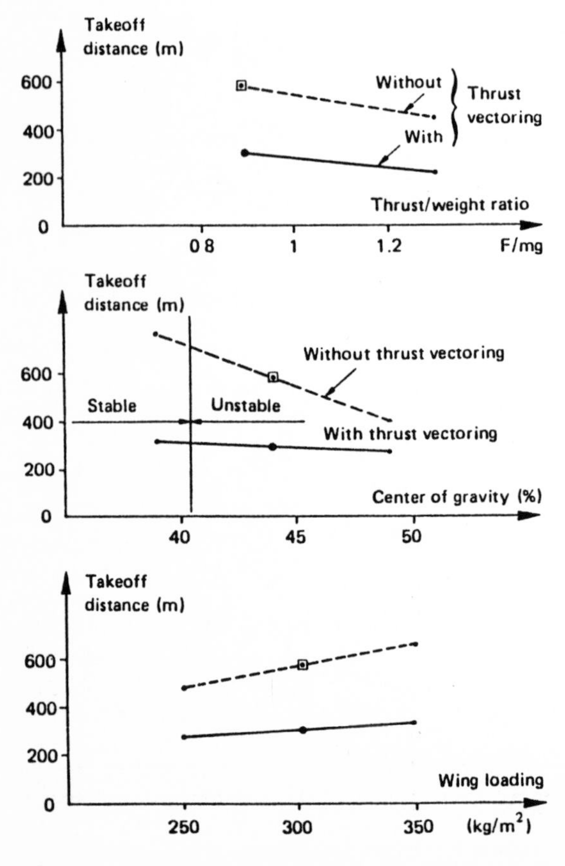

Fig. 1. **Advantage of thrust vectoring for takeoff distance.** The influence of the thrust-to-weight ratio, the center of gravity position and wing loading: thrust-to-weight ratio = 0.89, center of gravity = 44%, wing loading = 302 kg/m^2 (dry runway, altitude 0, temperature = 15°C). (Costes and Huynh, 182).

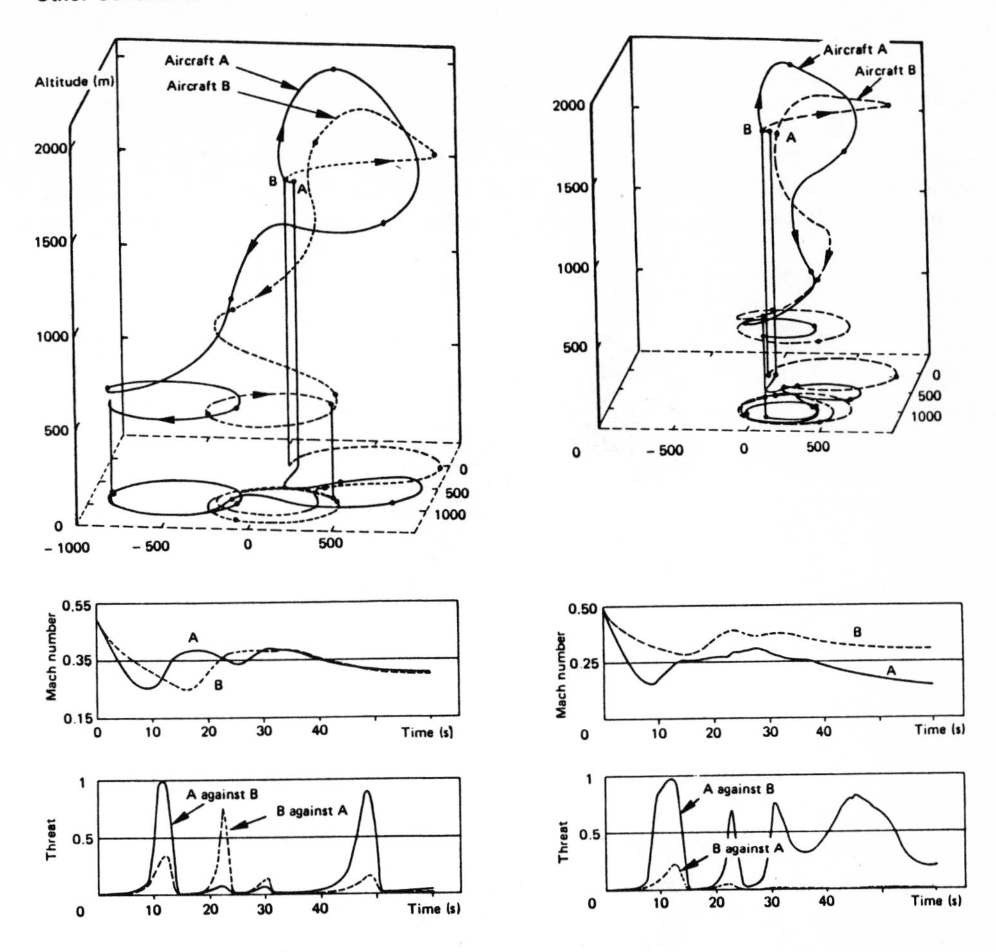

Fig. 2 (left) **Combat between two identical aircraft without thrust vectoring.** Aircraft A flight path optimization (the flight paths are monitored every 10 s) (Initial conditions: altitude 1,500 m, speed 150 m/s). (Costes and Huynh, 182)

Fig. 3 (right). **Air combat between two aircraft.** Flight path optimization of aircraft A with thrust vectoring (the flight paths are monitored every 10 s). (Costes and Huynh, 182)

The advantage of thrust vectoring was evaluated by comparing the "optimum" flight paths of two versions of the model; with and without deflectable engine nozzle. Optimization was achieved by a numerical code using a projected gradient method developed at ONERA. This method allows the time histories of the aircraft control surfaces (elevator, thrust deflector, canards, etc.), to be determined to minimize the takeoff distance, while complying with the various constraints inherent in the aircraft, such as control surface excursion, maximum AoA, etc.

With (pitch-only) jet-deflection restricted to maximum 20 degrees, a reduction of more than 50% of the takeoff distance was obtained, delimited by clearing an obstacle 15 m high (Fig. 1). While the takeoff distance was linearly reduced with increasing thrust-to-weight ratio – T/W – of the simulated fighter, the (takeoff) advantage of vectored over conventional fighter was found to be almost unaffected by T/W. However, that margin increased with increasing wing loading.

The advantage of thrust vectoring for air combat was evaluated by comparing the "optimum" maneuvers of two versions of the aircraft, with and without a deflectable (pitch-only) nozzle, against the same adversary, who adopts an aggressive closed-loop pursuit-law, attempting to cancel the off-boresight angle between the fuselage axes.

The "optimum" maneuvers are aimed at maximizing a performance criterion based on winning and survival probabilities computed along the flight-path on the basis of the respective instantaneous threats (cf. Ref. 182, and Table 1 in the Introduction).

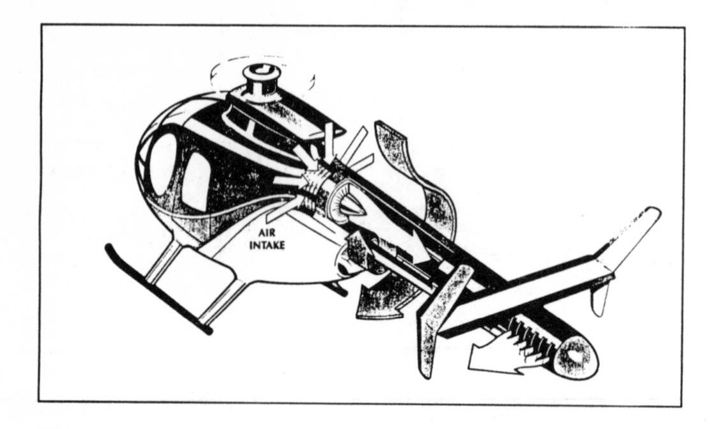

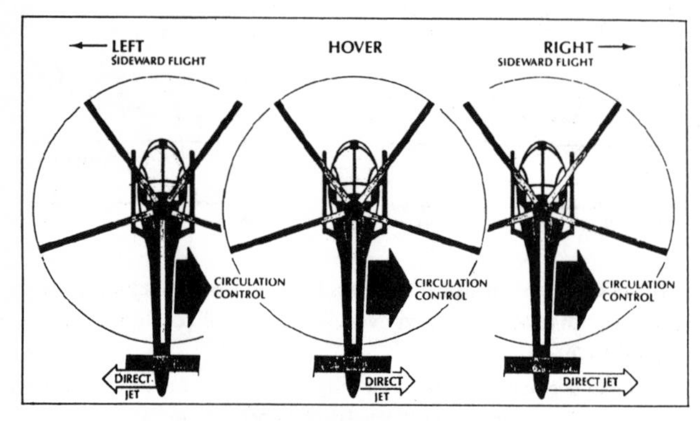

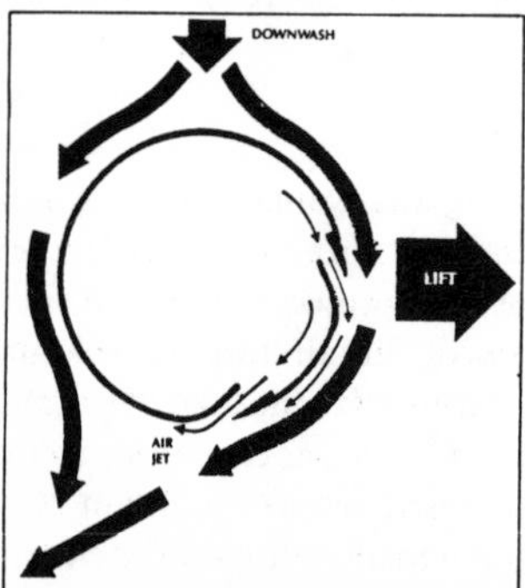

Fig. 4. **Partially-vectored helicopters may employ efficient directional/tail thrust-vectoring control and "circulation control" as depicted.** Note the air-jet blowing orientation into the downwash air stream.

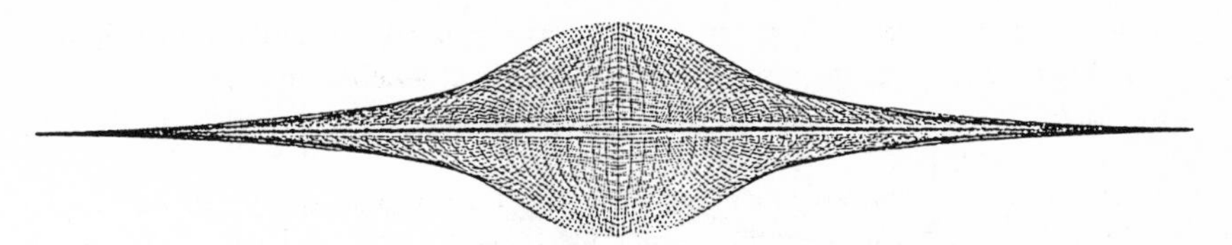

Figs. 5. **Front view of a pure vectored aircraft RPV model.** A computer-generated model made by one of the students at the Jet Propulsion Laboratory, TIIT, (1986). This topic will be discussed in Volume II.

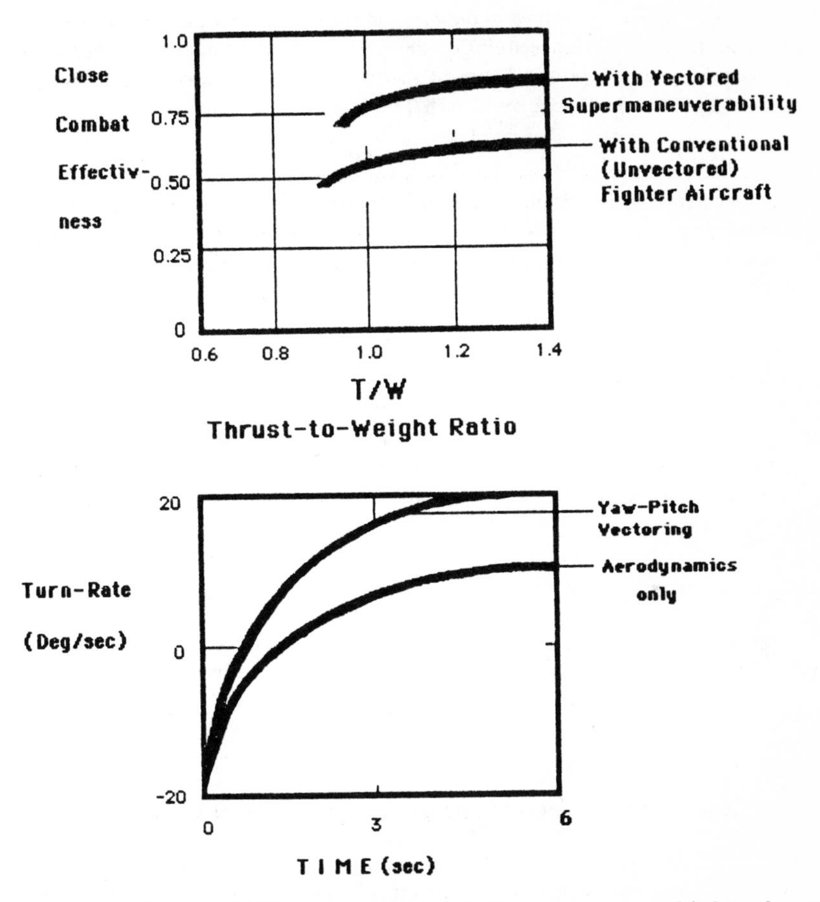

Fig. 6. **According to Herbst of MBB, all-round jet deflection generates speed-independent control moments in the yaw-pitch planes, thus allowing safe maneuvering in the stall and post-stall regimes.** This figure schematically shows the simulation of a scissoring maneuver and the associated turning rates which may be obtained with and without all-round vectoring flight/propulsion control at constant AoA = 40 degrees.

Consequently, with a jet-vectoring range of only 10 degrees, the roll turn-around time, i.e., the roll agility at high angles-of-attack, may be easily doubled. This effect becomes more prominent as AoA is increased, i.e., as the aerodynamic effect becomes weaker. The close-combat effectiveness was estimated on the basis of simulations of one-on-one combat against MiG-29. Cf. Introduction. (After Herbst, 188).

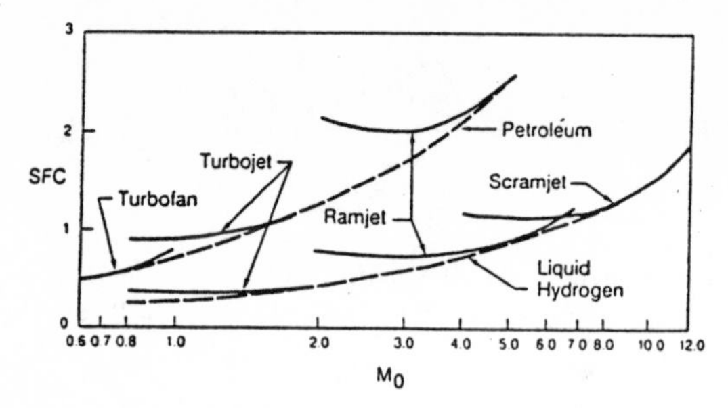

Fig. 7. **Thrust vectoring is not limited to turbojet and turbofan engines, nor to air-breathing engines using petroleum fuel.** In fact thrust vectoring is extensively used in boating and in rockets. The next step is to introduce this technology to PST-ramjets fueled by petroleum-based fuel, or by hydrogen. STOL benefits may also be imperative in such TV systems.

These simulations have demonstrated, for a variety of initial conditions, that even a pitch-only thrust-vectoring is significantly advantageous in terms of combat effectiveness.

For instance, in a one-on-one gunfight between two identical aircraft without thrust vectoring, maximizing the winning probability (over a duration of 60 s) leads to the well-known yo-yo maneuver, which gives a tactical advantage over the adversary (at approximately 10 s, see Fig. 2), thereby increasing the winning probability. However, if the pilot does not exploit this advantage in time, the evolution could become unfavorable (near 22 s, Fig. 2), especially in target-rich situations, due to a greater loss in speed.

The most interesting results of this study are shown in Fig. 3 for aircraft with thrust vectoring (aircraft A). Costes and Huynh noted, however, that the results depicted in Fig. 3 had been calculated with a pitch-down nozzle deflection of up to 60 degrees (and assuming that trimming of the aircraft remained possible). They also stress the need for (flight-testing) confirmation of their analysis and assumptions.

G.2 Vectored Helicopters

Fig. 4 demonstrates the main design features of the newly-introduced partially-vectored helicopters which employ directly vectored jet-force, as well as the effects of "circulation control" induced by the rotor down-wash.

G.3 Stealth/Vectored RPVs

As a general rule: Properly designed TV systems reduce signatures.

Unclassified material concerning this subject is to be published in Volume II. The configuration depicted in Fig. 5 may only represent a single set of "unit operations" required in the early conceptual design philosophy, and during laboratory and flight tests of such vehicles, as schematically depicted in the Preface to this Volume.

APPENDIX H

THOUGHT-PROVOKING AND THOUGHT-DEPRESSING QUOTATIONS

"The USA is relying heavily on thrust vectoring and stealth to restore its diminished technological superiority."

The Editor (JWRT)

Jane's All The World's Aircraft
1987–88, p. 43.[175]

(This quotation is included here for its first part. This inclusion does not mean that the author concurs with the second part of the quotation).

* * *

"NATO would be unwise to believe that it still has the clear superiority in conventional air power that it has enjoyed for most of the period since the Second World War."

The Editor (JWRT)

Jane's All The World's Aircraft
1987–88, p. 43.[175]

* * *

"No style of thinking will survive which cannot produce a usable product when survival is at stake."

Thomas Favill Gladwin

* * *

"If a professor thinks what matters most,
Is to have gained an academic post,
Where he can earn a living, and then,
Neglect research, let controversy rest,
He's but a petty tradesman at the best,
Selling retail the work of other men."

Kalidasa

REFERENCES

Let us come now to references to authors, which other books contain and yours lacks. The remedy for this is very simple; for you have nothing else do do but look for a book which quotes them all from A to Z, as you say. Then you put this same alphabet into yours... And if it serves no other purpose, at least that long catalogue of authors will be useful to lend authority to your book at the outset.

Miguel de Cervantes

1. B.L. Berrier, "A Review of Several Propulsion Integration Features Applicable to Supersonic Cruise Fighters". NASA TMX-73991, 1976.
2. A.D. Wolfe and A.E. Fanning, "Advanced Nozzle Technology", AGARD Fighter Aircraft Design, June 1978.
3. J.P. Werdner and W.A. Vahl, "A Preliminary Assessment of the Impact of Two-Dimensional Exhaust Nozzle Geometry on the Cruise Range of a Hypersonic Aircraft. Top Mounted Ramjet Propulsion", NASA TM-81481, Sept. 1980.
4. D.L. Bowers and J.A Laughrey, "Integration of Advanced Exhaust Nozzles", AGARD-CP-301, Sept. 1981.
5. R.J. Glidewell and R.E. Warburton, "Advanced Exhaust Nozzles Technologies", AGARD-CP-301, Sept. 1981.
6. B.L. Berrier, J.L. Palcza and G.K. Richey, "Nonaxisymmetric Technology Program — An Overview", AIAA 77-1023, Aug. 1977.
7. J.C. Gue, "Advanced Technology Thrust Vectoring Exhaust Systems", AIAA 73-1304, Nov. 1973.
8. E.H. Miller and J. Protopapas, "Nozzle Design and Integration on an Advanced Fighter", AIAA 79-1813, Aug. 1973.
9. W.W. Hinz and E.H. Miller, "Propulsion Integration of a Supersonic Strike Fighter", AIAA 79-0100, Jan. 1979.
10. F.J. Capone, "The Nonaxisymmetric Nozzle — It Is for Real", AIAA 70-1810, Aug. 1979.
11. P.E. Hiley and D.L. Bowers, "Advanced Nozzle Integration for Supersonic Strike Fighter Applications", AIAA 81-1441, 1981.
12. D.W. Speri and J.T. Blozy, "Development of Exhaust Nozzle Internal Performance Prediction Techniques for Advanced Aircraft Application", AIAA 81-1490, 1981
13. D.L. Bowers and J.L. Laughrey, "Application of Advanced Exhaust Nozzles for Tactical Aircraft", AIAA Proceedings, Vol. 1, p. 132, 141, 1982.
    "New Nozzles Design Aimed for the F-15, F-16 Aircraft", Aviation Week and Space Technology, Vol. 117, p. 67, 71, 73, Sept. 13, 1982.
14. Glidewell, R.J., "Installation Trades for Axisymmetric and Nonaxisymmetric Nozzles", AIAA 80-1084, 1980.
15. J.W. Paulson and J.L. Thomas, "Low Speed Power Effect on Advanced Fighter Configurations with Two-Dimensional Deflected Thrust", NASA TMX-74010, 1977.
16. J.P. Yip and J.W. Paulson, "Effects of Deflected Thrust on the Longitudinal Aerodynamic Characteristics of a Close-Coupled Canard Configuration", NASA TP-1090, 1977.

17. F.J. Capone, "Aerodynamic Characteristics Induced on a Supercritical Wing Due to Vectoring Twin Nozzles at Mach Numbers from .4 to .95", NASA-CR 78746, 1978.
18. B.L. Berrier and F.J. Capone, "Effect of Simulated In-Flight Thrust Reversing on Vertical Tail Loads of F-18 and F-15 Airplane Modes", NASA TP 1890, 1981.
19. W.C. Schnell and G.W. Ordonez, "Axisymmetric and Nonaxisymmetric Exhaust Jet Induced Effects on a V/STOL Vehicle Design", Part one, NASA CR 166146, 1981.
20. F.J. Capone and D.E. Reubusch, "Effects of Varying Podded Nacelle-Nozzle Installations on Transonic Aeropopulsive Characteristics of a Supersonic Fighter Aircraft", NASA TP 2120, 1983.
21. F.J. Capone and D.E. Reubusch, "Effect of Thrust Vectoring and Wing Maneuver Devices on Transonic Aeropropulsive Characteristics of a Supersonic Fighter Aircraft", NASA TP 2119, 1983.
22. F.J. Capone, "Supercirculation Effects Induced by Vectoring a Partial Span Rectangular Jet", Journal of Aircraft, Vol. 12, No. 8, pp. 633, 638, 1975.
23. A.I. Rilov, "Design of Supersonic Asymmetric Nozzles", Fluid Dynamics, Vol. 12. No. 3, pp. 414–420, 1978.
24. D.L. Bowers and F. Buchan, "An Investigation of the Induced Aerodynamic Effects of a Vectored Nonaxisymmetric Exhaust Nozzle", AIAA 78-1082, 1978.
25. W.P. Henderson *et al.*, "Canard Configurated Aircraft with Two Dimensional Nozzles", AIAA 78-1450, 1978.
26. W.C. Schnell and R.L. Grossman, "Vectoring Nonaxisymmetric Nozzle Jet Induced Effects on a V/STOL Fighter Model", AIAA 78-1080, 1978.
27. J.W. Paulson, J.L. Thomas and L.P. Yip, "Deflected Thrust Effect on a Close Coupled Canard Configuration", Jounral of Aircraft, Vol. 15, No. 5, pp. 287–292, 1978.
28. F.J. Capone, R.J. Re and E.A. Bar, "Thrust Reversing Effects on Twin Engine Aircraft Having Nonaxisymmetric Nozzles", AIAA 81-2365, 1981.
29. P.W. Banks, P.F. Quinto and J.W. Paulson, "Thrust Induced Effects on Low Speed Aerodynamics of Fighter Aircraft", AIAA 81-2612, 1981.
30. P.D. Wutten, L.W.Woodkey and J.E. Hamer, "Application of Thrust Vectoring for STOL", AIAA 81-2616, 1981.
31. R.E. La Froth, "Thrust Vectoring to Eliminate the Vertical Stabilizer", M.Sc. Thesis (AD-A079852) Avail NTIS HC A07/MFA01 CSCL 01/3, Dec. 1979.
32. D.L. Maiden, "Performance of an Isolated Two-Dimensional Variable Geometry Wedge Nozzle with Translating Shroud and Collapsing Wedge at Speeds up to Mach 2.01", NASA TN-D 7906, 1976.
33. R. Burley, "Flight Velocity Effects on Exhaust Nossle Installed on an Underwing Nacelle on an F-106 Airplane", NASA TMX 3361, 1976.
34. D.L. Maiden, "Two Dimensional Wedge/Translating Shroud Nozzle", NASA Case LAR-11919-1, U.S. Patent Appl. SN-672221, 1976.
35. D.E.Harrington and J.J. Schoemer, "Thrust Performance of Isolated Two Dimensional Suppressed Plug Nozzles with and without Ejectors at Mach 0, to 0.45", NASA TMX 3384, 1976.
36. D.L. Maiden, "Performance on an Isolated Two-Dimensional Wedge Nozzle at Fixed Cowl and Variable Wedge Centerbody at Mach Numbers up to 2.01", NASA TN-D 8218, 1976.
37. F. Capone, "Performance on Twin Two-Dimensional Wedge Nozzles Including Thrust Vectoring and Reversing Effects at Speeds up to Mach 2.2", NASA TN-D 8449, 1988.
38. G.T. Carson and M.L. Mason, "Experimental and Analitical Investigation of a

Nonaxisymmetric Wedge Nozzles at Static Conditions", NASA TD-1188, 1978. M.J. Harris and M.D. Falarshi, "Static Calibration of Two-Dimensional Wedge Nozzle with Thrust Vectoring and Spanwise Blowing", NASA TM-81161, 1980.

39. M.L. Mason and W.K. Abeyounis, "Experimental Investigation of Two Nonaxisymmetric Wedge Nozzles at Free Stream Mach Numbers Up to 1.2", NASA TO 2054, 1982.
40. D.L. Maiden and J.E. Petit, "Investigation on Two Dimensional Wedge Exhaust Nozzles for Advanced Aircrafts", AIAA 75-1317, 1975.
41. G.F. Goetz, J.H. Yong and J.L. Palcza, "A Two Dimensional Airframe Integrated Nozzle Design with Inflight Thrust Vectoring and Reversing Capabilities for Advanced Fighter Aircraft", AIAA 76-626, 1976.
42. J.E. Petit and F. Capone, "Performance Characteristics of a Wedge Nozzle Installed on a F-18 Propulsion Wind Tunnel Model", AIAA 79-1164.
43. J.L. Palcza, "Augmented Deflector Exhaust Nozzle (ADEN) for High Performance Fighter", AGARD Variable Geometry and Multicycle Engine, 1976.
44. R.J. Re and B.L. Berrier, "Static Internal Performance of Single External Ramp Nozzle with Thrust Vectoring and Reversing", NASA TP L962, 1982.
45. J.A. Lander and D.O. Nash, "ADEN Design for Future Fighters", AIAA 74-131, 1975.
46. D.O. Nash, T.G. Wakeman and J.L. Palcza, "Structural and Cooling Aspects of the ADEN Nonaxisymmetric Exhaust Nozzle", ASME 77-GT-110, 1977.
47. E.H. Miller, "Performance of a Forward Swept Wing Fighter Utilizing Thrust Vectoring", AIAA 83-2482, 1983.
48. D.M. Straight and R.R. Cullom, "Thrust Performance of a Variable Geometry, Nonaxisymmetric, Two-Dimensional, Convergent-Divergent Exhaust Nozzle on a Turbojet Engine at Altitude", NASA TP-2172, 1983.
49. J.D. Stevens. E.B. Thayer and J.F. Fullerton, "Development of a Multifunction 2D–CD Nozzle", AIAA 81-1491, 1981. Cf. NASA CR-145295, Cr-135252, 1978.
50. D.M. Straight and R.R.Cullom "Performance of a 2D/CD Nonaxisymmetric Exhaust Nozzle in a Turbojet Engine at Altitude", AIAA 82-1137, 1982.
51. G.H. McLafferty and J.H. Peterson, "Results of Test of a Rectangular Vectoring/Reversing Nozzle on a F-100 Engine", AIAA 83-1285, 1983.
52. B.L. Berrier and M.L. Mason, "A Static Investigation of Yaw Vectoring on 2D/CD Nozzles", AIAA 83-1288, 1983.
53. J. Pennington, "Simulation Study of the Effects of Thrust Vectoring on Combat Effectiveness of a Fighter Aircraft", NASA TM-X 3202, 1975.
54. F.C. Capone, "Static Performance of Five Twin Engine Nonaxisymmetric Nozzles with Vectoring and Reversing Capabilities", NASA TP-1224, 1978.
55. B.L. Berrier and R.J. Re, "Effects of Several Geometric Parameters on the Static Performance of Three Nonaxisymmetric Nozzle Concepts", NASA TP-1468, 1979.
56. F.J. Capone and B.L. Berrier, "investigation of Axisymmetric and Nonaxisymmetric Nozzles Installed on a 0.1 Scale F-18 Prototype Airplane Model", NASA TP-1638, 1980.
57. M.L. Mason, L.E. Putnam and R.J. Re, "The Effect of Throat Contouring on Two Dimensional Converging-Diverging Nozzles at Static Conditions", NASA TP-1704.
58. J.A. Yeter and L. Leavitt, "Effects of Sidewall Geometry on the Installed Performance of Nonaxisymmetric Convergent Divergent Exhaust Nozzle", NASA TP-1771, 1980.
59. P.J. Capone, "Aeropropulsive Characteristics of Twin Nonaxisymmetric Vectoring Nozzles Installed with Forward-Swept and After-Swept Wings", NASA TP-1778, 1981.
60. F.J. Capone, "Aeropropulsive Characteristics of Mach Numbers Up to 2.2 of Axisymmetric

and Nonaxisymmetric Nozzles Installed on an F-18 Model", NASA TP-2044, 1982.

61. P.E. Hiley, H.W. Wallace and D.E. Booz, "Study of Nonaxisymmetric Nozzles Installed on Advanced Aircrafts", AIAA 75-1316, 1975.
62. O.C. Pendergraft, "Comparison of Axisymmetric and Nonaxisymmetric Nozzles Installed on the F-15 Configuration", AIAA 77-842, 1977.
63. D.E. Berndt and A.P. Kuchar, "Nonaxisymmetric Nozzle Configuration Development for a Multimission Cruise Aircraft", AIAA 77-843, 1971.
64. C.M. Willard, F.J. Capone, M. Konarsky and H.L. Stevens, "Static Performance of Vectoring-Reversing Nonaxisymmetric Nozzles", AIAA 77-840, Journal of Aircraft, Vol. 16, No. 2, pp. 116–123, 1979.
65. W.C. Schnell, R.L. Grossman and G.E. Hoff, "Comparison of Nonaxisymmetric and Axisymmetric Nozzles installed on a V/STOL Fighter Model", SAE 77-0983.
66. P.E. Hiley, D.E. Kitzmiller and C.M. Willard, "Installed Performance of Vectoring/Reversing Nonaxisymmetric Nozzles", AIAA 78-1022, 1978.
67. F.J. Capone, N.S. Gowadia and W.H. Wooten, "Performance Characteristics of Nonaxisymmetric Nozzles Installed on the F-18 Aircraft", AIAA 79-0101, 1979.
68, R.A. Hutchinson, J.E. Petit, F.J. Capone and R.W. Whittaker, "Investigation of Advanced Thrust Vectoring Exhaust Systems for High Speed Propulsive Lift", AIAA 80-1159, 1980.
69. F.J. Capone, B.L. Hunt and C.E. Poth, "Subsonic/Supersonic Nonvectored Aeropropulsive Characteristics of Nonaxisymmetric Nozzles Installed on a F-18 Model", AIAA 81-1445, 1981.
70. H.L. Stevens, "F-15 Nonaxisymmetric System Integration Study Support Program", NASA CR-135252, 1978.
71. R.E. Martens, "F-15 Nozzle/Afterbody Integration", J. of Aircraft, Vol. 13, No. 5, pp. 327–333, May 1976.
72. P.E; Hiley, H.W. Wallace and D.E. Booz, "Nonaxisymmetric Nozzles Installed on Advanced Fighter Aircraft", Journal of Aircraft, Vol. 13, No. 12. pp. 1000–1006, 1976; AIAA-75-1316.
73. G.K. Richey, B.L. Berrier and J.L. Palcza, "Two-Dimensional Nozzle/Airframe Integration Technology — An Overview", AIAA 77-839, 1977.
74. J.L. Mace, E.B. Thayer and D. Bergman, "Nonaxisymmetric Nozzle Concept for an F-111 Test Bed", AIAA 77-841, 1977.
75. H.R. Wasson, G.R. Hall and J.L. Palcza, "Results of a Feasibility Study to Add Canards and ADEN Nozzle to the YF-17", AIAA 77-1227, 1977.
76. P.D. Whitten and R.W. Woodney, "Vectored Engine-Over-Wing Configuration Design", AIAA 77-1228, 1977.
77. G.F. Goetz, J.E. Petit and M.B. Sussman, "Nonaxisymmetric Nozzle Design and Evaluation for F-111 Flight Demonstration", AIAA 78-1025, 1978.
78. P.E. Hiley *et al.*, "An Air Vehicle Performance Evaluation Utilizing Nonaxisymmetric Nozzles", AIAA 79-1811.
79. D. Bergman, "Thrust Vectoring Applied to Aircraft Hang High Wing Loadens", AIAA 79-1812, 1979.
80. M.R. Robinson, "Enhanced Capabilities of Future Fighters as a Result of HiMat", AIAA 79-698, 1979.
81. L. Verhavert, "In Flight Thrust Vector Control", Ph.D. Thesis, Univ. Microfilm No. 75-211100, 1975.
82. G.B. Russel, "An Experimental Investigation of the Pressure Field Associated a Thrust Vector Control Device", M.Sc. Thesis, AL A021875, GAE/AE75D-17, 1975.

83. D.M. Davies and J.D. Hines, "The Effect of Thrust Vectoring on Aircraft Maneuvering", AD-A027367 USAFA-TR-76-9, 1976.
84. C.J. Yi, R.L. Humbold, R.J. Miller and E. Rachovitzky, "Flight and Propulsion Control Integration for Selected In-Flight Vectoring Modes", ASME 78-GT-79.
85. H.J. Kelley, E.M. Cliff and L. Lefton, "Thrust Vectored Differential Turns", Proceedings Vol. 1, (A 81 45502 21–63), 1980.
86. E. Lange, "Optimal Flight Path for Winged Supersonic Flight Vehicles Extension to the Case Where Thrust Can Be Vectored", Aeronautical Journal, Vol. 85, pp. 343–348, 1981.
87. J.E Pennington *et al.*, "Performance and Human Factors Results from Thrust Vectoring Investigations on Simulated Air Combat", Proceedings Vol. 1 (A81-4550221-63), Joint Automatic Control Conf., S. Francisco, Aug. 1980.
88. C.E. Robinson, "Exhaust Plume Thermodynamic Effects on Nonaxisymmetric Nozzle Afterbody Performance in Transonic Flow", AEDC-TR-78-24, 1978.
89. K.M. Gleason *et al.*, "Investigation of Infrared Characteristics of Three Generic Nozzle Concepts", AIAA 80-1160, 1980.
90. C.W. Chu *et al.*, "A Simple Nozzle Plume Model for I.R. analysis", AIAA 80-1808, 1980.
91. C.W. Chu and J. Der, "Modelling of a Two-Dimensional Nozzle Plume for an I.R. Signature Prediction Under Static Condition", AIAA 81-1108, 1981.
92. C.W. Chu and J., Der, "ADEN Plume Flow Properties for IR Analysis", Journal of Aircraft, Vol. 19, pp. 90–92, 1982.
93. B.G. Jaeggy, "A FORTRAN Program for the Computation of Two-Dimensional or Axisymmetric Nozzles", (in French), (ISL-R-110175), Available: HCA04/MFA01, 1975.
94. J.D. Hanck and N.O Stockman, "Computer Program for Calculating Two-Dimensional Potential Flow Through Deflected Nozzles", NASA-TM 79144m 1979,
95. P.D. Thomas, "Numerical Method for Predicting Flow Characteristics and Performance of Nonaxisymmetric Nozzles, Theory", NASA CR-3147, 1979.
96. P.D. Thomas, "Numerical Method for Predicting Flow Characteristics and Performance of Nonaxisymmetric Nozzles, Part Two: Applications", NASA CR-3264, 1980.
97. R.K. Tagirov and I.M. Shikhman, "Calculation of Supersonic in Flat and Axisymmetric Nozzles of a Given Geometry, Assuming Arbitrary Inlet Gas Properties", Fluid Mechanic-Soviet Research, Vol. 5, pp. 1–16, May, June 1976.
98. A.G. Mikhail, W. L. Hankey and J.S. Shang, "Computation of a Supersonic Flow Past an Axisymmetric Nozzle Boattail with Jet Exhaust", AIAA 78-993, 1978.
99. I.L. Osipov, "Numerical Method for Constructing Two-Dimensional Nozzles", Fluid Dynamics, Vol. 14, no. 2, pp. 312–318, 1979.
100. C.W. Chu, "Expedient Approach to Nonaxisymmetric Nozzle Performance Prediction", Journal of Aircraft, Vol. 17, pp. 127, 128, 1980.
101. R.C. Swanson, "Navier-Stokes Solutions for Nonaxisymmetric Nozzle Flows", AIAA 81-1217, 1981.
102. D.M. Straight, "Effect of Shocks on Film Cooling of a Full Scale Turbojet Exhaust Nozzle Having an External Expansion Surface", AIAA 79-1170, 1979.
103. B. Gal-Or, "2-D, Vectoring/Reversing Nozzle Cooling — An Experimental Study of the Temperature Fields in Sub and Supersonic Performance", a Technical Report of the Turbo and Jet Engine Laboratory, The Dept. of Aeronautical Eng., Technion – Israel Institute of Technology, Haifa, 1984.
104. H.D. Stetson, "Designing for Stability in Advanced Turbine Engines", International J. Turbo & Jet Engines, Vol. 1, No. 3, 1984, pp. 235–245.

105. C.C. Wu, *et al.*, "Study of an Asymmetric Flap Nozzle as a Vector-Thrust Device", AIAA-84-1360, 1984.
106. G.R. Barns, *et. al.*, "Vectoring Exhaust Nozzle Technology", AIAA-84-1175, 1984.
107. E. Hienz, "Requirements, Definition and Preliminary Design for an Axisymmetric Vectoring Nozzle to Enhance Aircraft Maneuverability", AIAA-84-1212, 1984.
108. J.A. Cohn, *et al.*, "Axisymmetric Thrust Reversing/Thrust Vectoring Exhaust System for Maneuver and Balanced Field Length Aircraft", AIAA-85-1466, 1985.
109. IDR, 1014, Aug. 1987.
110. K.E. Smith, *et. al.*, "Aircraft Control Integration Methodology and Performance Impact", AIAA-85-1424, 1985.
111. R.R. Popelewski, "Modified F-15 will Investigate Advanced Control Concepts", Aviation Week & Space Technology, Feb. 11, 1985, pp. 51–53. Also May 29, 1989, pp. 44–47.
112. L.D, Leavitt, "Static Internal Performance of a Two-Dimensional Convergent Nozzle with Thrust-Vectoring Capability up to 60°", NASA TP-2391, 1985.
113. M.M. Drevillon and R. Fer, "Tuyeres a pousee orientable pour avions de combat futurs", L'Aeronautique et l'Astronautique, No. 108, 1984–5, pp. 3–15 (in French).
114. J.R. Burley & J.R. Carlson. "Circular-To-Rectangular Transition Ducts for High Ar Nonaxisymmetric Nozzles", AIAA-85-1346, 1985.
115. Ramesh K. Agarwal and Jerry E. Deeze, "Euler Solutions for Airfoil/Jet/Ground-Interaction Flowfields". *J. of Aircraft*, Vol. 23, No. 5, May 1986.
116. Tavella D.A. and Karamcheti K., "Lift of an Airfoil with a Jet Issuing from Its Surface", *J. of Aircraft*, Vol. 24, No. 8, August 1987.
116. Tavella D.A. and Karamcheti K., "Lift of an Airfoil with a Jet Issuing from its Surface" *J. of Aircraft*, Vol. 24, No. 8, August 1987.
118, Sinha, K., Kimberlin R. and Wu J.M., "Equivalent Flap Theory: A New look at the Aerodynamics of Jet-Flapped Aircraft". AIAA 22nd Aerospace Sciences Meeting, Jan. 9–12, 1984, Reno, Nevada *(AIAA-84-0335).*
119. Gal-Or, B., "New Fighter Engines – a Review", Part I includes 103 references relevant to vectored engines, *Intern. J. Turbo and Jet Engines*, 1, 183–194 (1984).
120. Marcum, D.L. and Hoffman, J.D., "Calculation of Three-Dimensional Inviscid Flowfields in Propulsive Nozzles with Centerbodies", *AIAA-86-0449.*
121. Capone, F.J., "A Summary of Experimental Research On Propulsive-Lift Concepts in the Langley 16-Foot Transonic Tunnel", *AIAA-75-1315.*
122. Fair G. and Robinson, M.R., "Enhanced Capabilities of Future Fighters as a Result of Himat", AIAA-79-0698.
123. Sweetman, B., "ATF", *IDR*, 172–174 (1983); Gilson, C. "Future Fighters for the US Air Force", *IDR*, 165–171 (1983).
124. Joshi, P.B. and Hawthorne, C.A., "Approach and Landing Thrust Reverser Testing in Ground Effects", *AIAA-85-3075.*
125. Schultz, J., "Stealth Takes Shape", *Def. Elect.* 65, 1983.
126. Reneau, L.R., *et al.*, "Performance and Design of Straight, Two-Dimensional Diffusers", *J. of Basic Engineering*, 141, 1967.
127. Kowalski, E.J., "A Computer Code For Estimating Installed Performance of Aircraft Gas Turbine Engines", *NASA CR 159691*, 1979 (plus a few other NASA's computer programs and tapes purchased through COSMIC, USE).
128. Mello, J.F. and Kotansky, D.R., "Aero-Propulsion Technology For STOL and Maneuver", *AIAA-85-4013.*

129. Mihaloew, J.R., "Flight Propulsion Control Integration for V/STOL Aircraft", (NASA), *Intern. J. Turbo and Jet Engines*; In press.
130. Emami-Naeni, A; Anex, R.P.; Rock, S.M., "Integrated Control: A Decentralized Approach" Proceedings of the *IEEE Conference* on Decision and Control.
131. Hawkins, J.E., "YF-16 Inlet Design and Performance", *AIAA 74-1062.*
132. Guo, R.A. and Seddon, J., "Swirl Characteristics of an *S*-Shaped Air Intake With Both Horizontal and Vertical Offsets", *Aeronautical Quarterly*, May 1983.
133. Seddon, J. and Goldsmith, F.L., "Intake Aerodynamics", *Collins*, London, 1985.
134. Smith, Kenneth L.; Kerr, W. Bernie; Hartman, G.L.; Skira, charles, "Integrated Flight/Propulsion Control-Methodology, Design and Evaluation", *AIAA-85-3048*, 23p.
135. Bill Sweetman, "Steatlh Aircraft", *Airlife*, England, 1986.
136. Mctycka, D.L., "Determination of Maximum Expected Instantaneous Distortion Patterns From Statistical Properties of Inlet Pressure Data", *AIAA 76-705.*
137. Sedlock, D., "Improved Statistical Analysis Method for Prediction of Maximum Inlet Distortion", *AIAA-84-1274.*
138. Kuchar, A.P.,"Variable Convergent-Divergent Exhaust Nozzle Aerodynamics", Chart. 14 in Oates, Gorden, C., Editor: "The Aerothermodynamics of Aircraft Gas Turbine Engines"; University of Washington, Seattle, Wash. 1978, US-DOC AD/A-059 784.
139. Hiley, P.E., D.E. Kiotzmiller and C.M. Willard, "Installed Performance of Vectoring/Reversing Nonaxisymmetric Nozzles", AIAA-78-1022.
140. Joshi, P.B., "Unsteady Thrust Reverser Effects in Ground Proximity", *AIAA-85-4035.*
141. Cohn, J.A., Dusa, D.J., Wagenknecht, C.D. and Wolf, J.P., "Axisymmetric Thrust Reversing-Thrust Vectoring Exhaust System for Maneuver and Balanced Field Length", *AIAA-85-1466.*
142. Berndt, D.E., Glidewell, R. and Burnes, G.R., "Integration of Vectoring Nozzles in a STOL Transonic Tactical Aircraft", *AIAA-85-1285.*
143. Stromecki, D.J., "An Assessment of Gas Turbine Engine Augment or Technology and Needs for the 80's", *AIAA-80-1200.*
144. "Integrated Flight Propulsion Control (IFPC)" and related subjects: List of publications found in DIALOG (available in Technion's Jet Lab.).
145. Landy, R.J., Yonke, W.A.; Stewart, J.F., McDonnell Aircraft Co., St. Louis, MO, USA, "Development of Hidec Adaptive Engine Control Systems", *Journal of Engineering for Gas Turbines and Power*, Transactions of the ASME v109 N2 Apr. 1987, p. 146–151.
146. Gal-Or, L., "RCC Production/Processing By New Methods", Technion, 1988. A Research Contract Funded By USAF.
147. Carr, J.E., "Aerodynamic Characteristics of a Configuration With Blown Flaps and Vectored Thrust for Low Speed Flight", *AIAA-84-2199.*
148. Banks, D.W. and Paulson, J.W., "Approach and Landing Aerodynamic Technologies for Advanced STOL Fighter Configurations", *AIAA-84-0334.*
149. Franciscus, L.C. and Luidens, R.W., "Supersonic STOVL Aircraft With Turbine Bypass/Turbo-Compressor", *AIAA-84-1403.*
150. Zola, C.L.; Wilson, S.B. and Eskey, A., "Tundem Fan Applications in Advanced STOVL Fighter Configurations", *AIAA-84-1402.*
151. Seginer, A. and Salomon, M., "Augmentation of Fighter Aircraft Performance by Spanwise Blowing Over the Wing Leading Edge," *Tech. Mem. 84330, NASA Ames Res. Center*, 1983.
152. Englar, R.J. and Huson, G.G., "Development of Advanced Circulation Control Wing High

Lift Airfoils", David Taylor Naval Ship RD Center, Bethesda, Md. USA, *AIAA-83-1847.*

153. Woan, C.J., "A Three-Dimensional Solution of Flows Over Arbitrary Jet-Flapped Configurations Using a Higher-Order Panel Method", *AIAA-83-1846.*
154. Herbst, W.B., "Supermaneuverability", MBB/FEI/S/PUB/120, 7.10.1983; AGARD, FMP Conference on fighter maneuverability, Florence, 1981; AGARD CP-319.
155. Paulson, J.W. and Gatlin, G.M., "Trimming High Lift for STOL Fighters", *AIAA-83-0168.*
156. Kimberlin, R.D. and Sinha, A.K., "STOL Attack Aircraft Design Based on an Upper Surface Blowing Concept". *AIAA-86-1860.*
157. Glazer, A., Hugh, R.V. and Hunt, B.L., "Thrust Reverser Effects on the Tail Surface Aerodynamics of an F-18 Type Configuration", *AIAA-86-1860.*
158. Callahan, C.J., "Tactical Aircraft Payoffs for Advanced Exhaust Nozzles", *AIAA-86-2660.*
159. Banken, G.J., Cornetle, W.M. and Gleason, K.M., "Investigation of Infrared Characteristics of Three Generic Nozzle Concepts, *AIAA/SAE/ASME 16th Joint Proulsion Conference,* June 30–July 2, 1980, Hartford, Conn.
160. Norain, J.P., "Viscous-Inviscid Simulation of Upper Surface Blown Configurations", *AIAA-84-2200.*
161. Shaw, P.D., Blumberg, K.R. and Joshi, D.S., "Development and Evaluation of an Integrated Flight and Propulsion Control System", *AIAA-85-1423.*
162. Leavitt, L.D. and Burley, J.R., "Static Internal Performance of a Single-Engine Nonaxisymmetric-Nozzle Vaned-Thrust-Reverser Design With Thrust Modulation Capability", *NASA Tech. Paper 2519* (1985).
163. Imlay, S.T., "Numerical Solution of 2-D Thrust Reversing and Thrust Vectoring Nozzle Flowfields", *AIAA-86-0203.*
164. Sedgwick, T.A., "Investigation of Non-Symmetric Two-Dimensional Nozzles Installed in Twin-Engine Tactical Aircraft", USAF, Flight Dynamics Laboratory, Contract F3615-74-C-3051 to Lockheed, Burbank, California.
165. Lachmann, G.V., "Boundary Layer and Flow Control", *Pergamon Press,* N.Y., 1961.
166. Capone, F.J., "Lift Induced Due to Thrust Vectoring of a Partial-Span Two-Dimensional Jet at Mach Numbers from 0.4 to 1.30", *Langley Working Paper 116,* July 1973.
167. Yoshihara, H., "Transonic Performance of Jet Flaps on an Advanced Fighter Configuration", *USAF, Flight Dynamics Laboratory-TR-73-97,* 31 July, 1973.
168. Yonke, W.A, Terrel, L.A. and Myers, Larry P., "Integrated Flight/Propulsion Control: Adaptive Engine Control System Mode", *AIAA-85-1425.*
169. Smith, K.L., Kerr. W.B., Hartmann, G.L. and Skira, Charles, "Aircraft Control Integration-Methodology and Performance Impact", *AIAA-85-1424.*
170 Vizzini, R.W., "Integrated Flight/Propulsion Control System Considerations For Future Aircraft Application", *Journal of Engineering for Gas Turbines and Power,* Transactions ASME v107 N4 Oct. 1985, p. 833-837.
171. Yonke, W.A., Landy, R.J.; Cushing, J.M., "Integrated Flight/Propulsion Control. Hidec Modes", IEEE Proceedings of the 1984 National Aerospace and Electronic Conference V1, p. 472–478.
172. Emerson, L.D. and Davies, W.J., "Flight/Propulsion System Integration", *AIAA-83-1238.*
173. Landy, R.J. and Kim, D.B., "Multivariable Control System Design Techniques: An Appli-

cation to Short Takeoff and Landing Aircraft", *Proceedings of the AIAA/IEEE 6th Digital Avionics Systems Conference*, Dec. 1984, p. 132–139.

174. Gal-Or, B., "Vectored Aircraft For the 90's" *Intern. J. Turbo and Jet-Engines, 4*, 1 (1987).
175. "Jane's All The World's Aircraft", 1987–88 Ed., "RPVs", "Israel", "TIIT"; "Jane's Yearrbooks", Jane's Publishing Inc., New York, p. 826. Also *AW&ST*, p. 21, May, 1987.
176. Gal-Or, B., Ed., "A Critical Review of Thermodynamics", Mono Book Corp., Baltimore, Md. 1970, pp. 299, B-T Chue, Chapter on "Thermodynamics of Deformation of Solids".
177. Sobota, T.H. and Marble, F.E., "An Experimental and Numerical Investigation of Swirling Flows in a Rectangular Nozzle", *AIAA-87-2108.*
178. Von Glahn, U.H., "Secondary Stream and Excitation Effects on Two-Dimensional Nozzle Plume Characteristics", *AIAA-87-2112.*
179. Gal-Or, B., "vectored Propulsion"; Internal Reports, The Jet Propulsion Laboratory, Aero Eng.; Technion IIT, May, 1983; July, 1985; May 1986; May 1987; Aug., 1988 (see also Ref. 206).
180. Yugov, O.K., Selyvanov, O.D., Karasev V.N., and Pokoteeto P.L., "Methods of Integrated Aircraft Propulsion Control Program Definition", AIAA-88-3268 (Central Institute of Aviation Motors, Moscow, USSR).
181. Ogburn, M.E., Nguyen, L.T. and Hoffer, K.D., "Modeling of Large-Amplitude High-Angle-of-Attack Maneuvers", AIAA-88-4357-CP.
182. Costes P., "Thrust Vectoring and Post-Stall Capability in Air Combat", AIAA-88-4160-CP.
183. Beaufrere, H., "Integrated Flight Control System Design for Fighter Aircraft Agility", AIAA-88-4503.
184. Hodgkinson, J., Skow, A., Ettinger, R., Lynch, U., Laboy, O., Chody, J. and Cord, T.J., "Relationships Between Flying Qualities, Transient Agility, and Operational Effectiveness of Fighter Aircraft", AIAA-88-4329-CP.
185. Tamrat B.F., "Fighter Aircraft Agility Assessment Concepts and Their Implication on Future Agile Fighter Design", AIAA-88-4400.
186. Geidel, H.A., "Improved Agility for Modern Fighter Aircraft: PART II: Thrust Vectoring Engine Nozzles", MTU Motoren-und Turbinen-Union, München, Internal Report, 20.2.87.
187. Miller, S.C., "Future Trends – A European View" (Rolls-Royce), ISABE-87-7002.
188. Herbst, W.B., "Thrust Vectoring – Why and How?" ISABE-87-7061.
189. Adam, J.A., "How To Design an 'Invisible' Aircraft", IEEE Spectrum, April, 1988.
190. Knott, E.F. *et al.*, Radar Cross Section, Artech House, Dedham, Mass., 1985.
191. Sweetman, B., "UAVs", Interavia, 775, 8/1988. IDR, 787, 7/1988.
192. Nagabhushan, B.L. and faiss G.D., J. AIRCRAFT, Vol. 21, No. 6, 408, 1984.
193. Franklin, J.A., Hynes, C.S., Hardy, G.H., Martin, J.L. and Innis, R.C., J. GUIDANCE, 555, 1986.
194. Ransom, S., J. AIRCRAFT, Vol. 20, 59, 1983.
195. Foltyn, R.W., *et al.*, "Development of Innovative Air Combat Measures of Merit for Supermaneuverable Fighters", AFWAL-TR-87-3073, October 1987.
196. McAtee, T.P., "Agility – Its Nature and Need in the 1990s", Society of Experimental Test Pilots Symposium, Sept. 1987.
197. Bucknell, R.L., "STOVL Engine/Airframe Integration", AIAA-87-1711.

198. Tape, R.F., Glidewell, R.J., and Berndt, D.E., "STOL Characteristics of a Tactical Aircraft with Thrust Vectoring Nozzles", AIAA-87-1835.
199. Barnes, G.R., *et al.*, "Vectoring Exhaust Nozzle Technology", AIAA-84-1175.
200. Tape, F.F., *et al.*, "Vectoring Exhaust Systems for STOL Tactical Aircraft", ASME 83-GT-212.
201. Curry, S.G., *et al.*, "Exhaust Nozzle Concepts for STOL Tactical Aircraft", AIAA-83-1226.
202. Lewis, W.J. and P. Simkin, "A Comparison of Propulsion Systems for V/STOVL Supersonic Combat Aircraft", SAE-80-1141.
203. Adams, A. and S. Parkola, "Charting Propulsion's Future – The ATES Results", AIAA-82-1139.
204. Lynn, N., "Powered Lift", *Flight International*, p. 29, April 9, 1988.
205. Tape, R.F., Glidewell, R.J. and Berndt, D.E., "STOL Characteristics of a Tactical Aircraft with Thrust Vectoring Nozzles", Rolls-Royce Inc., 1987.
206. Janes "All-The-World-Aircraft", 'RPV';-TIIT, 1988/1989, AW&ST, May 18, 1987. See also Ref. 212.
207. Bare, E.A. and Reubush, D.E., "Static Internal Performance of a Two-Dimensional Convergent-Divergent Nozzle With Thrust Vectoring", NASA Tech. Paper, 2721, July, 1987.
208. Mason, M.L. and Berrier, B.L., "Static Performance of Nonaxisymmetric Nozzles With Yaw Thrust-Vectoring", NASA Tech. Paper 2813, May 1988.
209. Berrier, B.L. and Mason, M.L., "Static Performance of an Axisymmetric Nozzle With Post-Exit Vanes for Multiaxis Thrust Vectoring", NASA Tech. Paper, 2800, May 1988. See also AIAA-87-1834.
210. Richey, G.K., Surber, L.E. and Berrier, B.L., "Airframe-Propulsion Integration for Fighter Aircraft", AIAA-83-0084.
211. Well, K.H., DFVLR, Report A552-78/2, 1978.
212. Gal-Or, B., "The Principles of Vectored Propulsion", *International J. of Turbo and Jet-Engines*, Vol. 6, 1989.
213, Tamrat, B.F. and D.L. Antani, "Static Test Results of an Externally Mounted Thrust Vectoring Vane Concept", AIAA-88-3221.
214. Papandreas, E. "SCAT: A Small Low Cost Turbojet for Missiles and RPVs", AIAA-88-3248.
215. Eames, D.J.H. and Mason, M.L., "Vectoring Single Expansion Ramp Nozzle (VSERN) Static Model Test Program", AIAA-88-3000.
216. Miau, J.J., S.A. Lin, J.H. Chou, C.Y. Wei and C.K. Lin, "An Experimental Study of Flow in a Circular-Rectangular Transition Duct", AIAA-88-3029.
217. Humphreys, A.P., Paulson, J.W., and G.T. Kemmerly "Transient Aerodynamic Forces on a Fighter Model During Simulated Approach and Landing With Thrust Reversers" AIAA-88-3222.
218. Schneider, G.L. and G.W. Watt, "Minimum Time Turns Using Vectored Thrust", AIAA-88-4070-CP.
219. Sobel, K.M. and F.J. Lallman, "Eigenstructure Assignment for a Thrust-Vectored High Angle-of-Attack Aircraft", AIAA-88-4101-CP.
220. Klafin, J.F., "Integrated Thrust Vectoring On the X-29A", AIAA-88-4499.
221. Widdison, C.A., "Aircraft Synthesis with Propulsion Installation Effects", AIAA-88-4404.

222. VanOverbeke, T.J. and J.D. Holdman, "A Numerical Study of the Hot Gas Environment Around a STOVL Aircraft in Ground Proximity", AIAA-88-2882.
223. Oh, T.S. and J. A. Schatz, "Finite Element Simulation of Jets in a Crossflow with Complex Nozzle Configurations for V/STOL Applications", AIAA-88-3269.
224. Gal-Or, B., "The Fundamental Concepts of Pure Vectored Aircraft", ASME Conferences, Belgium, in press.
225. Gal-Or, B., "The Fundamental Concepts of Vectored Propulsion", J. Propulsion, in press.
226. Pavlenko, V.F., "Powerplants with In-Flight Thrust Vector Deflection", Moscow, Izdatel' stvo Mashinostroenie, 1987, 200p., 37 refs. In Russian.
227. Sweetman Bill, "B-2 Bomber for the 21st Century", Interavia, 1/1989, p. 22.
228. Cawthon, J.A., "Design and Preliminary Evaluation of Inlet Concepts Selected for Maneuver Improvement", AIAA-76-701.
229. Hawkins, J.E., "YF-16 Inlet Design and Performance", AIAA-74-1062.
230. Imfield, Wiliam F., "The Development Program for the F-15 Inlet", AIAA-74-1061.
231. Wiliams, J. and Butler, S.F.J., "Aerodynamic Aspects of Boundary-Layer Control for High-Lift at Low Speeds", J. of the Royal Aero. Soc. 67, No. 628 (April 1963), p. 201.
232. McLean, J.D., and Herring, H.J., "Use of Multiple Discrete Wall-Jets for Delaying Boundary-Layer Separation", NASA CR-2389, June 1974.
233. Banks, D.W., "Aerodynamics in Ground Effect and Predicted Landing Ground Roll of a Fighter Configuration With a Secondary-Nozzle Thrust Reverser", NASA TP-2834, Oct. 1988.
234. Bare, E.A., and O.C. Pendergraft, Jr., "Effect of Thrust Reverser Operation on the Lateral-Directional Characteristics of a Three-Surface F-15 Model At Transonic Speeds", NASA TP-2234, 1983.
235. Mattes, R. and W. Yonke, "The Evolution: IFPC to VMS", AIAA-89-2705.
236. Putnam, T.W. and R.S. Christiansen, "Integrated Control Payoff", AIAA-89-2704.
237. Tillman, K.D., T.I. Ikeler and R.A. Purtell, "The Pursuit of Integrated Control: A Real Time Aircraft System Demonstration", AIAA-89-2701.

INDEX

A number in brackets refers to a Reference number.
Otherwise a number refers to a page number.